Lecture Notes in Computer Science 1340

Edited by G. Goos, J. Hartmanis and J. van Leeuwen

Advisory Board: W. Brauer   D. Gries   J. Stoer

Springer
Berlin
Heidelberg
New York
Barcelona
Budapest
Hong Kong
London
Milan
Paris
Santa Clara
Singapore
Tokyo

Marc van Kreveld Jürg Nievergelt
Thomas Roos Peter Widmayer (Eds.)

Algorithmic Foundations
of Geographic
Information Systems

Springer

SELKIRK COLLEGE LIBRARY
CASTLEGAR, B.C.

Series Editors

Gerhard Goos, Karlsruhe University, Germany

Juris Hartmanis, Cornell University, NY, USA

Jan van Leeuwen, Utrecht University, The Netherlands

Volume Editors

Marc van Kreveld
Utrecht University, Department of Computer Science
P.O. Box 80.089, 3508 TB Utrecht, The Netherlands
E-mail: marc@cs.ruu.nl

Jürg Nievergelt
Thomas Roos
Peter Widmayer
ETH Zentrum, Department of Computer Science
CH-8092 Zürich, Switzerland
E-mail: {jn/roos/widmayer}@inf.ethz.ch

Cataloging-in-Publication data applied for

Die Deutsche Bibliothek - CIP-Einheitsaufnahme

Algorithmic foundations of geographic information systems / Marc
van Kreveld ... (ed.). - Berlin ; Heidelberg ; New York ; Barcelona ;
Budapest ; Hong Kong ; London ; Milan ; Paris ; Santa Clara ;
Singapore ; Tokyo : Springer, 1997
    (Lecture notes in computer science ; Vol. 1340)
    ISBN 3-540-63818-0

CR Subject Classification (1991): H.2, H.3, E.1, E.2,F.2, J.2

ISSN 0302-9743
ISBN 3-540-63818-0 Springer-Verlag Berlin Heidelberg New York

This work is subject to copyright. All rights are reserved, whether the whole or part of the material is
concerned, specifically the rights of translation, reprinting, re-use of illustrations, recitation, broadcasting,
reproduction on microfilms or in any other way, and storage in data banks. Duplication of this publication
or parts thereof is permitted only under the provisions of the German Copyright Law of September 9, 1965,
in its current version, and permission for use must always be obtained from Springer-Verlag. Violations are
liable for prosecution under the German Copyright Law.

© Springer-Verlag Berlin Heidelberg 1997
Printed in Germany

Typesetting: Camera-ready by author
SPIN 10647927   06/3142 – 5 4 3 2 1 0   Printed on acid-free paper

SELKIRK COLLEGE LIBRARY
CASTLEGAR, B.C.

Preface
Algorithmic Foundations of GIS

This volume aims to bring together the two lines of research whose interaction promises to have significant practical impact in the near future: the application-oriented discipline of Geographic Information Systems (GIS) and the technical discipline of geometric computation, or in particular, geometric algorithms and spatial data structures. GIS are complex systems consisting of a database part and a spatial data handling part. Spatial data handling in current GIS is less well-developed than the database aspects, and it is in relation to the spatial aspects that geometric algorithms and modeling can stimulate progress in GIS. GIS include many geometry-related problems such as spatial data storage and retrieval, visualization of spatial data, overlay of maps, spatial interpolation, and generalization.

Geometric Algorithms and GIS

Algorithms research has been an integral part of computer science since its beginnings. The goal is to design well-defined procedures that solve well-defined problems, and analyze their efficiency. Numerical algorithms were studied most thoroughly in the first phase of automated computation, and data management algorithms for sorting and searching became prominent in the 1960s. In the second half of the 1970s the systematic study of geometric algorithms gave birth to the discipline known today as computational geometry. The motivation for studying geometric algorithms comes from areas like robotics, computer graphics, automated manufacturing, VLSI design, and GIS.

The collaboration of GIS research and computational geometry has intensified in recent years, for several reasons. Firstly, the availability of geographic data in digital form is increasing rapidly. The acquisition of geographic data and the digitizing of maps are laborious, time-consuming tasks, and without appropriate digital data, GIS cannot operate effectively. Secondly, improved hardware now makes it possible to store and process large amounts of geographic data and to produce high-quality maps on computers. Thirdly, the research community in computational geometry, after having laid the conceptual and technical foundation, has recently shifted its attention toward more practical and more applied solutions.

Collaboration of GIS and Computational Geometry

The time is right for further development of the spatial components of GIS, as a result of joint efforts from computational geometry and GIS. What should such a collaboration look like? Many standard geometric problems and solutions, even those without any explicit reference to geography, are useful to GIS. Storing polygonal subdivisions on primary or background storage for efficient windowing queries is one example. Implementation and testing of data structures and query algorithms reveals their usefulness to GIS. Most of the basic algorithms developed in computational geometry, however, are not directly applicable to GIS. This is because standard geometric problems are often simplified to such an extent that they neglect important requirements of GIS. In addition, many problems arising in GIS have not yet been formalized in sufficient detail for the design of efficient algorithms.

As an example, consider generalization of maps. When the scale of a map is reduced, less information can be shown on the map, so it is necessary to remove or simplify certain objects on the map. However, it is far from clear which objects to remove or simplify. Different cartographers would produce different maps, if they were to generalize a map manually. A first step towards an automated map generalization method is the modeling of what should be generalized, when this is necessary, and how. These modeling questions cannot be answered by someone without cartographic knowledge. On the other hand, computational geometers are trained to abstract a problem into a form that is well defined. The joint knowledge of a cartographer and a geometer can result in a good model, a specification, for automated map generalization. Given such a specification, algorithms can be designed and implemented to compute the desired output.

There are several standard geometric structures and algorithms useful in GIS, with or without some modifications. These include topological data structures like the doubly-connected edge list or quad edge structure, spatial data structures like the quadtree and R-tree, the Voronoi diagram, the Delaunay triangulation, map overlay and buffer computation, and visualization algorithms. Most of these structures and algorithms are also used in other fields like computer graphics and robotics. They are well documented in textbooks or other standard texts.

Chapters in this Volume

This volume on algorithmic foundations of GIS contains a collection of survey papers on algorithms for spatial data. Standard texts on GIS and computational geometry are a starting point and not discussed in their generality here. But for completeness, references to those works are included.

Chapter 1 traces the history of the development of geometric computing, and illustrates some of its characteristic features. It explains how apparently simple algorithms pose difficult technical problems in the presence of degenerate data configurations. This fact turns the implementation of correct and robust algorithms, even if their logical structure is simple and clear, into a challenge that few programmers can meet successfully. Thus, basic geometric algorithms must be made available to application programmers in the form of program libraries developed by specialists. We present an example.

Chapter 2 discusses the use of Voronoi diagrams in GIS. It presents a general framework for handling fully dynamic and kinematic Voronoi diagrams for points and extended spatial objects. The Voronoi approach greatly simplifies some of the basic traditional GIS queries, such as map overlay and proximity queries, and even allows new types of higher level queries.

Chapter 3 contains a treatment of digital elevation models in GIS, with emphasis on TINs and algorithms. Although gridded elevation models are more widely used in current GIS than TINs, the advantages of TINs cannot be ignored. The grid is usually chosen for its simplicity, but this chapter shows that simple algorithms can solve the basic computations on the TIN model as well.

Chapter 4 gives a treatment of the visualization of terrains. Techniques for hidden surface removal like depth-sorting and visibility map computation are covered. The chapter also describes levels of detail in terrains to allow for a hierarchical multiresolution model.

Chapter 5 discusses the cartographically challenging problem of generalization. An overview is given of the types of generalization, the various approaches that have been attempted, and recent developments. Since generalization is a task that first needs a proper modeling, emphasis is placed on the modeling aspects.

Chapter 6 contains an extensive overview of spatial data structures that can be used on background storage. The design of such a structure obviously depends on the queries it should support efficiently. All structures are based on either an organization of the data objects or on an embedding of the underlying space. Design principles are usually based on common sense and simplicity.

Chapter 7 is dedicated to the design and analysis of space-filling curves in GIS. It presents an approach to theoretically evaluate the performance of range queries under a practical cost model that takes into account the number and locality of the data accesses.

Chapter 8 gives a treatment of algorithms on background storage. The assumption is that in GIS the amount of data that is input for an algorithm is often too large to fit in main memory. External memory algorithms attempt to minimize the number of block exchanges between main and external memory, since this affects the running time of the algorithm much more than the processing time itself. Several of these algorithms were based on common data structures for main memory, but have been adapted to work well on secondary storage.

Chapter 9 covers techniques for realizing precision and robustness in geometric computing. Standard geometric algorithms assume that real numbers can be stored and exact computation can be done. In practice, these assumptions are false, leading to false output and unexpected behavior of theoretically correct algorithms. There have been important developments recently in this practical aspect of geometric computing. To guarantee consistent output of a geometric algorithm, various advanced techniques may be needed.

As the editors of the book we take the opportunity to thank the contributing authors for their chapters, which in our opinion are excellent texts to be used for graduate student courses on GIS and applied geometric algorithms.

This book originated from the CISM Advanced School on the Algorithmic Foundations of Geographic Information Systems, which was held in Udine, Italy, from September 16 to 20, 1996. During the school, course material was distributed. Later, these course notes were revised and rewritten to form the chapters of this book. We thank the CISM for their support in the organization of the school, and their hospitality during the school. Finally, we acknowledge the support of the CGAL-project (ESPRIT IV LTR Project No. 21957), which has its homepage at http://www.cs.ruu.nl/CGAL/.

October 1997

Marc van Kreveld
Jürg Nievergelt
Thomas Roos
Peter Widmayer

International Centre for Mechanical Sciences
Centre International des Sciences Mécaniques

CISM Palazzo del Torso
Piazza Garibaldi 18
33100 Udine, Italy
E-mail: cism@hydrus.cc.uniud.it
WWW: http://www.uniud.it/cism/homepage.htm

Contents

3. Digital Elevation Models and TIN Algorithms

4. Visualization of TINs

Chapter 1. Introduction to Geometric Computing: From Algorithms to Software

Jürg Nievergelt

Dept. of Computer Science
ETH Zürich
Switzerland
jn@inf.ethz.ch

1   History of Geometry

Geometry started as a practical art for "measuring the earth". Soon it became the most formal of disciplines that served to define the standard of mathematical rigor. Later it spawned some of the most abstract mathematical disciplines to the extent of disavowing its original visual intuition.

Geometry as a computational tool re-emerged in the early days of computer graphics and computer-aided design. Computational geometry followed as a theoretical discipline that created hundreds of elegant and efficient algorithms. It came as a surprise, however, that "simple" operations such as intersection, when applied to degenerate configurations, cause conceptual and numerical problems that make it difficult to certify programs as correct and robust. The current challenge is to make proven algorithms generally available in the form of portable library programs. We describe the XYZ GeoBench and Program Library as an example of a system designed for teaching computational geometry.

1.1   Geometry: A Tool for Processing Spatial Data

Much of the "data" mankind has had to deal with throughout its existence is spatial in nature: the physical environment with its natural and man-made objects. Nature responded to man's need for spatial information by evolving the visual nervous system, a marvelous information-processing engine whose performance remains unrivaled by any man-made artifact. Man's quest to reason

1

about space gave rise to the oldest among the mathematical sciences, geometry, the science of "earth measurement".

The development of science admits many different approaches to the same problem. The long history of geometry shows all of the styles a mathematical science can assume, from pragmatic utility to extreme formalism. Let us illustrate this point with a few early examples of geometry in action.

Computational Geometry during the reign of King Ahusser of the Hyksos $\sim$ 1700 B.C. The need to re-partition land after the yearly floods of the river Nile called for computationally simple and efficent techniques for measuring areas. The crowning jewel of ancient Egyptian geometry is the following famous formula for squaring the circle.

Extract from the Rhind Papyrus (a copy of earlier papyri dated $\sim$ 1900 B.C.), Problem 50: "Method for calculating (the area of) a circular piece of ground of diameter 9 rods. What is its area? Subtract the ninth part of the diameter leaving 8 parts. Multiply 8×8 which gives 64. Hence the area is 6 khâ and 4 setat."

This formula assigns to a circle of diameter 9 the area of a square of side-length 8. In modern notation it would be written as $A = (8/9 \cdot 2 r)^2 = 256/81\, r^2 \approx 3.16\, r^2$, where A is the area of a circle of radius r. This surprisingly accurate "experimental quadrature of the circle" approximates πr^2 with a relative error of about $1/2\%$, an accuracy that certainly exceeded any practical needs 4 millennia ago.

This formula was presented merely as an empirical fact, without any mathematical justification. But whoever was clever enough to have found the superb pair of integers $(8, 9)$ must have wondered whether there might be a "better" pair. The next pair that yields a more accurate approximation, $(23, 26)$, reduces the relative error only slightly to $.36\%$. Even if the ancient Egyptians had known this, they would have been justified in prefering their computationally much simpler formula based on $(8, 9)$ that provided useful results.

Proof-based Geometry: Pythagoras $\sim$ 582-497 B.C. Ancient Egyptian and Babylonian geometry knew not only individual facts (such as the ratio 8/9 serving to measure the area of a circle), but also general statements that apply to an infinite class of objects. As an example, the theorem about right triangles, "$a^2 + b^2 = c^2$", was known as an empirical fact. The more general a statement is, the harder it is to verify by mere inspection of instances, and the more it calls for mathematical proof — a concept added by the Greeks. Pythagoras' innovation was not the statement of the theorem that goes by his name; rather, it was a geometric proof independent of numerical verification.

Axiomatic deductive geometry: Euclid $\sim$ 325-? B.C. The concept of proof soon raised the question as to what deductive steps are admissible. Euclid's main contribution was the first attempt at a rigorous formalization of the proof process. He showed how all of geometry then known could be derived from

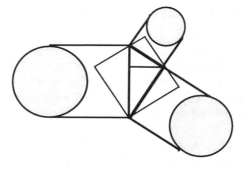

Fig. 1. Any three similar figures erected over the sides of a right triangle obey Pythagoras' theorem.

axioms, postulates, and definitions. Examples from Euclid's 'Elements':

Axioms:

- Things equal to the same thing are equal
- If equals are added to equals, the sums are equal
- If equals are subtracted from equals, ...
- Things which coincide with one another are equal to one another
- The whole is greater than the part

Postulates:

- A straight line can be drawn from any point to any other point
- A finite straight line can be drawn continuously in a straight line
- A circle can be described with any point as center, and with a radius equal to any finite straight line drawn from the center
- All right angles are equal to each other
- Given a straight line and any point not on this line, there is, through that point, one and only one line that is parallel to the given line.

Definition 1: A point is that which has no part.

Although Euclid's aim of formalizing proofs was clear, his tool, natural language, was inadequate to reach his ambitious goal. It took more than 2 millennia until mathematical logic succeeded in substantially improving on Euclid's technique.

The examples above illustrate three drastically different approaches to geometry, ranging from utilitarian pragmatism to abstract formality. In the course of several millennia, different branches of geometry have returned many times to these different styles. Computational or algorithmic geometry, the most recent branch, shares the concerns of ancient Egyptian geometry for practicality, simplicity and computational efficiency. These characteristics were decisive for the ancient Egyptian surveyor, and they remain decisive for the programmer who implements algorithms as robust programs.

3

2 Geometry Enters the Computer Age

Geometry found its way into computing in the late 1950s with the advent of computer graphics, when Ivan Sutherland created the first graphics system, "Sketchpad". Pictures that appear on the screen are usually generated "on the fly" from geometric models of 2-d or 3-d objects. Visibility algorithms that compute the picture as seen from a given viewpoint ("hidden line and surface removal") were among the earliest geometric algorithms studied in depth. The second major application to rely on geometric computation was computer-aided design, CAD. Here picture generation is less important than determining various geometric relationships among objects located in space (e.g. touching, inside, separates) and performing geometric operations on them. In "constructive solid geometry", for example, complex objects are defined by means of boolean (set-theoretic) operations on simple building blocks. As an example, Fig. 2 shows a perforated slab obtained as the set-difference of a box-shaped solid and a cylinder.

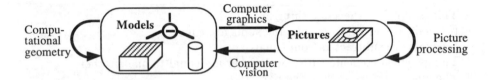

Fig. 2. Four disciplines related to geometric computation and pictures.

Geometry also enters into two other specialty disciplines that involve pictures. Picture processing typically operates on an array of pixels (gray-scale or colored dots) and modifies this array in some way (e.g. contrast enhancement) or extracts important features (e.g. boundary detection, medial axis transform). Computer vision, as used in robotics, attempts to build a 3-d model of a spatial configuration of objects from one or more pictures.

Computational geometry provides the technical know-how that guarantees an effective (correct, reliable, efficient) use of geometry in computer applications of all kinds. In its outlook on the types of problems to tackle and solutions to aim at, computational geometry has returned to the roots of its history. It studies simple practical problems, such as computing intersections of straight lines or answering proximity queries among points. It seeks simple, robust, efficient algorithms, just like the Egyptians' formula for squaring the circle. But of course the methods brought to bear on these problems are modern and powerful: the know-how developed over half a century of research in designing and analysing algorithms and data structures applies directly to geometric problems. Today we expect provably correct algorithms and programs, and mathematically rigorous error bounds.

Developing mathematically rigorous methods is not easy. Geometric software of the early days was characterized by the most straightforward techniques that appeared to solve a given problem. This often led to unpleasant surprises. Fig. 3 shows a notorious bug in a polygon-painting routine. In scan conversion, the electron beam is turned on and off as it scans the lines of the screen. Parity is the simplest approach to keeping track of the inside-outside state of the current beam position: every time the beam crosses a boundary segment of the polygon, the on-off state of the beam must change. For this parity technique to work, both line segments incident to a vertex must be recognized and counted as the beam visits this vertex. Because of round-off errors in evaluating the linear expression that determines the intersection between a line segment and the scan line, this operation is not as trivial as it may appear.

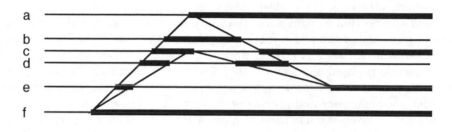

Fig. 3. The parity approach to area filling works for scan lines b and d, but may fail for a, c, e, and f.

For a polygon defined by line segments, a vertex is an exceptional boundary point: a segment ends and another one starts, so the boundary tracing algorithm must switch from one formula to another. A vertex is a locally degenerate configuration for a data set of line segments. Two line segments in general position either don't intersect, or they intersect in a point which is interior to both segments. But a vertex forces them to intersect at a common endpoint, thus causing a degeneracy.

Degeneracy was again revealed as a source of trouble in the next major application of geometric computation, CAD. Technical designs are often described as boolean expressions over simple building blocks. For example, a window can be considered as a hole in a wall, and the resulting surface may be described as the difference (wall – window). If the window reaches all the way to the ground, the wall and the window may have a common edge, a degenerate configuration. Assume our building blocks are rectangles taken as closed point sets, i.e. all of its boundary points belong to an object. Fig. 4 shows the result of set operations on objects in degenerate position. The object resulting from the intersection has an infinitely thin spike — not a desirable model of a physical object, considering that this design may control a metal cutting machine. The result of a difference operation has lost part of its boundary. A sophisticated "calculus of regularization" had to be developed in order to obtain a consistent theory of "constructive solid geometry".

5

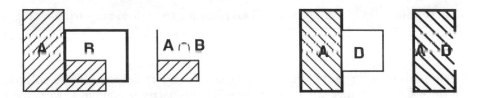

Fig. 4. Left: intersection figure has a dangling edge. Right: difference causes loss of a border.

Practitioners often tackled the difficulties caused by degeneracy by trial and error and clever ad hoc solutions. Theoreticians had to enter the field before a systematic attack on the problem of handling degeneracy in general could be started (see Section 5). But when theoreticians became attracted to computational geometry in the mid-seventies, inspired by the pioneering Ph.D. work of Michael Shamos (see, e.g. [SH75], where the sample problem we discuss in the next section is posed and solved), they first pursued another goal. They came up with a methodology for designing algorithms for geometric computation, analyzing their efficiency and proving optimality. They proved the validity of this approach by inventing hundreds of novel algorithms. This early phase is well documented in the first textbook dedicated to computational geometry [PS85]. The next section illustrates a simple example of this approach, before we return to the problems of degeneracy and robustness in Section 4.

3 The Interplay of Algorithm and Data Structure

In order to give the reader a flavor of computational geometry, we present a simple geometric problem and its elegant solution. The issues, concepts and techniques used in this example are typical of the entire discipline.

Access to spatial data is primarily guided by proximity relations among objects. Examples: when planning the motion of a robot arm one might ask for all objects within a certain distance of the robot's location; in a visibility computation one asks for the object nearest to the current viewpoint in a given direction. In order to gain insight into the great variety of proximity problems it is useful to understand the simplest instance, the "closest pair problem": given n points $p_1, \ldots, p_n$ in the plane, determine a pair (p_i, p_j), that minimizes the euclidian distance $d(p_i, p_j) = d_{ij}$, and compute $\delta = \min_{ij} d_{ij}$. We consider the evaluation of $d(p_i, p_j)$ to be an elementary operation to be charged unit cost. Thus, the minimum over all $n(n-1)/2$ distances d_{ij} identifies a pair (p_i, p_j) of closest neighbors and the minimal distance δ.

An algorithm designer will immediately ask questions about the complexity of the closest pair problem and the efficiency of the "naive" algorithm just described: Is it optimal, i.e. is it really necessary to compute $\binom{n}{2}$ distances? Or is it possible that some distance evaluations can be omitted or replaced by cheaper operations?

6

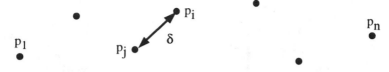

Fig. 5. Input data and result of a closest pair problem.

A rough sketch of a better algorithm [HNS88] shows that $5\,n$ distance evaluations always suffice to determine a closest pair. This remarkable reduction from $\Theta(n^2)$ geometric operations down to $\Theta(n)$, however, is not entirely free. It requires more complex data management, such as sorting, in order to carefully keep track of all the relevant information that can be deduced from the few distance evaluations actually performed. As we will see, the resulting trade-off "less geometry at the cost of more book-keeping" is beneficial. It leads to an $O(n \log n)$ algorithm which is a clear improvement over the naive $\Theta(n^2)$ algorithm.

The algorithm to be presented is of a type called plane sweep or line sweep, a paradigm that works well for many 2-dimensional problems. A vertical line, the "moving front", sweeps across the data from left to right, without ever backing up, processing "events" as it "sees" or encounters them. Sweep algorithms rely on an invariant characteristic of incremental algorithms: at all times, the front has accumulated the answer to the problem defined by the subset of data already seen. Once it has scanned all the data, the answer to the entire problem is at hand.

This algorithm schema is tailored to the closest pair problem as follows. The n points $p_1, \ldots, p_n$ are the events to be processed, sorted by x-coordinate. The sweep invariant I_k, for $2 \le k \le n$, states: when the front has processed the events $p_1, \ldots, p_k$ we know a closest pair (p_i, p_j) among the subset $\{p_1, \ldots, p_k\}$ and its distance $\delta = d_{ij}$. The invariant I_2 is initialized with $\delta = d_{12}$. At termination, $k = n$, I_n solves the closest pair problem. The essence of a sweep algorithm, as in a proof by induction, is the step from k to $k + 1$.

Fig. 6 illustrates how an event p is processed. The minimal distance δ among the k points to the left of the front is known. The current event, the $(k + 1)$-th point p, either changes the value of δ, or doesn't, depending on the data configuration in its immediate vicinity. Specifically, a new closest pair and a new δ come into existence iff there exists a point r within a semi-circle to the left of the front with center p and radius δ. Imagine the plane covered by a sheet of paper into which a semi-circular hole of the right shape and size has been cut, positioned with its center over p. This window defines a semi-circular query. The response to this query, i.e. all the points visible through the window, must be examined. If there are none, the event p leaves the previous solution unchanged. If any point r is visible, a new closer pair and a new value of δ emerge.

We can now substantiate the claim stated earlier: "$5\,n$ distance evaluations always suffice to determine a closest pair". How many points can be seen through the semi-circular window? The sweep invariant implies that points to the left of the front must have a distance $\ge \delta$; and the window has radius δ. Clearly, only

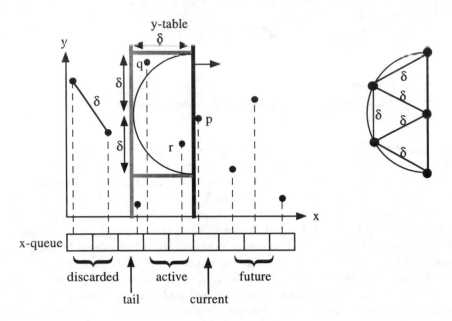

Fig. 6. A step of the sweep algorithm.

a small number of points can fit — on the right in Fig. 6 we see the maximum of 5 points packed into a semi-circle.

The practicality of the sweep algorithm just sketched hinges on our ability to answer semi-circular queries efficiently. The goal is clearly staked out. If the query requires time $\Theta(n)$, we get a useless $\Theta(n^2)$ algorithm that is more complicated than the naive $\Theta(n^2)$ algorithmus "compute all pairwise distances". If, on the other hand, we can answer the query in time $O(\log n)$, we obtain an efficient $O(n \log n)$ algorithm. The decisive factor is the existence of data structures that support semi-circular queries. Text books on algorithms and data structures describe dozens of standard recipes to answer region queries of various shapes; unfortunately, you won't find a solution ready-made for semi-circular queries. Thus, we have to think further before we can reduce the problem to known building blocks.

Orthogonal range queries bring us a step closer to our goal. We extend the semi-circular window to a rectangle of height 2δ and width δ. This approximation has the disadvantage of including points such as q in Fig. 6. After "unnecessarily" computing $d(p, q)$, q will be revealed as a "false drop" that we could have ignored had we used a semi-circular query. This nuisance is mitigated, however, by the fact that you can pack at most 6 points within a box of size 2δ by δ subject to the constraint of pairwise distances $\geq \delta$. Since one of these points is the current event p, the argument still holds that $5n$ distance evaluations suffice. The great advantage prevails that the rectangle approximation admits an efficient implementation, as follows.

8

A sweep algorithm maintains a queue of events, in this case the given points, sorted by x-coordinate. Thus, it is easy to continually discard points that trail the front by a distance $> \delta$. When answering a query, we consider only the "active" points, i.e. those that lie in a narrow strip of width δ to the left of the front. Depending on where the current event p lies on the front, any of the active points can form a new closest pair with p. We maintain the active points, sorted by y-coordinate, in a table that permits fast "dictionary operations": find, insert, delete, predecessor and successor.

This standard data type "dictionary" can be implemented using a dozen different data structures, resulting in different performance characteristics. We aim at an algorithm that runs in time $O(n \log n)$ in the worst case. Balanced trees implement dictionary operations with a worst case time bound of $O(\log k)$, where k is the number of entries in the dictionary — for our sweep algorithm, $k < n$. Predecessor and successor require only time $O(1)$. The asymptotic run-time analysis that follows does not depend on the specific choice of data structure but only on the existence of data structures that guarantee a time bound of $O(\log n)$ for each dictionary operation.

The current event p is inserted into the table according to its y coordinate $p.y$. Vertically, p lies in the middle of the 2δ-range of the rectangular query. Starting at p, we scan the active points in the table both upwards, applying the operation "successor" until we exit from the rectangle at coordinate $p.y + \delta$; as well as downwards, applying "predecessor" until we exit from the rectangle at $p.y - \delta$. Thus, we have visited each point q in the rectangle and submitted it to a distance computation $d(p, q)$. The total time required for processing an event p thus adds up to:

$$(\text{insert } p) +$$
$$(\# \text{ of points } q \text{ in the rectangle}) \times (\text{predecessor or successor} + \text{evaluate } d(p, q)).$$

Insert requires time $O(\log n)$; predecessor, successor and evaluate $d(p, q)$ need time $O(1)$. The number of points in the rectangle is ≤ 5. Thus, processing an event is dominated by the insertion of p into the dictionary and takes time $O(\log n)$. With a total of n events we have found a simple, efficient $O(n \log n)$ algorithm for the problem of the closest pair. Assuming that standard operations such as sorting and dictionary access can be obtained from a program library, this sweep algorithm takes at most one additional page of source code.

The success of sweep algorithms is due to a clever trick. By treating the x-axis as a time-dimension, a 2-d problem is reduced to a sequence of 1-d problems. For 1-d data, i.e. totally ordered domains, we know various efficient data structures. Thanks to these, sweep algorithms succeed in processing n 1-d problems, each of which may involve n data items, in time $O(n \log n)$ rather than in time $O(n^2)$. Unfortunately, this idea does not generalize efficiently to higher dimensions. If we sweep 3-d space with a plane, we transform a 3-d problem into a sequence of 2-d problems. The latter generates queries in a 2-d data space, and for these we rarely have data structures that answer queries in logarithmic time. We leave the discussion of spatial data structures to the survey in Chapter 7.

9

4 Degeneracy, Robustness, and the Quest for Perfection

The above description of the closest-pair sweep ignores a few details. Although these may appear trivial, a programmer cannot ignore them when aiming at a robust program, i.e. one that will work correctly for *all* data configurations, rather then merely for most of them. For example, plane sweep relies strongly on correctly sorted events. What happens if two or more events have exactly the same x-coordinate? What if two points in the input data turn out to have exactly the same x and y coordinates? Are they one point or two? If we treat them as two distinct points that happen to coincide, in what order do we process them? Will the algorithm correctly process a distance $\delta = 0$? What happens to the complexity analysis that relied heavily on the fact that at most 5 points can be located in a query rectangle of size 2δ by δ?

The closest pair sweep is such an exceptionally simple algorithm that it can easily be made immune against such nitpicking questions: events of equal x-coordinate can be processed in any order, and the algorithm terminates when a distance $\delta = 0$ has been computed or all events have been processed.

It came as a surprise, however, that "nitpicking questions" of the type above pose serious and technically very difficult problems for other geometric algorithms that also appear rather simple. Chapter 10, "Precision and robustness", treats this topic in depth. Here we merely aim to explain why the full extent of these difficulties was not recognized at first, and what the core cause of these difficulties is.

The concept of (small) numerical errors, unavoidable in numerical computation, has been extensively studied. The overriding concern of traditional numerical analysis is to minimize such errors, to study how they propagate, and to prove rigorous error bounds. An arsenal of techniques has been developed for this purpose, such as: using double or variable precision at critical steps; algorithms that adapt their discretization to the data; forward and backward error analysis; the concept of numerical stability; the distinction between well-conditioned and ill-conditioned problems, and many more. Thus, one might have expected that the techniques of error control developed by numerical analysts would apply to geometric computation as well. But those techniques were developed in the context of linear algebra and differential equations. They apply only to a limited extent to geometry, since they fail to address the central issue of correct geometric computation: topological consistency.

Topological consistency is a concept not commonly encountered in traditional applications of numerical analysis. It calls for computing certain relations among the data *exactly*, not merely approximately, in the presence of roundoff and other errors typical of finite number systems. Let us briefly address the issues: What is it? Why do we care? What concepts are involved? What questions arise? What can be achieved, and at what cost?

Everytime an algorithm executes a test, it asks a question about the data configuration and obtains an answer. The issue arises whether the answers obtained are consistent with each other, i.e. whether they correspond to a possible data configuration of the type the algorithm is supposed to handle. To understand the problem, think of M. Escher's famous drawings of impossible objects: each and every local snapshot is realistic by itself; but the way the local scenes relate to each other turns the aggregate into a picture of a physically impossible scene. This amusing artistic phenomenon is directly analogous to the problem of inconsistent answers in geometric computation.

Consider the tantalizing phenomenon of "The Braiding of Floating Point Lines" (L. Ramshaw). According to Euclid, two distinct straight lines in a plane intersect in at most one point. In view of roundoff errors, is it not surprising that several different points might simultaneously satisfy both equations that define the two lines, suggesting a cluster of intersection points. It is a bit surprising, though, to see that the straight lines $y = 4.3\,x/8.3$ and $y = 1.4\,x/2.7$, evaluated for various values of x using two decimal digit arithmetic, cross each other repeatedly as shown in Fig. 7, and as indicated numerically in the following table:

x	.78	.79	.85	.86	.92	.93	slope
$4.3\,x/8.3$	.39	.39	.43	.43	.46	.46	.5181
$1.4\,x/2.7$	.37	.40	.40	.44	.44	.48	.5185

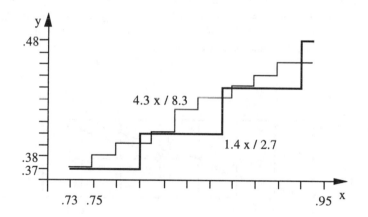

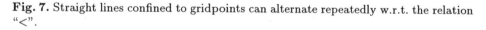

Fig. 7. Straight lines confined to gridpoints can alternate repeatedly w.r.t. the relation "$<$".

The numerical behavior observed, i.e. the repeated alternation of the relationship "above-below" as x grows monotonically, is inconsistent with the geometric properties of straight lines. Does it matter, given that the computed y-values are so close to each other?

That depends on what conclusions we draw from a few of these observations. Assume that, in a visibility computation of what can be seen from above, we are only interested in the question which of the two lines is above the other one in some interval of the x-axis. From the observed values at $x = .79$ and $x = .85$ we might wish to conclude that $y = 4.3\,x/8.3$ will be above $y = 1.4\,x/2.7$

11

for all $x \geq .85$, and thus save further evaluations. But in this case, the small roundoff errors that could be ignored if we were only interested in computing an intersection point mislead us into a global error about the vertical ordering of the lines. To see that we have declared the wrong line as uppermost, compare the slopes: $4.3/8.3 \approx .5181 < 1.4/2.7 \approx .5185$.

The problem of the braiding lines is of course not due to our use of only 2 decimal digits in the mantissa — for any fixed precision arithmetic examples of the same phenomenon can be found. Perhaps the problem can be avoided by using variable precision arithmetic, i.e. as many digits as the data requires — but that is costly. Can we invent a consistent geometry wherein any pair of straight lines defines three intervals: Line L above line L', an intersection interval where L and L' coincide, followed by line L' above line L?

There are other problems where inconsistencies of some kind simply cannot be avoided, as in the following example. In computer graphics points are usually restricted to integer grid points, as is the case for p_1, p_2 and the endpoints of segment L in Fig. 8. Intermediate results are often computed in higher precision and rounded to the integer grid when necessary. Our example assumes a procedure Whichside(p: point, L: line) that takes a grid point p and a real line L as arguments. If p_1 and p_2 both lie on the same side of L, we might expect that any convex combination $q = c\,p_1 + (1 - c)\,p_2$, $0 \leq c \leq 1$, also lies on this same side. But if q must be a grid point, Fig. 8 shows that even with correct rounding q may end up on the "wrong" side of L.

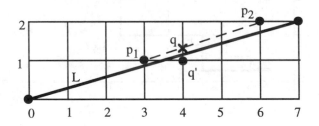

Fig. 8. The point $q = 2/3\,p_1 + 1/3\,p_2$ is rounded to a grid point q' on the wrong side of the line segment L defined by its endpoints $(0, 0)$ and $(7, 2)$.

What can an algorithm designer do to make a program robust against logical inconsistencies due to unavoidable small numerical errors? Half a dozen approaches to robustness are available today, but none offers a complete solution. Each and every algorithm, and its implementation, must be carefully crafted and analyzed. The long history of geometry is about to be extended by new "digital geometries" wherein seemingly familiar objects have new properties and obey new laws.

5 The XYZ GeoBench: A Program Library for Algorithm Animation

Computational geometry has now enjoyed two decades of rapid progress. It turned a field characterized by trial and error into a discipline where no programmer can work competently in ignorance of theory. But in order to make the advances of computational geometry accessible to application programmers, standard algorithms must be turned into correct, robust, efficient, well-tested, portable programs. It has become clear that the development of professional software for geometric computation calls for specialists with a broad range of experience that ranges from algorithm design and analysis to numerics and program optimization. Chapter 11, "Implementation and library design of geometric algorithms", discusses these issues and describes the project CGAL that aims at a library of production-quality geometric programs.

The implementation of a sophisticated geometric algorithm is an arduous endeavor if attempted without the right tools, such as: A library of abstract data types (e.g. dictionary, priority queue) and corresponding data structures (e.g. balanced tree, heap), reliable geometric primitives (e.g. intersection of 2 line segments), and visualization aids. What the applications programmer needs, but can rarely find today, are reliable and efficient reusable software building blocks that perform the most common geometric operations. Geometric modelers, the core of CAD systems, do not address this problem — they are typically monoliths from which application programmers cannot extract any useful part to use in their own programs.

In order to provide insight into the functions and structure of a geometry software system, this section presents an example aimed primarily at education. The XYZ (eXperimental geometrY Zurich) GeoBench and Program Library [NSdL+91]is used as a programming environment for rapid prototyping and visualization of geometric algorithms. The GeoBench is a loosely coupled set of modules held together by a class hierarchy of geometric objects and common abstract data types. The program library is a collection of well-tested, practical, efficient algorithms for basic geometric problems. The XYZ software (written in Object Pascal for the Macintosh) and documentation is available via anonymous ftp from ftp://ftp.inf.ethz.ch/pub/software/xyz/ or via the World Wide Web page http://wwwjn.inf.ethz.ch/geobench/XYZGeoBench.html.

5.1 Architecture and Components of the GeoBench

The main goal of the GeoBench and program library is to make available to the user a loosely coupled collection of carefully crafted software packages that serve both for the rapid prototyping of new algorithms and for the experimental testing and evaluation of existing programs. Fig. 9 presents the structure of this software system, where an arrow indicates the relationship 'is_based_on'.

Fig. 10 shows the backbone of the GeoBench. The class hierarchy defines the common data types and serves as interface between the components. This

13

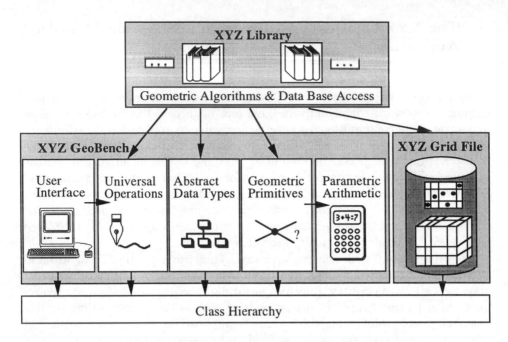

Fig. 9. The relationship among the XYZ components

tree describes the 'is_a' relationship among the classes. Algorithms are methods associated with the class on which they operate. For example, the Voronoi-algorithm is a method in the class 'pointVector' that yields an object of type 'voronoiDiagram'. Inheritance ensures that all methods for a given class are also available to its descendants. Thus, the method 'Voronoi-diagram' is also applicable to 'polyLine', 'polygon', and 'convexPolygon', for each of which it may have its own implementation.

5.2 User Interface and Algorithm Animation

Algorithm animation is useful in teaching and debugging. Whereas a static understanding of an algorithm based on invariants appeals to our rational mind, a dynamic visual trace of sample program runs adds intuitive insight. Algorithm animation is a creative art. The animation programmer has to design an intuitively understandable visual representation of an abstract concept: the current state of a program and its data.

Fortunately, geometry admits obvious visual representations of its data, spatial objects. This makes it possible to automate the production of animation code to a fair degree. Display routines are associated with each class of geometric objects, to be called at those moments in a program run when the picture needs updating.

Using conditional compilation to include animation code makes it possible for the same source code to serve for both production and animation runs. The

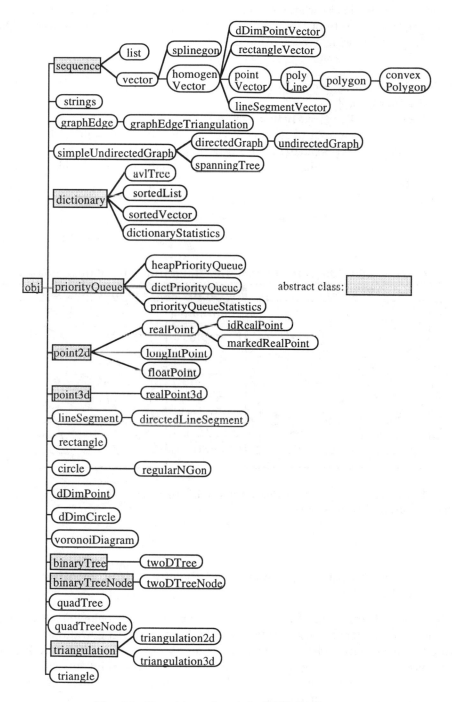

Fig. 10. Class hierarchy of the XYZ GeoBench

animation code checks whether animation is turned on and, if so, updates the currently visible state of the algorithm and waits for the user to proceed. This code has the following general structure.

```
...
{ Geometric algorithm changing internal state. }
{$IFC myAlgAnim }
    if animationFlag[myAlgAnim] then
        { Update graphical state information, usually draw some objects. }
        waitForClick(animationFlag[myAlgAnim]);
        { Update graphical state information, usually erase some objects. }
    end;
{$ENDC}
...
```

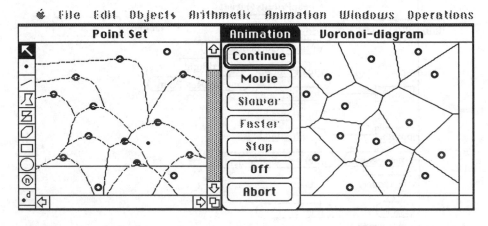

Fig. 11. Screendump of the GeoBench while animating the computation of a Voronoi-diagram using Fortune's sweep

5.3 Interchangeable Arithmetic and Parameterized Floating Point Arithmetic

The choice of arithmetic has a significant impact on the behavior of an algorithm, in particular for degenerate or nearly degenerate configurations. In order to experiment with different number systems, the user can specify the kind of arithmetic to be used (e.g. integer, long integer, floating point in various radix systems). In the best case, no line of code of an implementation must be modified in order to try out a different number system. Since points are the basic building blocks of all geometric objects, we achieve this goal by defining an abstract class 'point2d'; it has no instance variables for the coordinates, but it specifies

16

an interface that includes access procedures to the coordinates and to various geometric primitives.

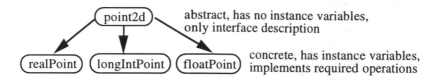

Fig. 12. The abstract class 'point2d' and its descendants

From this abstract 'point2d' we derive concrete point objects with instance variables and geometric primitives in their respective arithmetic systems. Algorithms that use only the functions and procedures specified by the abstract type 'point2d' can run in any of the three kinds of arithmetics currently supported. Whereas 'realPoint' uses the built-in floating point arithmetic, the object 'floatPoint' calls upon a software floating point package that uses any base and precision specified by the user. The latter can be used both to simulate high precision arithmetic, as well as low precision in order to accentuate rounding errors and thus facilitate testing.

5.4 The Program Library: Experimental Performance Evaluation

The XYZ program library contains carefully programmed and tested algorithms for several dozen standard geometric problems. The presence of several algorithms for solving the same problem reflects our concern for experimental assessment and comparison. As an example, consider the well-known problems of finding the closest pair and all-nearest-neighbors in a set of n points given in the plane. These problems date from the early days of computational geometry [SH75] and admit several optimal algorithms [PS85]. Convinced that plane-sweep yields the practically best algorithms for many simple 2-d problems, we developed new algorithms for these well-known problems [HNS92] and compared them to other algorithms implemented on the GeoBench, using comparable coding techniques (see also [Sch92]).

As is the case for many proximity problems, closest pair and all-nearest-neighbors can be solved easily in linear time after the powerful $O(n \log n)$ preprocessing step of computing the Voronoi diagram. But this diagram contains much more information than required, and this costs time. Vaidya's box-shrinking algorithm [Vai89] is another general approach that works in any number of dimensions. In contrast, plane-sweep yields direct solutions tailored to 2-d point proximity problems whose code complexity, memory requirements, and run times compares favorably with its competitors, as Fig. 13 shows. All algorithms are implemented as part of the XYZ library. The box-shrinking algorithm is optimized for d = 2; the Voronoi-based algorithm uses the efficient Fortune's sweep [For86]. The graph with logarithmic scales shows run times in seconds for these three algorithms on a Macintosh IIfx, implemented using the same floating point arithmetic. The test data are random point sets uniformly distributed in a square. Assuming a formula of the form $cn \log_2 n$ for the run times, the table shows the experimentally determined constants for all three algorithms and the maximum size of a configuration that can be handled in a 4 MByte partition.

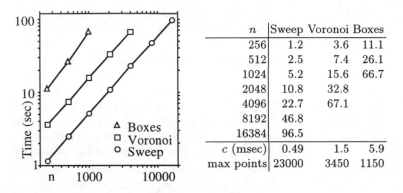

n	Sweep	Voronoi	Boxes
256	1.2	3.6	11.1
512	2.5	7.4	26.1
1024	5.2	15.6	66.7
2048	10.8	32.8	
4096	22.7	67.1	
8192	46.8		
16384	96.5		
c (msec)	0.49	1.5	5.9
max points	23000	3450	1150

Fig. 13. Performance measurements for three all-nearest-neighbors algorithms

This experiment shows clearly that "all of the tested algorithms may be optimal, but some are more optimal than others". The constant factors hidden by an asymptotic formula $O(n \log n)$ can easily differ by an order of magnitude, as they do in the example above. A system that facilitates experimental performance assessment is an essential tool for the development of program libraries.

6 Conclusion

This introduction to geometric computation has touched lightly on four topics: its historic development, the nature of its algorithms, the difficult problem of handling degenerate configurations correctly, and the practical problem of building program libraries in order to make production-quality geometric software available to application programmers.

This latter concern is urgent. An increasing number of applications, such as geographic information systems, process spatial data. Software for geometric computation is the necessary technical ingredient for spatial data processing. Only a few specialists have the deep knowledge in algorithms and software production required for creating top quality programs for geometric computation. Only if their products are packaged in portable libraries or other systems accessible to application programmers and end users, will the community reap the fruits of their labor [Nie94].

Computational geometry is now a well-established academic discipline, accessible to anyone with a standard background in algorithms and data structures. Textbooks of different coverage and levels of depth include [dBvKOS97], [Kle97], [Meh84], [Mul94], [NH93], [PS85].

18

Acknowledgments

Peter Schorn is the main architect of the XYZ GeoBench; Christoph Ammann, Michele De Lorenzi, Adrian Brüngger and many students have made valuable contributions. Thanks to Nora Sleumer for helping with the production of this chapter.

References

[dBvKOS97] M. de Berg, M. van Kreveld, M. Overmars, and O. Schwarzkopf. *Computational Geometry*. Springer, 1997.

[For86] S. Fortune. A sweepline algorithm for Voronoi diagrams. In *Proc. 2nd Annu. ACM Sympos. Comput. Geom.*, pages 313–322, 1986.

[HNS88] K. Hinrichs, J. Nievergelt, and P. Schorn. Plane-sweep solves the closest pair problem elegantly. *Inform. Process. Lett.*, 26:255–261, 1988.

[HNS92] K. Hinrichs, J. Nievergelt, and P. Schorn. An all-round sweep algorithm for 2-dimensional nearest-neighbor problems. In *Acta Informatica*, volume 29, pages 383–394, 1992.

[Kle97] R. Klein. *Algorithmische Geometrie*. Addision-Wesley, 1997.

[Meh84] K. Mehlhorn. *Data Structures and Algorithms 3: Multi-dimensional Searching and Computational Geometry*, volume 3 of *EATCS Monographs on Theoretical Computer Science*. Springer-Verlag, Heidelberg, West Germany, 1984.

[Mul94] K. Mulmuley. *Computational Geometry: An Introduction Through Randomized Algorithms*. Prentice Hall, Englewood Cliffs, NJ, 1994.

[NH93] J. Nievergelt and K. Hinrichs. *Algorithms and Data Structures, with Applications to Graphics and Geometry*. Prentice-Hall, 1993.

[Nie94] J. Nievergelt. Complexity, algorithms, programs, systems: The shifting focus. In *Proc. ALCOM Workshop on Algorithms: Implementation, Libraries, and Use, J. Symbolic Computation*, volume 17, pages 297–310, 1994.

[NSdL$^+$91] Jürg Nievergelt, Peter Schorn, Michele de Lorenzi, Christoph Ammann, and Adrian Brüngger. XYZ: A project in experimental geometric computation. In *Computational Geometry — Methods, Algorithms and Applications: Proc. Internat. Workshop Comput. Geom. CG '91*, volume 553 of *Lecture Notes in Computer Science*, pages 171–186. Springer-Verlag, 1991.

[PS85] F. P. Preparata and M. I. Shamos. *Computational Geometry: An Introduction*. Springer-Verlag, New York, NY, 1985.

[Sch92] P. Schorn. The XYZ GeoBench for the experimental evaluation of geometric algorithms. In *DIMACS Workshop on Computational Support for Discrete Mathematics*, 1992.

[SH75] M. I. Shamos and D. Hoey. Closest-point problems. In *Proc. 16th Annu. IEEE Sympos. Found. Comput. Sci.*, pages 151–162, 1975.

[Vai89] P. M. Vaidya. An $O(n \log n)$ algorithm for the all-nearest-neighbors problem. *Discrete Comput. Geom.*, 4:101–115, 1989.

Chapter 2.   Voronoi Methods in GIS

Christopher M. Gold

Centre de Recherche en Géomatique
Université Laval
Canada
christopher.gold@scg.ulaval.ca

Peter R. Remmele

Dept. of Computer Science
ETH Zürich
Switzerland
remmele@inf.ethz.ch

Thomas Roos

Dept. of Computer Science
ETH Zürich
Switzerland
roos@inf.ethz.ch

1   Introduction and Historical Background

Traditional vector-based GIS organize basic objects of interest such as roads, rivers, towns, or houses in thematic multi-layered (polygonal) maps representing them as polygons, arcs, and nodes (see Corbett [6] for a summary). Much of the early work was based on the line-intersection model of space where the global detection of intersecting arcs or lines was the sole method of determining connectivity – see Dutton [10] and Chrisman et al. [5]. Relations between polygons, arcs, and nodes imposed a "topology" which linked the basic map objects.

At that time, GIS systems were designed to process polygonal thematic maps, in order to respond to simple queries of the type "Show me all the areas with Type 1 or Type 2 soil, not zoned as industrial, within 50m of a lake". It turned out that such queries can be answered using only three basic operations. The first operation, "Reclassify and Merge", reclassifies a polygonal map on the basis of each polygon's attributes and combines adjacent polygons of the same resulting class. The second operation, "Corridor" or "Buffer Zone", constructs a polygon set with boundaries at a specified distance around the source object set (lakes in our example). The third operation, "Polygon Overlay", combines two polygon layers into a single layer such that each polygon gets the properties from each original layer (e.g. Type 1 soil and industrial in our example). All operations are global and depend therefore on a complete rebuild of the topology after each operation.

Many years of experience allowed us to identify the weaknesses of this technique. The detection of intersections itself is both an expensive operation and prone to errors due to digitizing limitations; with the consequence that if an intersection is not found the result is an incomplete knowledge of topology. As adjacency is defined via intersection, disconnected features such as islands are difficult to handle correctly. As a result, there is no readily-available approach to locally modify the topology and hence no true dynamic system permitting the addition and deletion of individual map objects. Dissatisfaction has started to grow concerning these limitations, because even minor modifications to the map require a global rebuild. Although a variety of techniques can be used to modify a small patch and then to sew it back into the main map, subsequent synchronization problems and localization problems (finding locally modifiable patches) arise.

As the emphasis on the user interface continues to grow, it is clear that the current non-interactive spatial model will be superseded by something closer to the human interaction with a paper map using a pen – or even our interaction with real-world geographic space. This requires that our actions take effect immediately, both in spatial queries and spatial modifications of our map. If we can detect the adjacent objects to our moving pen at any time during map construction – so must the GIS; this provides the possibility of testing line segment intersections and snapping points to line segments. If we can detect polygons or other structures that fail to close exactly – so must the GIS. If we can locate the polygon containing our pen, an isolated data point, or an island without special processing – so must the GIS. We thus need a spatial model possessing the tiling and adjacency properties of the raster model but the direct relationship to real-world objects of the vector model.

While this has puzzled the GIS community for some years, at least one potential answer is readily available from Computational Geometry: the Voronoi diagram [32], a universal data structure for representing proximity (see, e.g., the early works by Shamos and Hoey [29] and by Green and Sibson [21]). It supports a multitude of nearest neighbor queries (details can be found in the textbooks by Preparata and Shamos [27] and Okabe et al. [26]). As we will see, the Voronoi diagram for points and line segments with its dual, the Delaunay multigraph [7], provides the basic spatial adjacency properties between map objects. This not only resolves the basic difficulties with the line-intersection model, but allows us to maintain the topology of a GIS.

Nowadays, the Voronoi diagram and its variants are well known in many areas of science (see Aurenhammer [2] for a survey) and many algorithmic paradigms of Computational Geometry apply to compute them efficiently (see [3, 11, 22, 29]). In Geomatics and GIS, Gold [14, 15, 16] discovered the Voronoi diagram as a fundamental tool for representing and maintaining topology in a map. Generalizations towards dynamic operations [9] and kinematic motions [4, 8, 23, 31] have been investigated for about one decade.

22

This paper gives a survey of static, kinematic, and dynamic Voronoi diagrams as basic tools for Geographic Information Systems (GIS). We present a method that allows the insertion, deletion, and translation of points and line segments in a Voronoi diagram of n generators. All elementary operations are available in $O(\log n)$ expected time and linear expected storage space. The Voronoi approach also greatly simplifies some of the basic traditional GIS queries and allows even new types of higher level queries. The concept of a persistent, locally-modifiable spatial data structure that is always complete provides an important new approach to spatial data handling that is not available with existing systems.

The paper is organized as follows. Section 2 provides the Computational Geometry basis by an introduction into static and kinematic Voronoi methods; we show how both can be combined towards a fully dynamic approach. In Section 3, we look at the results from a practical point of view and show how the presented methods can be applied to obtain efficient algorithms for GIS.

2 Voronoi Diagrams

A brief summary of the elementary definitions and properties of Voronoi diagrams of points and line segments in the Euclidean plane is given in the following. Let d denote Euclidean distance and let

$$S := \{l_1, \ldots, l_n\}$$

be a finite set of $n \geq 3$ disjoint line segments (so-called generators) in the Euclidean plane. Line segments are allowed to degenerate to points. Later, we will also allow line segments to share endpoints, in which case we split off such endpoint from all incident line segments leaving behind open[1] line segments. Our results still apply, as we can leave an arbitrarily small gap between the line segments and the endpoint. We use the current assumption of disjoint line segments in the following, as it provides a more intuitive understanding.

In order to simplify the presentation, we assume the line segments of S to be in *general position*: we claim that there is no point in the Euclidean plane having the same distance to four different line segments and that there is no line in the plane tangent to three line segments, such that they all lie on the same side of this line.

2.1 Static Voronoi Diagrams

The static Voronoi diagram of S partitions the plane according to the nearest neighbor rule: each generator is associated with the region closest to it. The *bisector*

$$B_{ij} := \{x \in \mathbb{R}^2 \mid d(x, l_i) = d(x, l_j)\}$$

---

[1] We call a line segment *open* if one or two endpoints have been split off.

23

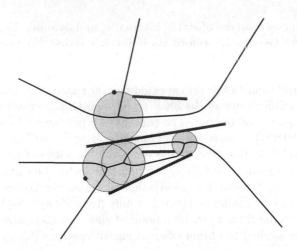

Fig. 1. Voronoi diagram and empty-circle property.

of two line segments l_i and l_j is a differentiable curve consisting of line and parabola segments. The *Voronoi region* or *Voronoi cell*

$$v_i := \{x \in \mathbb{R}^2 \mid d(x, l_i) \leq d(x, l_j) \text{ for all } j\}$$

of l_i is generalized-star-shaped [25] with respect to l_i. The "vertices" of the Voronoi regions, i.e. the non-differentiable points on the boundary ∂v_i, are called *Voronoi points* and the bisector parts on the boundary are called *Voronoi edges*. Finally, the *Voronoi diagram* $V(S)$ is the collection of its Voronoi polygons. Figure 1 shows a Voronoi diagram of six line segments and points. Using Euler's formula, it is easy to see that the number of Voronoi vertices and edges is bounded by $O(n)$; the space complexity needed to store the Voronoi diagram is therefore linear in the number of line segments.

Three line segments $l_i, l_j, l_k \in S$ form a Voronoi point in $V(S)$ iff there exists a circle of contact touching these three generators and no other line segment intersects the interior of this circle[2] (cf. Figure 1). This fact is well-known under the term *empty-circle property* for Voronoi diagrams.

When talking about topology and neighborhood relations of geometric objects, the dual multigraph $D(S)$ of the Voronoi diagram $V(S)$ is often more intuitive. In $D(S)$, each generator of S corresponds to a node and each Voronoi edge of $V(S)$ corresponds to an edge; this is done in such a way that the cyclical order of the dual edges around a dual node corresponds to the ordered sequence of Voronoi edges of the corresponding Voronoi region (see Figure 2). In general position, the dual multigraph $D(S)$ can be extended to form a complete tri-

---

[2] Notice that if three line segments of S generate two circles of contact, these circles have different orientation with respect to the cyclical order of the generators on their boundary.

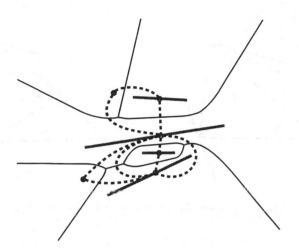

Fig. 2. The dual multigraph.

angulation by adding a point at infinity (c.f. [28]); this extension is called the *topological structure* of the Voronoi diagram.

There exist several algorithmic techniques for constructing Voronoi diagrams of points and line segments in $O(n \log n)$ time and $O(n)$ space (see [9, 11]).

2.2 Kinematic Voronoi Diagrams

We now extend the notion of the Voronoi diagram to continuously moving generators (see [1] for a recent survey). For our purpose, it is sufficient to consider one point p moving along a straight line within a static scene S of line segments (compare [28]).

As point p moves, the Voronoi diagram $V(S \cup p)$ changes continuously, but at certain critical instants in time, topological *events* occur that cause a change in the topology. At each instant of time, the topology of the Voronoi diagram is completely determined by the Voronoi points in $V(S \cup p)$ and by the unbounded Voronoi edges. Therefore, the topological structure of $V(S \cup p)$ can only change when a Voronoi point appears/disappears, or when a change on the boundary of the convex hull $\partial CH(S \cup p)$ occurs.

It is well-known that in the dual multigraph $D(S \cup p)$, both types of changes can be described as *swaps* of the diagonal edge of two neighboring triangles (see Figure 3). As p alone is moving while all other generators in S are fixed at their current position, a topological event can only occur when p runs into or out of a circle of contact or when p enters or leaves the boundary of the convex hull $\partial CH(S \cup p)$ (see [18] for more details). During the motion of p a list of swaps can be maintained in a priority queue which uses $O(\log k)$ time per event in the worst case; here, k is the current degree of p in the dual multigraph. The swap itself can be performed in $O(1)$ time when using the quad-edge data structure by Guibas and Stolfi [24] (see [28] for more details).

SELKIRK COLLEGE LIBRARY
CASTLEGAR, B.C.

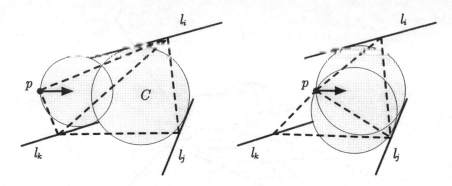

Fig. 3. The quadrilateral before and after the swap.

2.3 Dynamic Voronoi Diagrams

We now generalize our concept to allow *dynamic* operations such as insertions and deletions of points and line segments in a Voronoi diagram; we present a surprisingly simple and general framework to handle insertions and deletions in $O(\log n)$ expected time. For inserting and deleting a point p into the generator set S, we adopt the framework of Devillers et al. [9] using a slightly modified definition of the conflict region by Boissonnat et al. [3].

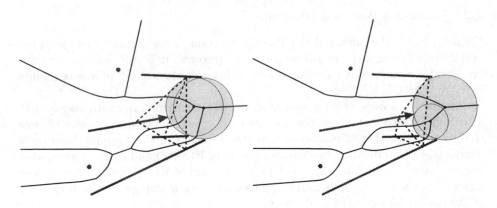

Fig. 4. A Voronoi diagram during an expansion.

For handling line segments, we add a tool for expanding and shrinking line segments using kinematic Voronoi methods. More precisely, we *insert* a line segment into a Voronoi diagram of line segments by first inserting one of its endpoints into the Voronoi diagram. Afterwards, we expand the generator by moving the point to the other end of the line segment (cf. Figure 4). In this way, a line segment is generated by a rubber band pinned down at one end and dragged to the other. In this setting, the topological events are even simpler

26

SELKIRK COLLEGE LIBRARY
CASTLEGAR, B.C.

than in the general case since the moving point – the free endpoint of the line segment – only runs into (but never out of) circles of contact corresponding to Voronoi points (see [19, 28]). We *delete* a line segment in a Voronoi diagram of line segments by reversing this operation: we first shrink the line segment to a single point and remove this point from the Voronoi diagram.

Using input randomization, one can prove that the expected time for expanding/shrinking a line segment is bounded by $O(\log n)$ (see [18] for details).

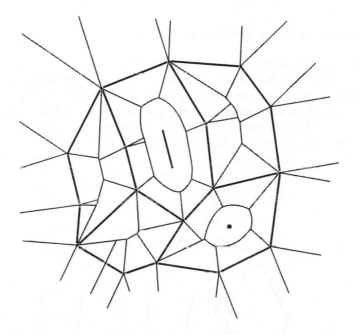

Fig. 5. A Voronoi diagram of points and open line segments.

3   GIS – A New Approach

Traditional GIS normally are coordinate-based systems, which means that a uniform grid will span the geographic environment and this grid will define the environment model's resolution. Objects of interest such as mountain peaks, corner stones, or even towns (depending on the resolution) will be represented by a point on the model's grid. In order to incorporate extended objects such as rivers and estates, it is useful not only to employ single points but also line segments and polygons. Besides a variety of information attached to these basic objects, a GIS has to store some kind of neighborhood relationship to allow local updates and to answer queries on locality. Therefore, a data structure designed as the base of a GIS must be able to handle at least these objects within its neighborhood relations.

27

The Voronoi diagram and its dual, the Delaunay multigraph, exactly fulfill these properties; the Voronoi diagram allows us to locate the nearest generator to a given query point whereas the dual gives us fast access to the neighbors of a generator stored in the GIS. Manipulations on the data such as insertions, deletions, or translations can easily be handled as we have seen before. In order to design the complex objects of a polygonal map, we have to allow line segments that share endpoints (as described in Section 2). Figure 5 shows a Voronoi diagram of a polygonal map.

3.1 Traditional GIS Operations

Within the Voronoi diagram, a map polygon creates Voronoi cells in its interior. *Labeling a polygon* consists of selecting any of these cells (by using the proximal query) and performing a flood-fill to collect all adjacent cells (see Figure 6). Attributes are then associated with the polygon label. When reclassification gives the same class on each side of a line segment, this segment is removed using the line deletion process described above.

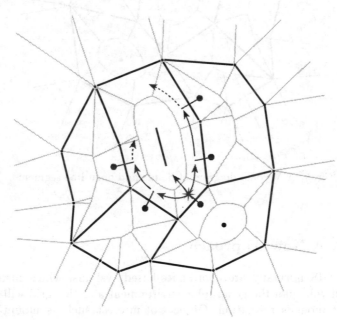

Fig. 6. A flood-fill of a polygon.

Buffer-zone generation is performed by inserting the target object set, e.g. the boundaries of a set of lakes, into a Voronoi layer (see Figure 7). For each Voronoi cell the intersection of the corridor boundary (at the specified distance from the generating object) may be calculated – if it is present at all. This is a complex task within a traditional GIS, but is easily available with the Voronoi diagram [14], being an expression of the "wavefront" or "prairie fire" analogy.

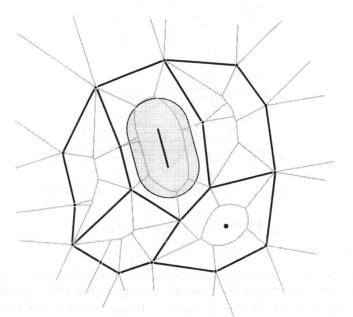

Fig. 7. Buffer zone generation.

Polygon overlay in the Voronoi context may be performed by drawing the second polygon set, line by line, onto the Voronoi diagram of the first polygon set. The attributes associated with each line segment are preserved. A form of flood-fill is then used to ensure that labeling is correct in the resulting map, even in the case of islands.

3.2   Separating the Topology from the Map Data

There are three main data types stored in a Voronoi system: coordinates, Delaunay triangles (containing pointers to the three adjacent triangles and to the three map objects forming the vertices), and the map objects themselves, which may be points or open line segments (and have pointers, as required, to the coordinate records and the matching line segment). Equivalently, the quad-edge data structure of Guibas and Stolfi [24] may be used. Unlike the DCEL structure (see, e.g., [27]), no pointers are associated with the map objects themselves; instead, all topology is contained within the Delaunay triangle records. This is an extremely desirable property in a GIS, as the same object may then be inserted in several layers simultaneously. This eliminates object duplication in the attribute database, as well as avoiding having different coordinate representations of the same object in various layers – eliminating *sliver polygons* when overlaid.

29

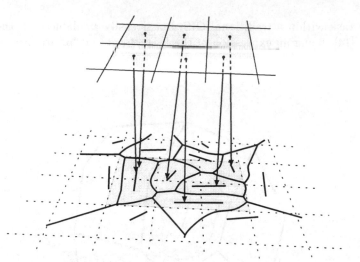

Fig. 8. Point location by a grid.

As a consequence of this design decision, there remains the basic question of how to access the portion of the triangulation associated with a particular map object, or with any specified (x, y) location. In practice, a simple walk through the triangulation from some previously accessed triangle (see, e.g., [24]) has been found quite sufficient to date, despite its poor theoretical efficiency. The next object to be processed is nearly always close to the previous object. Where necessary, a grid-based point location process may be employed (see [26] and Figure 8), at the cost of a more elaborate structure. However, it should be noted that even a simple grid, containing a pointer to some object falling within the grid cell, followed by a simple walk, will significantly improve the performance.

3.3   The Intersection Problem

We showed how a line could be generated as the locus of a moving point and how to create and delete points and line segments. This works satisfactorily when the end points are known in advance. For real-world digitizing, connections are formed when the line being drawn crosses a previously-defined line. In this case, intersections must be detected in advance of the moving point. Gold [13] showed that a line segment can only be intersected by an expanding line segment if both have been Delaunay neighbors shortly before. So, if a line segment becomes a new Delaunay neighbor of an expanding line segment, an intersection test is made and, if necessary, the intersection point calculated, and both the intersected line segment and the expanding line segment are split off at the intersection point as described above. This allows us to split and merge Voronoi diagrams along a common line (as shown in Figure 9).

30

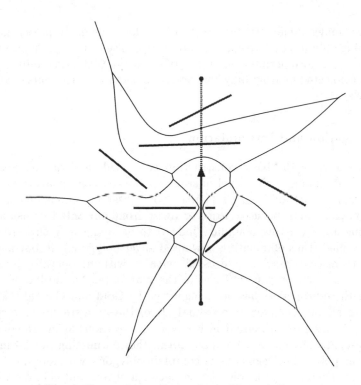

Fig. 9. Separating and merging a Voronoi diagram.

This gives a fully dynamic algorithm for the creation and deletion of points and line segments, for point movement and for intersection and the snapping together of lines during digitizing.

3.4 Update Management

As the topological structure is a triangulation, both global and local searching techniques are directly achievable as we have seen before (compare [30]). Because the update process is incremental, the addition and deletion of objects is immediate, and queries may be asked at once. This is of great value in identifying objects to be joined, for example, as it is only necessary to point within the Voronoi cell of an object in order to select it. This proximal query is basic to all interaction and is equivalent to normal human gestures for specifying objects. The objective, as with much recent work on graphic interfaces, is to conform as far as possible to natural human gestures. Also, a fully dynamic map construction opens various ways of managing the history of a map; this appears to be an interesting approach to spatio-temporal GIS (compare Frank et al. [12]).

Finally, spatial operations not normally associated with the topology may be implemented using a single spatial data structure. These include robot navigation, interpolation, fluid flow simulation, dynamic network analysis, etc. (see [17]

31

for some examples). The structure gives a fully local means of preserving topology, allowing fully interactive map update in response to the user's graphical actions (see [13]). It also permits the mixing of many types of data – fully connected polygons, connected hydrography, discrete points and line segments – within the same overlay.

3.5 Navigation and Interpolation

The local nature of the basic construction commands and queries gives great flexibility in designing higher level queries. As the topology is always complete, robot-navigation methods using a "cursor" or "observer" embedded within the Voronoi diagram may be used to steer away from unwanted collisions, or to follow along a pre-existing boundary judged to be sufficiently close to the trajectory specified (thus eliminating unwanted sliver polygons). Robot-navigation problems themselves may be addressed, with or without operator interaction, and with or without ongoing changes in the map data. An outline of a marine GIS using these properties has been suggested by Gold and Condal [17].

Because all objects have a proximal zone, the area-stealing interpolation model may be used to estimate field values (e.g. elevation) at intermediate locations. These map objects may be of any form, thus permitting precise interpolation between points, line segments, or complete polygons, with any specified level of continuity, eliminating the distinction between object and field data (see [20] for details).

3.6 Map Partitioning and Parallel Processing

When maps become very large – maybe even global – additional issues become critical, e.g., partitioning the map into portions that are manageable in memory becomes a significant issue. Traditional GIS, with map sheet boundaries, do not have a good answer. Some suggestions for partitioning maps along line segment boundaries or constrained triangulation edges are given in [33]. This has the added attraction that such a partitioning may be used to control access to particular portions of the map, either so that several operators may simultaneously be working on map update within assigned regions, or else to allow parallel processing to build the whole map more quickly without danger of conflict. Splitting and merging Voronoi diagrams as described above are fundamental in this process.

4 Conclusions

The Voronoi diagram embodies a form of "automated" topology, which – used as a basic data structure in GIS – may be the tool for the appropriate problems in the domain of Geomatics, and also the basic framework for algorithmic correctness and efficiency.

The concept of an always-complete, persistent, locally-modifiable spatial data structure provides an important new approach to spatial data handling. We have implemented the concept to a level that validates the proposed system design. However, analysis of future GIS needs leads to significant implementation issues where the field of Computational Geometry could contribute to a systematic resolution of an important field of spatial analysis. Clarification and simplification of problems concerning spatial or topological relationships could be the gain of a collaboration of Geomatics with Computational Geometry.

Acknowledgements

The funding of the first author for this research was made possible by the foundation of an Industrial Research Chair in Geomatics at Laval University, jointly funded by the Natural Sciences and Engineering Research Council of Canada and the Association de l'Industrie Forestière du Québec. The second and third author gratefully acknowledge the support by the Swiss National Science Foundation (SNF) under grant 21-39328.93 and 20-45407.95.

The authors would like to thank Jack Snoeyink and the anonymous referees for their helpful comments. An extended version of this chapter will appear in the special issue of Algorithmica on Cartography and Geographic Information Systems [18].

References

1. G. Albers, L.J. Guibas, J.S.B. Mitchell, and T. Roos, *Voronoi diagrams of moving points*, Tech. Rep. 235, ETH Zürich, 1995, to appear in Intl. J. Comp. Geom. & Appl.
2. F. Aurenhammer, *Voronoi diagrams: A survey of a fundamental geometric data structure*, ACM Comput. Surv., Vol. 23, pp 345–405, 1991.
3. J.-D. Boissonnat, O. Devillers, R. Schott, M. Teillaud, and M. Yvinec, *Applications of random sampling to on-line algorithms in Computational Geometry*, Discrete & Computational Geometry, Vol. 8, pp 51–71, 1992.
4. P. Chew, *Near-quadratic bounds for the L_1 Voronoi diagram of moving points*, Proc. 5th Canad. Conf. Comp. Geom. *CCCG'93*, pp 364–369, 1993.
5. N.R. Chrisman, J.A. Dougenik, and D. White, *Lessons for the design of polygon overlay processing from the Odyssey Whirlpool algorithm*, Proc. 5th Intl. Symp. on Spatial Data Handling *SDH'92*, Charleston, pp 401–410, 1992.
6. J.P. Corbett, *A general topology for spatial reference*, SORSA Report, 1985.
7. B.N. Delaunay, *Sur la sphere vide*, Bull. Acad. Science USSR VII: Class. Science Math., pp 793–800, 1934.
8. O. Devillers and M. Golin, *Dog bites postman: Point location in the moving Voronoi diagram and related problems*, Proc. 1st Annual European Symp. on Algorithms *ESA'93*, LNCS 726, pp 133–144, 1993.
9. O. Devillers, S. Meiser, and M. Teillaud, *Fully dynamic Delaunay triangulation in logarithmic time per operation*, Comp. Geom. Theory & Appl., Vol. 2, pp 55-80, 1992.

10. G. Dutton (Ed.), *First International Advanced Symposium on Topological Data Structures for GIS*, Harvard University, Cambridge, MA, Vol. 8, 1978.

11. S. Fortune, *A sweepline algorithm for Voronoi diagrams*, Algorithmica, Vol. 2, pp 153–174, 1987.

12. A.U. Frank, I. Campari, and U. Formentini (Eds.), *Theories and Methods of Spatio-Temporal Reasoning in Geographic Space*, LNCS 639, 1992.

13. C.M. Gold, *An object-based dynamic spatial model, and its application in the development of a user-friendly digitizing system*, Proceedings, 5th Intl. Symp. on Spatial Data Handling *SDH'92*, Charleston, pp 495–504, 1992.

14. C.M. Gold, *Problems with handling spatial data – the Voronoi approach*, CISM Journal, Vol. 45, No. 1, pp 65–80, 1991.

15. C.M. Gold, *Spatial data structures – the extension from one to two dimensions*, In: L.F. Pau (Ed.), Mapping and Spatial Modelling for Navigation, NATO ASI Series F, No. 65, Springer-Verlag, Berlin, pp 11–39, 1990.

16. C.M. Gold, *The interactive map*, In: M. Molenaar and S. de Hoop (Eds.), Advanced Geographic Data Modelling and Query Languages for 2D and 3D Applications, Netherlands Geodetic Commission, Publications on Geodesy, No. 40, pp 121–128, 1994.

17. C.M. Gold and A.R. Condal, *A spatial data structure integrating GIS and simulation in a marine environment*, Marine Geodesy, Vol. 18, pp 213–228, 1995.

18. C.M. Gold, P.R. Remmele, and T. Roos, *Fully dynamic and kinematic Voronoi diagrams in GIS*, Special Issue on Cartography and Geographic Information Systems, Algorithmica, to appear.

19. C.M. Gold, P.R. Remmele, and T. Roos, *Voronoi diagrams of line segments made easy*, Proc. 7th Canadian Conference on Computational Geometry *CCCG'95*, Laval University, Quebec City, pp 223–228, 1995.

20. C.M. Gold and T. Roos, *Surface Modelling with Guaranteed Consistency – An Object–Based Approach*, Proc. Int. Workshop on Advanced Research in GIS *IGIS'94*, LNCS 884, pp 70-87, 1994.

21. P.J. Green and R. Sibson, *Computing Dirichlet tessellations in the plane*, The Computer Journal, Vol. 21, pp 168–173, 1978.

22. L.J. Guibas, D.E. Knuth, and M. Sharir, *Randomized incremental construction of Delaunay and Voronoi diagrams*, Proc. 17th Intl. Colloquium on Automata, Languages and Programming *ICALP'90*, LNCS 443, Springer, pp 414 – 431, 1990.

23. L.J. Guibas, J.S.B. Mitchell, and T. Roos, *Voronoi diagrams of moving points in the plane*, Proc. 17th Intl. Workshop on Graph Theoretic Concepts in Computer Science *WG'91*, Fischbachau, Germany, LNCS 570, pp 113–125, 1991.

24. L.J. Guibas and J. Stolfi, *Primitives for the manipulation of general subdivisions and the computation of Voronoi diagrams*, ACM Transactions on Graphics, Vol. 4, pp 74–123, 1985.

25. D.T. Lee and R.L. Drysdale, *Generalization of Voronoi diagrams in the plane*, SIAM J. Comput., Vol. 10, No. 1, pp 73–87, 1981

26. A. Okabe, B. Boots, and K. Sugihara, *Spatial tessellations – concepts and applications of Voronoi diagrams*, John Wiley and Sons, Chichester, 1992.

27. F.P. Preparata and M.I. Shamos, *Computational Geometry: An introduction*, Springer, New York, 1985.

28. T. Roos, *Dynamic Voronoi diagrams*, PhD thesis, University of Würzburg, Germany, 1991.

29. M.I. Shamos and D. Hoey, *Closest point problems*, Proc. 16th Annual IEEE Symp. on Foundations of Computer Science *FOCS'75*, pp 151–162, 1975.

30. M. Teillaud, *Towards dynamic randomized algorithms in computational geometry*, LNCS 758, Springer-Verlag, Berlin, 1993.
31. T. Tokuyama, *Deformation of merged Voronoi diagrams with translations*, TRL Research Report TR87-0049, IBM Tokyo Research Laboratory, 1988.
32. G.F. Voronoï, *Nouvelles applications des paramètres continus à la théorie des formes quadratiques*. Premier Mémoire: *Sur quelques propriétés des formes quadratiques positives parfaites*, J. Reine Angew. Mathematik, Vol. 133, pp 97–178, 1907. Deuxième Mémoire: *Recherches sur les parallélloèdres primitifs*, J. Reine Angew. Mathematik, Vol. 134, pp 198 – 287, 1908 and Vol. 136, pp 67–181, 1909.
33. W. Yang and C.M. Gold, *Managing spatial objects with the VMO-Tree*, Proc. 7th Intl. Symp. on Spatial Data Handling *SDH'96*, Vol. 2, Delft, The Netherlands, pp 11B-15 to 11B-30, August 1996.

Chapter 3. Digital Elevation Models and TIN Algorithms

Marc van Kreveld

Dept. of Computer Science
Utrecht University
The Netherlands

marc@cs.ruu.nl

1 Introduction

Two of the most important types of map are the *choropleth map* and the *isoline map*. A choropleth map is basically a subdivision into regions, where the boundaries separate regions with a different attribute or property. For example, on a map showing countries the boundaries always have different countries to the two sides. An isoline map is also a subdivision into regions, but now the boundaries show all locations where an attribute has a fixed value. A precipitation map showing the lines of 750 mm, 800 mm, 850 mm, and 900 mm of precipitation is an example.

Isoline maps are a way to visualize elevation data, which is represented in a GIS as a *digital elevation model*. Mathematically, an elevation model is a continuous function in two variables. A digital elevation model simply is a finite representation of an elevation model. The most well-known example of an elevation model is height above sea level; therefore, one often uses the term *terrain*, or *digital terrain model* for digital elevation model.

In the GIS literature, papers abound on the application and representation digital elevation models. Survey papers have been written as well (most notably, by Weibel and Heller [125]), and textbooks on GIS and automated cartography also deal with digital elevation models [4, 11, 69, 73, 84, 111, 128]. But no survey has been written with the emphasis on algorithms on terrains (for brevity, we often use the term terrain as a shorthand for digital elevation model). It is almost impossible to produce a complete survey on all terrain algorithms, or even a nearly complete bibliography. Instead, this survey highlights the most important concepts and problems on terrain data, and discusses a few algorithms more thoroughly. The emphasis is on the efficiency of the algorithms, but it appears that 'model' and 'algorithm' cannot be seen as separate issues. Two algorithms that compute the drainage network on a terrain generally use a different model for the drainage network, so the algorithms that perform the computation cannot really be compared on efficiency: the algorithms don't compute the same thing. Even worse, the algorithms may be based on different terrain models.

Since the choice of a model and an algorithm go hand in hand, this survey will deal with both. We concentrate on the triangulated irregular network model for representing terrains, but sometimes we also deal with the other common model, the grid. The techniques underlying the algorithms have been developed both by computational geometers and by GIS researchers.

Another issue of importance is how the efficiency of algorithms should be analyzed. Computational geometers usually consider the worst possible inputs and make sure that the algorithm works well even in these cases. GIS researchers often don't analyze their algorithms, or give timings of an implementation. We'll adopt an intermediate view: if the worst case efficiency is of the same order as the typical efficiency for real-world inputs, then we'll use worst case analysis. If there seems to be an important difference, we'll try to track down why the worst case analysis is too pessimistic in practice, and try to motivate a more realistic efficiency statement.

This survey certainly doesn't include all aspects of terrain modelling and algorithms. Not included are data compression for terrains [43, 45], surface networks [96, 122, 127], and dealing with errors and uncertainty [5, 6, 76]. Other aspects are treated only briefly, like hierarchical terrain modelling, viewshed analysis, path planning, and statistics. Also, standard algorithms that have been described in all textbooks on computational geometry [18, 91, 98] are omitted. These include the computation of subdivision intersections (or: map overlay), Voronoi diagrams, and Delaunay triangulations.

2 Terrain Models and Representation

2.1 The Regular Square Grid

The regular square grid—or simply the grid—is a structure that specifies values at a regular square tesselation of the domain, see Figure 1 (a). In the computer it is stored as a two-dimensional array. For every square in the tesselation, or entry in the array, exactly one elevation value is specified. There are different interpretations of the grid. Firstly, the elevation stored can be thought of as the elevation for every point inside the square. In this case the digital elevation model is a non-continuous function. Secondly, the elevation stored can represent only the elevation at the center point of the square, or the average elevation inside the square. In these cases, an interpolation method is necessary to get a digital elevation model that specifies the elevation of every point. A possible interpolation method for any point p different from a square center is using the weighted average of the elevations of the four centers surrounding the point p, where the weight depends on the distances to the centers. But there are other possibilities too, like interpolation by splines.

12	12	13	13	14	15	14	14	14
12	12	12	13	14	15	14	14	13
12	12	12	13	13	14	14	13	12
12	12	12	13	13	13	13	12	12
12	12	12	12	13	13	13	12	12
12	12	12	12	12	13	12	12	12
12	12	11	12	12	13	12	12	11
11	11	12	12	12	12	12	11	10
12	12	12	12	12	12	12	10	9

(a) grid model   (b) contour line model

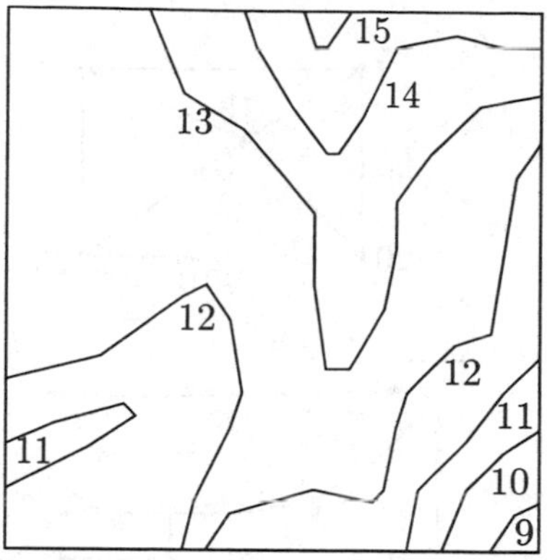

Fig. 1. Two models for elevation.

2.2 The Contour Line Model

In the contour line model to represent elevation, some set of elevation values is given, and every contour line with one of these elevations is represented in the model. So a collection of contour lines is stored along with its elevation, see Figure 1 (b). Sometimes the term isoline or isocontour is used when the elevation model represents something else than height above sea level. Throughout this survey we'll continue to use contour line in all situations.

A contour line is usually stored as a sequence of points with its x- and y-coordinates. It then represents a simple polygon or polygonal chain of which the elevation is specified. Since the contour line model specifies the elevation only at a subset of the domain, an interpolation method is needed to determine the elevation at other points. Since any point lies in a region bounded by contour lines of only two elevations, one usually only uses the bounding contour lines for the interpolation.

The contour lines can be stored in a doubly connected edge list [18, 97, 98] or quad edge structure [59]. An alternative to this representation is by means of the *contour tree* [46, 120] (or *topographic change tree* [71] or *Reeb graph* [112]). An contour line subdivision is a special type of planar subdivision with no vertices of degree three or greater. In theory they could exist if a contour line contains a pass, but such a situation would be coincidental and undesirable in mapping. So one generally assumes that each contour line is either a cycle of edges (polygon) or a chain between two points on the boundary of the elevation model. Suppose that every contour line corresponds to a node in a graph, and two nodes are connected by an arc if they bound the same region. This graph is easily seen to be a tree, the *contour tree*. A contour line model can therefore be stored by storing the contour tree, and with every node the cycle or chain of edges that together form the contour line.

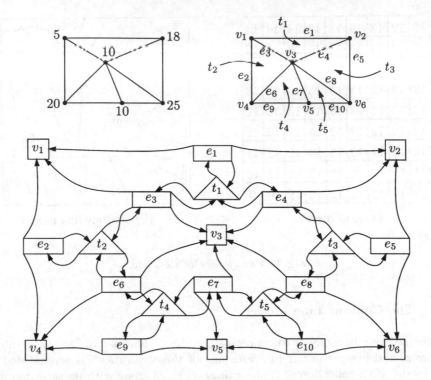

Fig. 2. A TIN and the network structure for it. The three values and the list of each vertex are not shown.

2.3 The Triangulated Irregular Network Model

In the triangulated irregular network model, usually abbreviated to TIN, a finite set of points is stored together with their elevation. The points need not lie in any particular pattern, and the density may vary. On these points a planar triangulation is given. Any point in the domain will lie on a vertex, an edge or in a triangle of the triangulation. If the point doesn't lie on a vertex, then its elevation is obtained by linear interpolation (of 2 points if it lies on an edge, and of 3 points if it lies in a triangle). So the model is a piecewise linear model that—in 3-dimensional space—can be visualized as a simply connected set of triangles. A TIN is continuous but not differentiable in the whole domain. The TIN model for terrains has been used since the seventies [50, 93, 95].

A possible storage scheme of a TIN is the doubly connected edge list [97, 98] or the quad edge structure [59]. Both are ways of representing the topology of any planar subdivision. These representations allow for all necessary traversal operations in an efficient manner.

An alternative representation can be used for a triangulation, since it is a special type of planar subdivision, see also Figure 2. This alternative saves storage space and makes traversal and other operaties a bit simpler. For every triangle t, edge e, and vertex v, there is a record (or object) for that feature. The record of a triangle t has three fields with pointers. These pointers are directed to the records of each of the three edges incident to t. The record of an edge e has four fields with pointers. Two of the pointers are directed to the records of the two incident triangles, and the other two pointers are directed to the records of the incident vertices. The record of a vertex v has three fields with values. These are the x- and y-coordinates and the elevation of the vertex.

The topological network structure just described allows for finding—for every triangle—the elevations of its vertices in constant time, finding the adjacent triangles for a given triangle in constant time, and more. This allows us, for instance, to walk through the triangulation along a straight line efficiently, which is necessary to determine a profile of a terrain. Variations on the structure are possible, for instance by storing a list of pointers at the vertex records to the incident edges.

2.4 Hierarchical Models

A hierarchical terrain model is a terrain model that represents a terrain in various levels of error, or, to make it sound less badly, various levels of imprecision. Most approaches of this type are based on TINs. Generally, TINs with more vertices have less error than TINs with fewer vertices. On the other hand, TINs with many vertices are more expensive to compute on. So if it is okay to have some error in an application, it may be better to work with a TIN with fewer vertices. A hierarchical terrain model allows the user to choose a terrain with the appropriate precision for each task.

Issues of importance of hierarchical terrain modelling are:

- The storage required by the model. Explicitly storing a terrain at many levels of detail leads to redundancy and excessive use of storage.
- The model may incorporate an efficient search structure automatically. For instance, locating a point on a terrain may be possible by first locating it on the coarsest level, and then locating it at repeatedly finer levels of detail.
- The triangulations should preferably be well-shaped, for instance, using the Delaunay triangulation.
- The model may allow a mixture of different detail levels in different parts of the terrain. This is useful in flight simulation, where the terrain close by must be shown with more detail than parts far away.
- It may be important that the model is consistent with respect to morphologic features. For instance, if there is a significant peak in a terrain at some detail level, one would wish that the same peak also exists on every finer detail level.

The list of issues already indicates that there probably isn't one best solution to all applications. Many hierarchical terrain models have been suggested [20, 21, 23, 38, 65, 103, 121]; the list of references is far from complete. A survey of the topic also exists [36].

41

3 Access to TINs

3.1 Traversal of a TIN

There is an old but good method to traverse a TIN and visit all its vertices, edges, and triangles in a simple way [51, 53, 55]. The nice thing about the method is that hardly any additional storage is needed: no mark bits, no stack, just one access point to the TIN.

Let T be a TIN stored in the structure just described, and let v be the bottom left vertex of T. For any triangle, we will give names to the incident edges. This implies that one edge receives two names, one for each incident triangle. Let t be

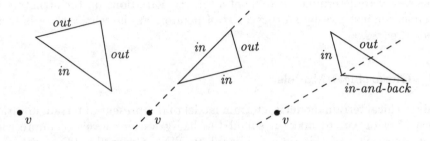

Fig. 3. Types of edges for the traversal algorithm.

any triangle of the TIN, we name an incident edge to be *in* if the line containing that edge separates the vertex v and the interior of the triangle t. If the vertex and the interior of t lie to the same side, the edge is *out*. If the line supporting the edge contains the vertex v we have a special case. The edge is *out* if the interior of t lies left of the supporting line, otherwise it is *in*, including the case of horizontal supporting lines. This is well-defined for TINs with a rectangle as the boundary, since we chose v to be the bottom left vertex. Notice that any interior edge of the TIN is *in* for the one incident triangle and *out* for the other.

All triangles have either one *in* edge and two *out* edges, or two *in* edges and one *out* edge. The *in* edges will represent the edge through which the triangle will be entered during a traversal. To make sure that we enter a triangle only once, consider any triangle with two *in* edges. These two *in* edges share a vertex, and the line through this vertex and through v separates the two *in* edges. The *in* edge that lies above, or left of this line, or is supported by the line, will be the real *in* edge. The other *in* edge will be *in-and-back*: the traversal enters the triangle through it, but will go back through that same edge immediately after. Similarly, for triangles that have two *out* edges, one will be the first *out* edge and the other the last *out* edge. Consider the line through v and the vertex incident to both *out* edges, the first *out* edge is the one left of this line. Note that for any triangle, we can determine easily, in constant time, which of its edges are *in* and *out* distinguish the real *in* and the *in-and-back* edges, and distinguish the first and last *out* edges.

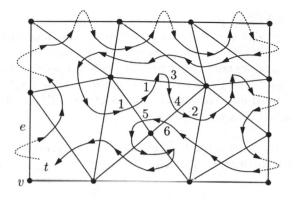

Fig. 4. Traversing a TIN; numbers at arrows correspond to the algorithm.

The algorithm starts at the triangle that has the bottom left vertex v as a vertex, and has one edge on the left side of the bounding rectangle. Let that triangle be t, and let e be the edge on the left side. The algorithm proceeds from triangle to triangle by crossing edges, and make the decision on which edge to cross next based solely on the type of edge through which it was just entered, and the type of triangle it is currently in. Since there are two types of triangles, and each has three edges, the algorithm distinguishes six cases to decide how the traversal should proceed, see also Figure 4. Edge e will be the first edge crossed in the traversal. When the algorithm reaches e a second time, the traversal is completed.

1. If e is the only *in* edge of t, then let e' be the first *out* edge of t, and let t' be the triangle on the other side of e. Repeat the algorithm with $t := t'$ and $e := e'$.

2. If e is the real *in* edge of t, then let e' be the *out* edge of t, and let t' be the triangle on the other side of e'. Repeat the algorithm with $t := t'$ and $e := e'$.

3. If e is an *in-and-back* edge of t, then let t' be the triangle on the other side of e. Repeat the algorithm with $t := t'$ and e.

4. If t has two *out* edges and e is the first one, then let e' be the last *out* edge of t, and let t' be the triangle on the other side of it. Repeat the algorithm with $t := t'$ and $e := e'$.

5. If t has two *out* edges and e is the last one, then let e' be the *in* edge of t, and let t' be the triangle on the other side of it. Repeat the algorithm with $t := t'$ and $e := e'$.

6. If e is the only *out* edge of t, then let e' be the real *in* edge and t' the triangle on the other side of it. Repeat the algorithm with $t := t'$ and $e := e'$.

If there appears to be no triangle on the other side of the edge we have just crossed, then we return to the previous triangle through the same edge we just

43

left immediately. Then we proceed as usual: the current triangle has just been entered through an edge that happens to be part of the bounding rectangle.

The algorithm visits every triangle exactly three times, once through each of its edges. At all times, we need only know the current triangle, the edge through which it was accessed, and the vertex v. We can report a triangle when we visit it for the first time. If all edges or all vertices of the TIN should be reported, some simple adaptations are needed.

The idea described above has been extended to the traversal of other subdivisions than just triangulations [19, 26].

3.2 Efficient Access to a TIN

In a regular square grid structure there is direct access to every part of the terrain. If one wants to know the elevation at a point with coordinates x and y, these coordinates can simply be rewritten to index values in the two-dimensional array. In a TIN this is not so easy, because a TIN is a pointer structure. If one wants to know the elevation at a point given its coordinates, one could test each triangle to see if it contains the point. This is rather inefficient, obviously. We briefly review three methods to gain access to the TIN at a specific point.

Access using quadtrees. Quadtrees and other spatial indexing structures like R-trees can be used to get access to a TIN at a specific query point efficiently. A quadtree is a rooted search tree of degree four, meant to store 2-dimensional data. Its nodes represent a recursive decomposition of a big square (associated with the root of the quadtree) into four subsquares (the children of the root). Each of the four subsquares is decomposed into four yet smaller squares. Further decomposition stops as soon as the part of the TIN that falls in the subsquare is simple enough. For example, when only a few vertices of the TIN lie inside the square. A leaf node is created that represents this square, and at the leaf node we store a pointer to the triangle record of the triangle that contains the center of the square. From that triangle record we can walk in the topological structure to locate the triangle record of the triangle that contains the query point.

Another possibility is to take the smallest enclosing axis-parallel rectangle of each triangle of the TIN. The resulting set of rectangles may overlap, but one can expect that at each point there won't be many rectangles containing it. A quadtree, R-tree, or other 2-dimensional search tree can be used to store the rectangles, see for instance the book by Samet [100]. For a query point, we determine all rectangles that contain it, and then find the triangle enclosed by one of the located rectangles that really contains the query point.

Access using planar point location. In the computational geometry field several efficient methods have been developed to determine which region of a subdivision contains a given query point. This problem is known as point location. For a TIN with n triangles, it is possible to construct an $O(n)$ size data structure that allows for point location in $O(\log n)$ time. Among the many results, the most simple and efficient ones are by Sarnak and Tarjan [101], Clarkson and Shor [12], Kirkpatrick [66] and Seidel [105]. Descriptions can also be found in textbooks on computational geometry [18, 91, 98].

Jump-and-walk strategy. The simplest method to locate a point in a TIN among these three is the jump-and-walk strategy. It also requires very little additional storage, so it may well be the best choice in practice. Suppose that access to the TIN structure is provided by one pointer to some feature. Rather than traversing all triangles until we find one that contains the query point, we can also traverse the TIN in a straight line from the access point to the query point. We typically encounter much fewer triangles on the way than if we would traverse the whole structure. The query time will be even better to take more than one starting point. Among those we'll choose as the real starting point the one closest to the query point, in the Euclidean sense.

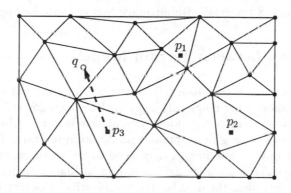

Fig. 5. Jump-and-walk with three starting points.

There are two natural schemes for choosing starting points. They can be placed in a regular, grid-like pattern, and stored in a 2-dimensional array with a pointer from each entry into the TIN structure. The pointer gives access to the triangle record that contains the point of the regular grid. This first scheme is good in most practical situations. If there are parts of the TIN where many small triangles occur, then the scheme is not so good. Either we don't have fast access to region with small triangles, since we must walk through many triangles before reaching the point to be located, or we must store a large array. A second scheme that overcomes this problem is the following. Choose m points $p_1, \ldots, p_m$ from the set of vertices of the TIN at random, and store a copy of those points in an unsorted list. By choosing the points from the TIN vertices, we generally have that regions that contain many triangles will also have several access points. With each point p_i we store a pointer to one of the triangle records incident to the point. When we query with a point q, we first determine the access point

45

p_i closest to q in $O(m)$ time, and then we start tracing the line segment $\overline{p_i q}$, starting at the triangle containing p_i. It appears that the choice $m = c \cdot n^{1/3}$ for a constant c is good. There is also some theoretical justification for this [7, 88]

3.3 Windowing

When a user is interested in a part of a terrain, then this part must be extracted from the whole terrain. We assume that the selection of the interesting part is defined by some rectangle, the window, and all triangles intersecting that rectangle should be located. Again we take the TIN as the elevation model to which windowing is applied.

The algorithm consists of two steps. First, the position of the window on the terrain should be located, and second, the TIN triangles inside the window can be traversed. We can locate for instance the upper left corner of the window, and start traversing from there; we already discussed ways to locate a point in a TIN. Traversal can be done in time linear in the number of triangles intersecting the window. The ideas of traversal without mark bits can also be applied here [19].

4 Conversion between Terrain Models

Terrain data can be entered into a GIS in various formats. Often, contour line data is entered when paper maps with contour lines are digitized by hand. Also quite often, gridded data is the input format, for instance when the data is acquired by remote sensing or automatic photo-interpretation. There are various reasons why data in one format may need to be transformed into another. Gridded data usually is huge in size, resulting in high memory requirements and slow algorithms when the data is processed further. Contour line data often needs to be interpreted anyway before anything useful can be done with it. In several cases it may help to store and compute with TINs instead. Conversion from TINs into grids or contour lines may also be useful. The former problem is algorithmically quite straightforward and we won't discuss it here. The latter problem shows up when a TIN is visualized as a contour map; we deal with that issue later in Section 6.

This section gives a few algorithms and references for converting to TINs. First we discuss how point sample data can be converted to a TIN, then grid-to-TIN conversion is handled, and then we go from contour lines to a TIN.

4.1 From Point Sample to TIN

Suppose a set P of n points in the plane is given, each with an elevation value. To convert this information into a TIN, one could simply triangulate the point set. In fact, the triangulation is an interpolation of a region based on the points of P. It is common to use the Delaunay triangulation because it attempts to create well-shaped triangles. Efficient algorithms for the Delaunay triangulation have been known for a while. See for instance Lee and Schachter [74], Guibas and Stolfi [59], or any textbook on computational geometry [18, 91, 98].

When the interpolation provided by the Delaunay triangulation is not appropriate, one can use different, more advanced interpolation methods like natural neighbor interpolation [54, 99, 106], weighted moving averages [4], splines [40], or Kriging [4, 123]. It is possible to combine the advantages of the TIN with the quality of these more advanced interpolation methods as follows. Suppose we are given the point set P together with an interpolation function and a maximum error. The idea is: construct a TIN based on P and the interpolation function, such that at any point, the interpolated elevation and the elevation given by the TIN differ by at most the maximum error. Now it may not suffice to use only the points of P as TIN vertices, and we need to select more points. Ideally, we select as few points as possible, the optimization problem that shows up is very difficult. Heuristics can be used to choose additional points as vertices in the TIN, and hopefully not too many additional ones. It may also be the case that only a subset of the points of P is needed to represent the interpolation function with the desired accuracy. Or perhaps a completely different set of points is best. In any case, finding a TIN with minimum number of vertices will be difficult. It is known that the problem of computing a TIN with the minimum number of vertices, given a TIN and a maximum allowed error, is an NP-hard problem, implying that efficient algorithms are unlikely to exist.

4.2 From Grid to TIN

Grid-to-TIN conversion can be seen as a special case of the conversion of sample points to a TIN. Also, grid-to-TIN conversion can be seen as a special case of TIN generalization: reducing the number of vertices of a TIN to represent a terrain. A grid can simply be triangulated to a fine regular triangulation. Various algorithms have been proposed in the literature; see for instance the survey by Garland and Heckbert [49], see also Lee [78]. Most of the methods have the following distinguishing features: (1) selecting which grid points to keep or discard, and (2) deciding when to stop selecting or discarding.

One method decides which grid points to keep or discard by initially assigning an importance to each grid point [8]. The importance is determined by comparing the elevation of a grid point with the interpolated elevation at the grid point based on the elevations of the eight neighbors. Only the grid points where the difference is greatest are kept. These points can be triangulated, for instance using the Delaunay triangulation, and become the TIN vertices.

A second method differs from the previous one by discarding grid points incrementally instead of using a precomputed importance [77, 78]. In a sense, the importance computation is postponed until the point is really discarded. A more detailed description follows later.

Thirdly, there are methods that start out with a coarse triangulation of only the four corner grid prints, and keep on refining the triangulation by adding more points in the triangles [34, 49, 61, 108]. Refining a triangle further stops when the triangle approximates the grid points that lie in it sufficiently well.

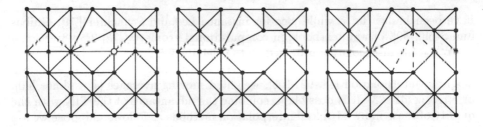

Fig. 6. Left, a TIN with one vertex indicated by a circle. Middle, the polygon that appears when the indicated vertex is removed. Right, a Delaunay triangulation of the polygon.

Another method start out by detecting surface specific features on the grid like peaks, pits, saddle points, ridges, and valley lines [41]. Then they complete these retained points and line segments to a TIN. Extraction of surface specific features is discussed later in this survey.

We describe two methods in more detail, namely, the drop heuristic method by Lee [77, 78] of the second type, and a method by Heller [61] (see also Fjällström [34] and Garland and Heckbert [49]) of the third type.

The drop heuristic. Lee's drop heuristic method takes a TIN as its input, and iteratively discards one vertex at a time to obtain a TIN with fewer vertices. Obviously, it also applies to grids as input if we consider it to be a triangulated regular grid. If a vertex is discarded, the incident edges are also removed and a polygon appears in the subdivision. To get back to a triangulation, the polygon is triangulated using the Delaunay triangulation. This will ascertain that if the algorithm starts with a Delaunay triangulation, then after every iteration we'll still have a Delaunay triangulation.

To decide which vertex should be discarded, each vertex is temporarily removed and the appearing polygon is triangulated—see Figure 6. Then we determine the vertical distance between the removed point and the new, simplified TIN. The removed vertex lies in one of the new triangles in the polygon, so this is easy to do. This vertical distance can be viewed as the error introduced by the deletion. Once we know the error that would be introduced, we add the removed vertex back to the TIN and temporarily remove another vertex. After we have done so for all vertices, we select the one for which the computed error is smallest and really discard it. The process continues until the created error is more than the prespecified allowed error.

It should be noted that the error at a vertex after it is discarded can become bigger when more vertices are discarded. So there is no guarantee that the error at all of the discarded vertices really is within the prespecified error. The drop heuristic method completely forgets about vertices that are discarded, although,

48

at the expense of more computation, a variant of the method could still consider them. Below we'll analyze the typical running time of the standard algorithm.

A straightforward implementation of the algorithm requires $O(n \log n)$ time per iteration on a TIN with n vertices. Discarding the vertices temporarily to determine their error and retriangulation can be done in $O(n \log n)$ time in total. This is true because the total complexity of all polygons to be triangulated is linear in n. Unfortunately, the vertical distances may have to be recomputed after an iteration, because the deletion of some vertex may result in a change of introduced error of other vertices. It is possible to construct an example of a TIN with n vertices where the algorithm has to recompute the errors many times for many vertices, but this is not a typical case. Observe that if a vertex v is removed in some iteration, then only the vertices of the TIN adjacent to v can have a change in introduced error. For all other vertices, the error resulting from their removal stays the same. So the question is how many neighbors a vertex in the Delaunay triangulation has. In the worst case this number may be $n - 1$, all other vertices, but on the average, a vertex has degree at most six. We'll analyze the typical case where any vertex that is removed has constant degree. Under this assumption we can design a variation of the given algorithm that will run in $O(n \log n)$ time. We describe this variation below.

1. For each vertex v in the TIN:
   - Temporarily remove it v.
   - Compute the Delaunay triangulation of the appearing polygon. Determine the vertical distance $error(v)$ of v to the new TIN.
   - Add the removed vertex back to the TIN.

   Store $error(v)$ for each vertex v sorted in a balanced binary tree $\mathcal{T}$. At each node of $\mathcal{T}$ storing some $error(v)$, store a pointer to the vertex v in the TIN. At v, we store a pointer back to the corresponding node in $\mathcal{T}$.

2. Consider the node with smallest $error(v)$ in $\mathcal{T}$. If it is greater than the prespecified maximum error, the algorithm stops. Otherwise it proceeds with the next step.

3. Remove the node storing the smallest $error(v)$ from $\mathcal{T}$. Remove the corresponding vertex v from the TIN structure. Let $w_1, \ldots, w_j$ be the vertices adjacent to v. Retriangulate the polygon defined by $w_1, \ldots, w_j$ using the Delaunay triangulation.

4. For every vertex $w_i \in \{w_1, \ldots, w_j\}$:
   - Remove the node that stores $error(w_i)$ from $\mathcal{T}$.
   - Recompute the vertical distance to the terrain if w_i were removed as we did in the first step.
   - Insert the new $error(w_i)$ in $\mathcal{T}$.

   Continue at step 2.

If all vertices in the TIN have constant degree, then each iteration requires only $O(\log n)$ time. The polygon to be retriangulated only has a constant number of vertices, and only for these a new vertical distance need be computed. The insert and delete operations on the tree $\mathcal{T}$ take $O(\log n)$ time each. So the algorithm takes only $O(n \log n)$ time in typical cases.

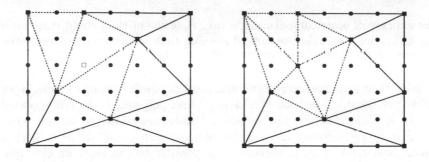

Fig. 7. The right TIN shows the situation if the square grid point on the left is the one with maximum error.

Incremental refinement. The algorithm to be descibed next takes a grid and a maximum allowed error ϵ as the input. Unlike the drop heuristic, the algorithm to be described really does guarantee that the final TIN has error at most ϵ. The approach is to start with a coarse TIN with only a few vertices, and keep adding more points from the grid to the TIN to obtain less error.

1. Let P be the set of midpoints of grid cells, with their elevation value. Take the four corner points and remove them from P, and put them in a set S under construction.
2. Compute the Delaunay triangulation DT(S) of S.
3. Determine for all points in P in which triangle of DT(S) they fall. For points on edges we can choose either one. Store with each triangle of DT(S) a list of the points of P that lie in it.
4. If all points of P are approximated with error at most ϵ by the current TIN then the TIN is accepted and the algorithm stops. Otherwise, take the point with maximum approximation error, remove it from P and add it to S. Continue at step 2.

If we assume a simple and slow implementation of the algorithm, we observe that at most n times a Delaunay triangulation is computed. For each one, the points in P are distibuted among the triangles of DT(S). This requires $\Theta(n^3)$ tests of the type point in triangle, if a linear number of points are added to S.

A much faster implementation has a worst case performance of $O(n^2 \log n)$ time, and in typical situations even better: typically $O(n \log n)$ time. The algorithm resembles incremental construction of the Delaunay triangulation to some extent [18, 57, 12]. Our algorithm, however, must also distribute the points of P and find the one with maximum approximation error. We'll show that these steps can be done efficiently.

Assume that $p \in P$ has been determined as the point with maximum error, and p must be removed from P and added to S. Then we locate the triangle t of DT(S) that contains p, and we find the vertices that will become neighbors of p in DT($S \cup \{p\}$). This update step of the Delaunay triangulation is the same as in the incremental construction algorithm. To distribute the points of $P \backslash \{p\}$

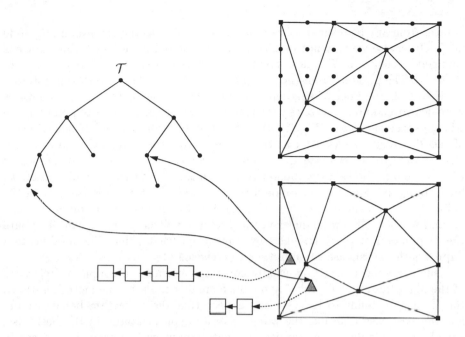

Fig. 8. The situation for a TIN with vertices shown as small squares (top right), and the corresponding structure with a few of the pointers between triangle records, list elements, and tree nodes.

over the triangles of DT($S \cup \{p\}$), observe that only the triangles of which p is a vertex in DT($S \cup \{p\}$) have changed. So for all triangles of DT(S) that don't exist in DT($S \cup \{p\}$), we collect the associated lists of points. These points are distibuted among the new triangles and stored in new lists.

The problem that remains is locating the point with maximum error. It is solved as follows. For each triangle of the TIN we determine the point of P inside it with maximum error. These points are stored in the nodes of a balanced binary tree $\mathcal{T}$ sorted on error. This allows us to locate the point p with maximum error efficiently; it is in the rightmost leaf of $\mathcal{T}$. Before p is moved from P to S, the Delaunay triangulation must be changed accordingly. To find the triangle in DT(S) that contains p we'll use a pointer from the node in $\mathcal{T}$ to the triangle record in the TIN structure; such pointers are shown as dashed lines with arrows in Figure 8. The triangle records are shown as grey triangles. After updating the TIN to be DT($S \cup \{p\}$) we move p from P to S.

Then we reorganize the lists that were stored with the triangles. When p was added to the Delaunay triangulation, some triangles were destroyed. The point of P inside each one that had maximum error is deleted from $\mathcal{T}$. The lists of points of the destroyed triangles contain p and the points that must be distributed among the new triangles, and stored in new lists. For each of the new lists we must find the point that realizes the maximum error in the cooresponding triangle, and store it in $\mathcal{T}$. For efficiency reasons it is a good idea to use a cross-pointer from any list element that stores a point of P to the corresponding node in $\mathcal{T}$. Otherwise we may not be able to locate the points of which the error has changed efficiently in $\mathcal{T}$. These pointers are shown as solid lines with arrows in Figure 8. The pointers from the triangle records to the lists are shown dotted.

If k is the number of neighbors of p in $DT(S \cup \{p\})$, then $k - 2$ triangles were destroyed and k new ones were made. Let m be the number of points in the triangles incident to p in $DT(S \cup \{p\})$. Then the iteration that added p as a vertex of the TIN requires $O(k + \log n)$ time for updating the Delaunay triangulation, $O(km)$ time to redistribute the m points over the k triangles, and $O(k \log n)$ time to update the balanced binary tree $\mathcal{T}$. In the worst case, m and k are both linear in n, giving an worst case performance of $O(n^3)$. But redistribution of the points can also be done in $O(k + m \log m)$ time by sorting the m points by angle around p. Since all new triangles in the TIN are incident to p, we can distribute the m points over the k triangles by using the sorted order. The modification improves the worst case running time to $O(n^2 \log n)$.

One can expect that k is usually constant, and after a couple of iterations of the algorithm, m will probably be much smaller than n. The more iterations, the smaller m tends to be. One can expect that the algorithm behaves more like the best case than like the worst case, for typical inputs. In the best case, k will be constant, and every list of points stored with a triangle reduces in length considerably each time it is involved in a redistribution. This means that later iterations in the algorithm go faster and faster, since m decreases from linear in n to a constant. If k is assumed to be a constant, we needn't use the modification to distribute the points, but simply spend $O(km) = O(m)$ time. Using an amortized analysis technique, one can show that the whole algorithm will take $O(n \log n)$ time under the assumptions given.

4.3   From Contour Line to TIN

Contour line to TIN conversion algorithms are useful because elevation data is often obtained by digitizing contour line maps. A contour line map is already a vector data structure, in fact, a planar subdivision where the vertices and lines are assigned the elevation of the contour line they are on. To convert the contour lines to a TIN, the obvious thing to do is triangulate all regions, that is, triangulate between the contour lines. Each region can be seen as a polygon with holes, and there are standard triangulation algorithms known for this problem in computational geometry [18, 91, 98].

Instead of using any triangulation it is a good idea to use one that gives nicely shaped triangles, like the Delaunay triangulation. However, the input to the triangulation algorithm is a polygon, not a point set. There exists a triangulation that follows the Delaunay triangulation as closely as possible, given some given set of edges must be present. It is called the *constrained Delaunay triangulation* [9, 22].

In the GIS literature, a couple of approaches to triangulate between contour lines have been described [10, 28, 47, 102]. One of the problems with the constrained Delaunay triangulation and some of the other methods is that they may create horizontal triangles. This side effect of the triangulation is known as the wedding cake effect. It is especially undesirable when visualizing the terrain with the use of hill shading. Several of the known methods avoid such horizontal triangles. Of course the choice of a suitable triangulation comes down to choosing a particular type of interpolation function between the contour lines.

5 Mathematical Computations on Terrains

In many applications it is useful to do things like adding or subtracting the elevation data in two terrains, or squaring the elevation data of a terrain. For example, suppose the data of two terrains represent the height above sea level, and the depth from the surface to the groundwater. Then the subtraction of the latter data set from the former one yields the height of the groundwater above sea level. Similarly, if the depths from the surface to two types of soil data is stored in terrains, then the thickness of the soil in between can be obtained by subtraction.

As an example where it is useful to square and cube terrain data, consider wind erosion [85]. Particles of a certain size can be lifted from the earth's surface by the wind, transported, and deposited again. It has been shown that the detachment capacity of wind varies with the square of the wind velocity, and the transporting capacity with its cube. To model erosion by wind, we need data on wind velocity at the surface, which can be seen as elevation data and modelled by a grid or TIN. Squaring this elevation data gives a model for the detachment capacity of the wind that can be used in further computations and simulations.

5.1 Adding and Subtracting Terrains

Assume that two TINs T_1 and T_2 are given, and we wish to add up the elevation data in them. Subtracting would be the same after placing a minus sign in front of the elevation values of the terrain to be subtracted. The addition of two TINs can be determined exactly and stored into a new TIN, because the addition of piecewise linear functions (which TINs represent) again yields a piecewise linear function. The addition is done by performing an overlay of T_1 and T_2. There are several algorithms known for computing the overlay [30, 58, 81, 89]. After computing the overlay—the refinement of each of the TINs—we obtain a subdivision where all faces have three, four, five, or six edges. It is trivial to triangulate and obtain a proper TIN again. We now must fill in the height information for the vertices of the overlay. Every vertex originally in T_1 receives its height plus the interpolated height in T_2, and the analogous thing holds for the vertices of T_2. The vertices in the overlay that come from the intersection of two edges are assigned the height that is the sum of the interpolated heights on those two edges. The easiest and most efficient overlay algorithm of the ones mentioned above is the one by Guibas and Seidel [58], which is based on a topological plane sweep of the two TINs. It requires $O(n + k)$ time, where n is the number of vertices in the TINs T_1 and T_2, and k is the number of vertices in

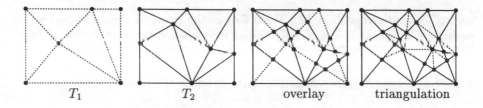

T_1 T_2 overlay triangulation

Fig. 9. The overlay of two TINs and its triangulation.

their overlay. The value of k can be as large as quadratic in n, but more often it will be linear or close to linear in n.

5.2 Squaring a Terrain

Suppose we want to compute and represent a function in two variables x and y that is the square of another function, represented by a terrain T. The square of T will obviously be a piecewise quadratic function, so a TIN can never represent the square of T without introducing error. The problem we'll discuss is representing the square of a TIN in another TIN but with a guaranteed maximum allowed error ϵ at any point. What would happen if the square of T were computed simply by squaring the elevation of each vertex, and represented by a TIN $\hat{T}$ with the same topological structure? The TIN $\hat{T}$ will always overestimate the true square T^2 of T. The error of $\hat{T}$ as a representation of T^2 is $\max(\hat{T} - T^2)$, maximized over all points (x, y) on the two terrains.

Let's consider one edge of T, where the lower vertex has elevation a and the higher vertex has elevation b. Then $\hat{T}$ will represent this edge as the linear interpolation from a^2 to b^2, whereas the true square of the edge will be a quadratic function from a^2 to b^2. The maximum error over the edge always occurs exactly in the middle of the edge, and the error itself has the value $\frac{1}{4}(b-a)^2$. So the error is not dependent on the position or length of the edge, only on the difference in elevation of the incident vertices. We conclude that the maximum error of $\hat{T}$ always occurs on an edge, and never interior to a triangle.

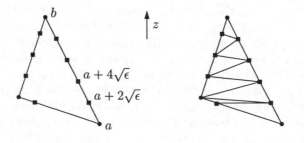

Fig. 10. Refining a triangle.

54

Suppose that the maximum allowed error ϵ is given. To compute a TIN $\tilde{T}$ that represents the square of T with error at most ϵ, we'll refine the edges of T so that none of them has an elevation difference more than $2\sqrt{\epsilon}$. For any edge spanning the elevations from a to b, the number of points needed to refine that edge is $\lceil \frac{1}{2}(b-a)/\sqrt{\epsilon} \rceil$. We place these points at $a + 2\sqrt{\epsilon}$, $a + 4\sqrt{\epsilon}$, $a + 6\sqrt{\epsilon}$, and so on, until the last one is at elevation $b - 2\sqrt{\epsilon}$ or higher. We do so for every edge of the TIN, and then triangulate every triangle with the additional points as flat as possible. This can be done without introducing any edges that span an elevation more than $2\sqrt{\epsilon}$. Then we square the refined TIN to obtain $\tilde{T}$. From the discussion in the previous paragraph, the error of $\tilde{T}$ is at most ϵ.

6 Computation of Contour Lines

One of the most useful structures that can be obtained from a digital elevation model are the contour lines. Contour lines are probably the most common and natural way to visualize elevation data. Other applications lie in site planning. When a new construction site must be determined, one of the requirements may be that the site lie on an elevation below 1000 meters. Or a spatial query done by a user of a GIS may request all geographic objects of a certain type that have at least a certain elevation. For example, the parliament of a country may consider to partially fund an irrigation system for all crop fields that receive less than 250 mm percipitation annually. To estimate how much this will cost, the total area of these crop fields must be determined. This in turn requires the contour lines of 250 mm on an elevation model representing the annual percipitation.

In this section we use the term *contour line* for one connected set of line segments with a given elevation. We use the plural term *contour lines* for all connected sets of line segments with the given elevation. We next consider two methods for determining contour lines on a TIN. The first method simply scans the TIN to determine the contour lines, while the second method uses preprocessing to be able to find the contour lines more efficiently. This is particularly useful in interactive situations. The last issue treated in this section is the choice of elevations for which the contour lines are selected for display. It is a form of classification.

6.1 Direct Computation of Contour Lines

When considering the contour lines on a TIN, observe that all vertices of the contour line lie on edges or vertices of the TIN, and all segments of the contour line lie on triangles or horizontal edges of the TIN. We assume that there are no horizontal triangles on the elevation of which we want the contour line. This can be enforced as follows. Suppose the contour lines of elevation Z are needed, and at some moment a horizontal triangle t with elevation Z is located. Then we only take the edges of t for which the other incident triangles have a vertex higher than Z. This basically comes down to tracing a contour line a very small amount higher than Z. With this enforcement we can from now on forget about whole triangles on the contour line. In a similar way we can forget about complications introduced by saddle points at elevation Z on the TIN (saddle points are vertices that have four or more incident segments of a contour line). If a contour line doesn't contain saddle points, it must be a simple polygon (closed) or a simple polygonal line between two points on the boundary of the TIN.

Given a TIN and an elevation value Z, there is a very simple way to find the contour lines of elevation Z: Traverse the whole TIN and for every triangle, determine if it contains a segment of the contour lines. If so, report it. This algorithm requires $O(n)$ time for a TIN with n triangles. One shortcoming of this algorithm is that it gives the segments on the contour lines in an arbitrary order. Sometimes it is necessary that each contour line be returned as a separate sequence of segments, for instance when smoothing should be performed.

There are two ways to obtain the contour lines in a structured form, as sequences of segments. The first way is by postprocessing the segments that were found by the trivial algorithm. Sort all endpoints of the segments lexicographically on the coordinates. Then all endpoints that are shared among two segments become adjacent in the order. This allows us to structure the separate segments to sequences, each of which is one contour line. If the contour lines together contain k segments, then the postprocessing step takes $O(k \log k)$ time. So in total, the method takes $O(n + k \log k)$ time. Since k is expected to be much smaller than n on real data, proportional to $\sqrt{n}$ is often argued, the overhead of $O(k \log k)$ time is no big deal.

The second way to obtain the contour lines in structured form is by tracing each contour line directly on the TIN. If the TIN is stored in a topological structure like the one described in Subsection 2.3, the traversal of one contour line from a starting point can easily be done in time linear in the number of segments of the contour line (there is a small catch if the contour line passes through a vertex of high degree; a possible solution [118] won't be discussed here since it won't be worthwhile in practice). It remains to find all starting points from which to start tracing. If the TIN structure has mark bits stored with the edge records or the triangle records, the following method can be used. Initially all mark bits are reset. For each triangle of the TIN, determine if its mark bit is reset and it contains a segment of the contour lines. If so, start tracing the contour line and set the mark of all triangles that are traversed. The tracing can stop if the boundary of the TIN is reached or a cycle has been completed. After the tracing has stopped, we continue with the next triangle. After all triangles have been tested, all mark bits must be reset again to allow a next request for contour lines. The whole algorithm clearly takes $O(n)$ time. A disadvantage is that mark bits are required in the structure.

6.2 Preprocessing for Contour Lines

The brute-force contour line extraction approach described above is unsatisfactory especially when the number of triangles that cross the elevation Z is much smaller than the total number of triangles in the TIN. A more efficient solution can be obtained in situations where preprocessing is allowed. Then we can build

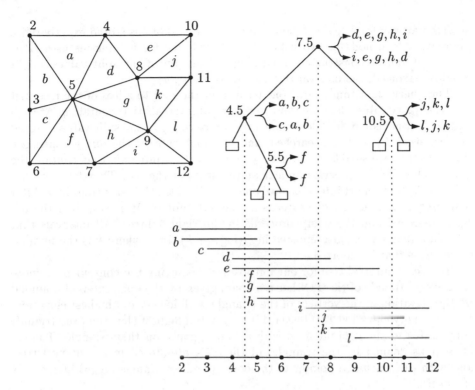

Fig. 11. An example of a TIN and the corresponding interval tree. The split value and the two lists are shown with each node, where L is the upper list and R the lower list.

a data structure and query with the elevation of which the contour lines are requested. This idea has been the basis of two different approaches, described by De Floriani et al. [37] and by the author of this survey [118]. The method is also used in the visualization of isosurfaces, the higher-dimensional counterpart of contour lines [1, 79].

We describe the *interval tree*, a geometric data structure that stores a set of intervals of the real line. It was developed by Edelsbrunner [25] and McCreight [83] independently. Here we give a brief description—see Figure 11.

Let I be a set of open intervals of the form (a, b), where $a, b \in \mathbb{R}$ and $a < b$. The interval tree for I has a root node δ that stores a split value s. Let I_{left} be the subset of intervals (a, b) for which $b \leq s$, let I_{right} be the subset of intervals (a, b) for which $a \geq s$ and let I_δ be the subset of intervals for which $a < s < b$. The subsets $I_{left}, I_{right}, I_\delta$ form a partition of I. The subset I_δ is stored in two linear lists that are associated with node δ. One list L_δ stores I_δ on increasing value of the left endpoint, and the other list R_δ stores I_δ on decreasing value of the right endpoint. If I_{left} is not empty, then the left subtree of δ is defined recursively as an interval tree on the subset I_{left}. The right subtree of δ is defined in a similar way for I_{right}. It follows that any interval of I is stored exactly twice (namely, at one node in two lists). An interval tree for n intervals uses $O(n)$ storage, it can be constructed in $O(n \log n)$ time and if the split values s split roughly balanced, the interval tree has depth $O(\log n)$.

The query algorithm follows one path from the root to a leaf of the tree. Let q be the query value, thus, we want to report all intervals that contain q. At each node δ that is visited, it is determined by comparing q to the split value s stored at δ whether L_δ or R_δ is searched, and in which subtree the query continues. If $q < s$, then we search in the list L_δ and report all intervals that contain the query value. These intervals appear at the start of the list. Therefore, we can traverse L_δ and report intervals until one is reached that doesn't contain q. After searching in L_δ, the query proceeds in the left subtree. If $q > s$, then the list R_δ is searched and the query proceeds in the right subtree. All intervals that contain a query value are reported in $O(\log n + k)$ time, where k is the number of intervals that is reported.

To use an interval tree for our purposes of retrieving the contour lines, note that every triangle of the TIN has a z-span, given by the open interval bounded by the elevation of the vertices of the triangle with lowest and highest elevation. For any query elevation Z between this lowest and highest elevation, the triangle contributes to the contour lines with a line segment on that triangle. The set of z-spans defined by the triangles of the TIN are stored in an interval tree, and with each z-span a pointer to the corresponding triangle record in the TIN structure.

Not only triangles, but also horizontal edges of the TIN can contribute to the contour lines with a line segment. The z-span of a horizontal edge is the closed interval containing a single elevation, the elevation of that edge. The interval tree can easily be adapted to store these closed intervals. Given the query elevation Z, the search in the interval tree retrieves all triangles that lie partially below and partially above Z, and all edges with elevation Z. The line segments of elevation Z on these triangles and edges together form the contour lines for elevation Z. The query time is $O(\log n + k)$, where k is the number of segments in the contour lines.

We conclude that the contour lines of any elevation on a TIN can be found in only $O(\log n + k)$ time, if we are allowed to do preprocessing and use linear additional storage.

The method that was just described computes the contour lines in unstructured format. We continue by considering how the contour lines can be found as sequences of segments, and still use the interval tree to have fast query time. We can combine the methods of the previous subsection with the interval tree just described. By postprocessing the segments of the contour lines, we can get them in structured form in additional $O(k \log k)$ time. This makes the total query time $O(\log n + k \log k)$. This time, the additional term may be significant, because in practical cases the $k \log k$ term is likely to be significantly larger than $\log n + k$.

The method with mark bits can also be combined with the interval tree. Since we stored with each interval a pointer to the corresponding triangle record in the TIN structure, we have immediate access to start tracing. One subtlety is the following. We cannot reset all mark bits after a query by traversing the whole TIN; that would blow up the query time to linear in n again. Instead, we repeat the query interval tree with the only objective to reset the mark bits. So we trace each contour line on the TIN structure again and reset all mark bits. This will double the query time but no more than that. So, to conclude we have seen that the contour lines of a query elevation Z can be obtained in structured form in $O(\log n + k)$ time, where k is the number of segments in the contour lines.

6.3 Classification

Classification is the operation of determining the elevation values of contours that are appropriate for mapping. These elevations that bound regions of the terrain in different classes can be chosen in several different ways. To mention a few, the elevations can be chosen at fixed intervals, such as 0, 500, 1000, 1500, and 2000. The classes induced are: up to 0, from 0 to 500, from 500 to 1000, and so on. One could also classify elevation data by choosing class boundaries using statistical measures, for instance at $\mu - 1.6\sigma$, $\mu - 0.8\sigma$, μ, $\mu + 0.8\sigma$, and $\mu + 1.6\sigma$, where μ is the mean elevation and σ is the standard deviation. A third way of classification is to compute class boundaries such that each class receives an equal amount of area on the map, given the number classes that can be used. Evans gives a good overview of types of classification [4, 27].

Several types of classification make use of the *density function*. It is well-known that a finite population of interval data can be described by a histogram. For continuous interval data, the density function—or frequency distribution— is the corresponding descriptive statistic. It shows how frequent each elevation occurs in the data. We study the computation of the density function of a TIN. Note that it is more appropriate to compute class intervals based on the density function than on the elevations of the vertices of the TIN. These vertices are generally not spread randomly, because large and nearly level regions are represented by only a few vertices. The elevations of these regions would be underrepresented by the set of elevations of all vertices, and an unfair classification would result (see e.g. [27, 62]). The following algorithm to compute the density function and the equal area clasification is by the author of this survey [119].

We begin with a useful observation and a straightforward algorithm. Consider just one triangle Δ in 3-space with vertices u, v, w. Assume for simplicity that $h(u) > h(v) > h(w)$, where $h(..)$ denotes the elevation of a vertex. Then the density on the triangle for a given elevation t is $l \cdot \cos(\alpha)$, where l is the length of the intersection of the triangle Δ with the plane $z = t$, and α is the angle between the normal of Δ and any horizontal plane. The density is zero for all elevations t with $t > h(u)$ or $t < h(w)$. It is given by a function f_{uv} depending linearly on t if $h(u) > t > h(v)$, and it is given by a different function f_{vw} depending linearly on t if $h(v) > t > h(w)$. So we have $f_{uv}(t) = a \cdot t + b$, where a and b depend only on the coordinates of u, v, w and thus are fixed. The same holds for f_{vw}, but with different a and b.

59

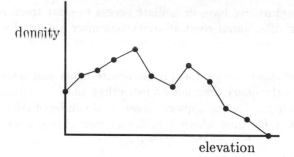

Fig. 12. Density function of a TIN.

For simplicity of exposition we assume that all vertices have different elevations. This restriction can be overcome without problems, but some care must be taken. Let $v_1, \ldots, v_n$ be the vertices of the TIN, and assume that they are sorted on decreasing elevation. This holds without loss of generality because we can simply relabel the vertices to enforce $h(v_1) > h(v_2) > \cdots > h(v_n)$. Consider the density for an elevation t, where $t \in (h(v_j), h(v_{j+1}))$. In such an open interval, the density is the sum of a set of linear functions, which is again a function linear in t. We denote the linear function that gives the density over the whole TIN in the interval $(h(v_j), h(v_{j+1}))$ by $F_j(t)$. So the linear functions $F_0, F_1, \ldots, F_n$ form the density function, where each function is only valid in its interval. By default we set $F_0(t) = 0$ and $F_n(t) = 0$ for the intervals $(h(v_1), \infty)$ and $(-\infty, h(v_n))$, because for these elevations the density is zero. One can show that the density function based on a TIN with n vertices is a piecewise linear continuous function with at most $n + 1$ pieces. The density function need not be continuous when there are vertices with the same elevation.

The straighforward algorithm to construct the density function on the TIN is the following. Sort the vertices by elevation, and for each interval $(h(v_j), h(v_{j+1}))$, determine the set of linear functions contributing to it. Then add up these linear functions to get one linear piece F_j of the density function. Since we have $O(n)$ vertices, we have $O(n)$ intervals and for each we can easily determine in $O(n)$ time which linear functions contribute. The total time taken by this algorithm is $O(n^2)$.

The efficient computation of the density function is based on the sweeping approach. We will exploit the fact that the linear function F_j can be obtained easily from the linear function F_{j-1} since the contributing f are for the larger part the same ones. We compute the summed linear functions F from top to bottom, which comes down to a sweep with a horizontal plane through the TIN. Throughout the sweep we maintain the density function of the current elevation. Using sweeping terminology, every vertex of the TIN gives rise to one event. The event list is a priority queue storing all these $O(n)$ events in order of decreasing elevation. With each event we store a pointer to the vertex of the TIN that will cause the event. The status structure is trivial: it is simply the summed linear function F for the current position of the sweep plane, and is stored in two reals.

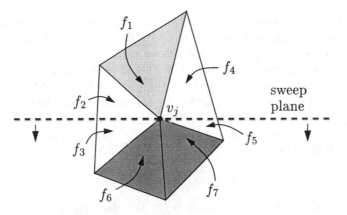

Fig. 13. Passing a vertex with the sweep plane.

The final ingredient to the sweep algorithm is handling the events. When considering how F_{j-1} should be changed to get F_j when the sweep plane passes the vertex with elevation $h(v_j)$, we must examine how the density function changes. The vertex v_j is incident to some triangles, for which it can be the highest vertex, the lowest vertex or the vertex with middle elevation. We update F_{j-1} to get F_j according to the following rules:

- For all triangles for which v_j is the lowest vertex (lightly shaded in Figure 13), we subtract from F the appropriate linear function (f_1 in Figure 13).
- For all triangles for which v_j is the highest vertex (darkly shaded), we add to F the appropriate linear function (f_6 and f_7).
- For all triangles for which v_j is the middle vertex (white in the figure), we subtract the one linear function (f_2 and f_4) and add the other (f_3 and f_5).

We don't need to precompute or store the linear functions f on each triangle to update F; the f can be obtained from the coordinates of the vertices on the TIN when the event at vertex v_j is handled. We have fast access to vertex v_j in the TIN; recall that an extra pointer was stored in the event list.

We also evaluate the function F_j at the event. The sequence of evaluations gives the breakpoints of the (piecewise linear) density function. These breakpoints are computed from right to left in Figure 12 since the sweep goes from high to low elevations.

Considering the efficiency of the algorithm, the initial sorting of the events takes $O(n \log n)$ time for a TIN with n vertices. Extraction of an event takes $O(\log n)$ time; for all events this adds up to $O(n \log n)$ time. Updating the status structure at an event v_j requires time linear in the number of triangles incident to

v_j. Summed over all vertices this is linear in n by Euler's formula. The evaluation to determine the breakpoints requires constant time per event. So in total the sweep algorithm requires $O(n \log n)$ time.

Once the density function is computed, the class intervals may be determined. Suppose as an example that the objective is to determine seven classes such that each class occupies an equivalent amount of area on an contour line map. We assume that the contour line map and the TIN have the same domain, otherwise we can clip the TIN with the domain of the contour line map before doing the sweep. The total area of the contour line map is the same as the total area under the density function and is denoted A. The area under the density function in the elevation interval $[a, b]$ is denoted $A(a, b)$. If $F(t)$ denotes the (piecewise linear) density function, then

$$A(a, b) = \int_a^b F(t)dt$$

The value of $A(a, b)$ is exactly the area for the class $[a, b]$ on the contour line map. We know the total area A and compute $A/7$, the desired area for each class. We then determine the lowest elevation such that $A/7$ of the area is below that elevation. This operation is easy by scanning over the known density function $F(t)$ from left to right and maintaining the area under $F(t)$ (this is also a kind of sweep). This gives the lowest class boundary. Continuing the scan gives all six boundaries of the seven classes in $O(n)$ time. In a similar way one can compute a non-fixed number of classes with the property that the within-class variance is less than or equal to a certain threshold, for each class. Finally, the density function can be used class interval selection by natural breaks in the data: They are the local minima of $F(t)$. We refer to Burrough [4] and Evans [27] for other classification schemes.

The sweep algorithm that was described for the density function requires linear working storage to store all the events. For most realistic terrains, the working storage can be reduced considerably. We make the following simple observation. Every vertex except the local maxima—the peaks—have a higher neighbor in the TIN. So we can initialize the event list with the local maxima only. When the event at a vertex v is handled, we insert all lower neighbors of v in the event list. This guarantees that every event is present in the event list when the sweep plane reaches it. The storage required by the algorithm is linear in the sum of the number of local maxima and the number of edges in the largest complexity cross-section.

7 Topographic Features

Geomorphologists study the shape of the land, and what processes influence it [56, 60, 104]. The quantification of the shape of the land is necessary in order automatically recognize certain features of shape. This on its turn may lead to a partition of the land into regions where for instance erosional processes have the same behavior.

Terrain features can be zero-, one-, or two-dimensional. We discuss the most important ones in the next subsections. Then we treat slope and aspect defined on a terrain.

7.1 Points on Terrains

Any point on a terrain has a certain elevation. When we also consider the neighborhood of a point, the slope and aspect at it can be defined. The slope (also called gradient) at a point is the maximum ratio of change in elevation and change of position in the xy-plane at that point. Mathematically, it is the maximum value of the directional derivative at that point (maximized over the direction). Note that the slope of an elevation model is an elevation model itself. Therefore it can be visualized, for example, as a contour line map by classification of the slopes.

The aspect or exposure of a point in an elevation model is the compass direction in which the directional derivative is maximum. With an aspect map it is easy to see which hill sides face to the south, for instance. The aspect of an elevation model is not an elevation model. Instead, it is a bivariate function that maps $\mathbb{R} \times \mathbb{R}$ to the circular scale $(-\pi, \pi]$. It is undefined at points that lie on a horizontal part of the terrain. The combination of slope and aspect is needed to produce hill shading on maps.

On a terrain there are certain special points that are more important or characteristic than others. These are the peaks, the pits, and the passes. The latter are also called saddles. A peak is a point such that in some neighborhood of it, there is no higher point. Similarly, in some neighborhood of a pit there is no lower point. A pass is a point where locally, four (or more) different parts of the contour lines meet. These definitions don't specify what neighborhood should be taken, and what should be considered a peak when there is a whole region of equal elevation points. Choices of this type have to be taken depending on the application.

Peaks, pits, and passes are elements that are used to describe terrain form. They are the basis of so-called surface networks and their relatives [96, 122, 127]. One such relative, the Warntz network, can be obtained by first identifying all passes, and from there, traverse the terrain in the directions of steepest ascent and steepest descent until peaks or pits are reached. The paths traversed together define a partition of the terrain into regions, of which one can hope that they have similar geomorphological features. What should be considered a pass for this idea to work well has considerable influence on the output [126].

7.2 Valleys and Ridges

Valleys on a TIN are 1-dimensional features. They are usually defined as the edges for which the two incident triangles each have an outward normal vector whose vertical projection on the xy-plane is directed towards the valley edge, when it is also projected vertically [42, 114]. There are potential problems with this definition if the outward normal vectors are parallel to the edge, in the

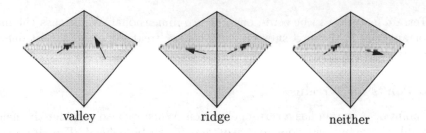

valley ridge neither

Fig. 14. Three times two adjacent triangles and their outward normal vectors.

projection. We could define an edge to be a valley edge if at least one incident triangle has its outward normal towards the edge, and the other one has its outward normal towards the edge or parallel to it. A similar definition can be made for ridge edges, where the outward normals lead away from the edge.

All of these definitions have the disadvantage that valley lines may be interrupted, even though the valley itself seems to be just one feature on the terrain. When we discuss drainage networks and basins, we'll see an alternative definition that can be used for valleys and ridges and avoids interruption as much as possible.

7.3 Curvature

The two-dimensional terrain specific features are obtained by considering *plan curvature* and *profile curvature*, being the curvature in a horizontal and vertical cross-section of the terrain. These curvatures specify whether the terrain is convex, flat, or concave in the cross-section. For smooth surfaces the convexity depends on the sign of the second derivative in the cross-section. One could partition the terrain into regions where the two curvatures are within certain boundaries, a type of classification [29, 48, 92, 110]. For instance, for profile convexity the bounds $-0.1\,°/m$ and $+0.1\,°/m$ are used. The terrain elements as in Figure 15 are the ones that can be obtained in such a classification. Profile curvature is related to the position on a hillside. Most hillsides are profile convex near the top and profile concave near the foot.

On gridded elevation models the curvature of a pixel can be determined by considering the 3×3 window of pixels, choosing a suitable interpolator for the nine pixels, and computing the plan and profile curvature of the interpolated surface at the center [48, 92].

On TINs a simple approach has been suggested: every edge of the TIN can be seen as flat, convex, or concave. Then triangles can be classified according to the type of incident edges. A convex region is one consisting only of triangles of which all incident edges are convex. A similar statement can be made for concave and flat regions. The triangles that are incident to triangles of different types are defined as saddle triangles. These definitions allow the regions of uniform

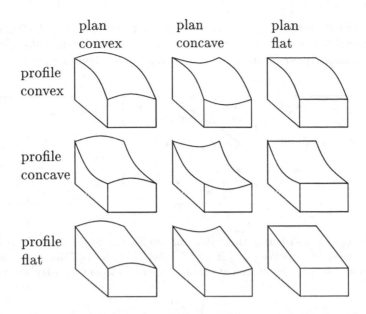

Fig. 15. Nine landform elements classified by plan and profile curvature.

curvature to be traced out on the TIN by straightforward graph traversal, in linear time [29].

A disadvantage of the TIN approach above is that it doesn't distinguish between plan and profile curvature. A region that is profile convex and plan concave is defined saddle, and so is a region that is profile concave and plan convex. It is possible to obtain a TIN curvature classification that includes plan and profile curvature. We define the plan curvature at a vertex v as follows. If v is a peak, the plan curvature is convex. If v is a pit, the plan curvature is concave. If v is a saddle, the plan curvature is undefined. In all other cases, the plan curvature of v is determined by the contour line through v. Vertex v is incident to two line segments s and s' on that contour line (usually across TIN triangles). When traversing the contour line with the higher terrain to the left and the lower terrain to the right, then v is plan convex if the contour line makes a left turn at v. If it makes a right turn, v is plan concave, and if it makes no turn (or a turn below some threshold), then it is plan flat. Note that saddle vertices have to be excluded because four line segments of the contour line meet at a saddle vertex.

We define the profile curvature at a vertex v as follows. If v is a peak, the profile curvature is convex. If v is a pit, the profile curvature is concave. If v is a saddle, the profile curvature is undefined. To define the profile curvature at another vertex v we must select a suitable vertical plane through v. Consider again the line segments s and s' on the contour line through v. We take the vertical plane through v that separates s and s', and makes an equivalent angle with them. This plane is in a sense perpendicular to the tangent of the contour

65

line at v, and therefore a reasonable choice. Next we consider the intersection of the vertical plane with the terrain at v, which is the profile. We can define concexity, concavity, and flatness in the profile in the obvious way.

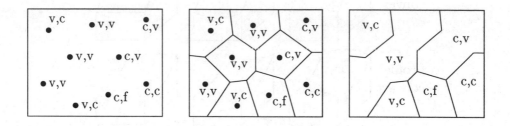

Fig. 16. Vertices labelled with their plan and profile curvature, where v is used for convex, c is used for concave and f is used for flat. The Voronoi diagram is shown in the middle, and boundaries of similar regions are erased to obtain a terrain partition.

Now we know for most of the TIN vertices their plan and profile curvature, see Figure 16. To obtain regions from this information we can use interpolation. For any point on a triangle or edge, we define its plan and profile curvature to be the same as the curvature of the nearest vertex (not a saddle). This nearest-neighbor interpolation approach induces a Voronoi diagram on the TIN vertices, excluding the saddle vertices. Adjacent regions that have the same curvature labels can be merged by erasing their common boundary. The result is a terrain partition into regions of uniform curvature both in plan and profile. Since the Voronoi diagram of n points can be computed in $O(n \log n)$ time [18, 91, 98], the terrain partition requires $O(n \log n)$ time to compute as well.

The approach can be supplemented with a scale-dependent parameter. For example, consider the plan curvature at a vertex v again. Instead of looking at the angle of the line segments on the contour line incident to v, we may locate two points p and p' on this contour line at a certain distance from v. This distance is the scale-dependent parameter. Then we determine the angle $\angle pvp'$, and decide upon the plan curvature. This refined approach may for instance cause small concavities in a convex region to be eliminated.

7.4 Drainage Information

Any terrain induces a more or less natural flow of water on it. For instance, water always flows downward, following the direction of gravity, and water collects into streams. These streams join and form rivers. The more downstream, the bigger a river becomes. It is possible to predict from an elevation model where the streams will be. The collection of all streams and rivers is called the *drainage network*. In this section we only consider how the form of the terrain influences the flow of water, and review a few possible definitions of the drainage network and related concepts. To find a definition that corresponds to the drainage network in reality is a problem that requires various types of data of the terrain. The area of hydrology also includes issues like surface permeability, subsurface flow, evaporation, and more [80, 85].

Generally, the drainage network can be seen as a group of connected acyclic networks (a forest of trees on the graph sense) of which the links are directed to the pits of the terrain, see Figure 17. Each connected network is also called a *river system*, and the part of the terrain that drains into some river system is called a *drainage basin* (or *basin*) of that system.

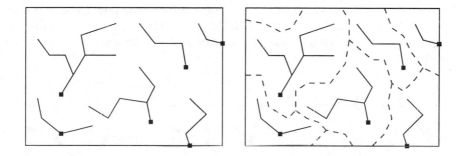

Fig. 17. Left, the drainage network as a forest of trees rooted at the pits. Right, the basins of each river system.

The drainage network on a grid. One of the first attempts to compute the drainage network on a terrain was by Peucker and Douglas [94]. Their algorithm works on a grid is extremely simple: slide a 2×2 window over the grid and flag the highest pixel in the window. After all subwindows have been treated, the unflagged pixels together form the drainage network. The maps produced by this method suffer from isolated dots and interrupted channels.

A more advanced approach was taken by O'Callaghan and Mark [90] and Mark [82], who also modelled the *accumulation* of water flowing in the terrain (they credit Speight [110] for this idea). Define for every pixel the drain neighbor to be one of the eight neighboring pixels to which the steepest descent is greatest. This drain neighbor is assumed to be unique. A pit doesn't have a descent direction and therefore no drain neighbor. Then assign every pixel one unit of water, and trace all units on the grid downward to the drain neighbors until they end in the pits. By maintaining counters to determine for every pixel how many units of water flow through it, the drainage network can be defined. It consists of all pixels for which the counter is higher than some well-chosen threshold. By treating the pixels in order of decreasing elevation, the method requires $O(n^2 \log n)$ time on an $n \times n$ grid (needed for the sorting). The method also requires quadratic additional storage.

The accumulation idea solves the problem of interrupted channels. If some pixel belongs to the drainage network because its counter exceeds the threshold, then the whole path along drain neighbors to a pit must also be part of the drainage network. The accumulation idea also helps to define drainage basins. Since the path from any pixel can be traced to a pit, it is possible to determine what pixels drain into any pit. So it is possible to outline the basins, the parts of the terrain drain into one single pit.

The drainage network on a TIN. On TINs, a definition of the drainage network has been suggested by Frank et al. [42]. They define the drainage network to consist of all valley edges of the TIN. This definition suffers from the possible interruption of streams, which can end in points other than pits, as observed by Theobald and Goodchild [114]. See for instance Figure 18. Furthermore, there is no concept of flow, so basins cannot be defined as an extension of the model for the drainage network.

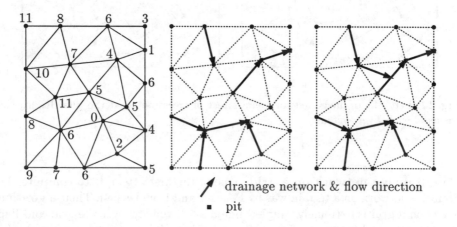

Fig. 18. Left, a TIN with elevations of the vertices. Middle, the drainage network by the definition of Frank et al. Right, the drainage network by the definition of Yu et al.

Yu et al. [129] showed recently that the idea of accumulation can be applied to TINs as well. On a TIN, there are no pixels to assign units of water to, and it isn't a good idea either to assign water to complete triangles. But the direction of flow can still be defined conveniently of a point on a TIN as the direction of steepest descent. In the interior of the triangles this direction is unique, but on edges and vertices we may need to choose a direction if there is more than one direction of steepest descent. Once the direction of flow is defined, flow paths can be traced and one can discover where flow paths join. It is natural to define for each point on the terrain the area of the region from which the flow paths go through that point. For many points on the TIN, this area is zero because there is no 2-dimensional region. But there are also points that receive water from a region with positive area. Define the drainage network to be those points on the terrain that receive water from a region whose area exceeds a certain threshold

68

area. We'll study the case where the threshold is set to 0. It is clear that the drainge network we obtain will include all drainage networks for larger threshold values.

One can show a number of properties of the drainage network defined this way. Most importantly, the drainage network consists of all valley edges, and furthermore, exactly of all flow paths from their lower vertices. This follows from the fact that water can only start accumulating at valley edges. The drainage network will have merge points where two or more streams join and continue together. These merge points are either vertices of the TIN, or points on valley edges. Since we assumed that at every point of the TIN the direction of flow is unique, streams cannot split.

When comparing the definitions of Frank et al. and Yu et al. we can easily observe that under the former definition, the drainage network has complexity at most linear in the number of edges of the TIN. It is considerably less obvious what the size of the drainage network is under the second definition. De Berg et al. [16] have shown that it is at most cubic in the number of edges of the TIN, and that the cubic bound is tight for some artificially constructed TINs. Whether the cubic worst case bound has any relevance in practice is doubtful. An emperical study on this issue has been done and the actual size on real terrain data appears to be roughly 20% more than under the definition by Frank et al. The tests were done on six different terrains represented by TINs with up to 12,000 vertices [115].

To compute the drainage network by the definition of Yu et al., we first identify all valley edges, and then follow the flow paths of their lower vertices to the pits. Since flow paths can merge, we can stop tracing any flow path from a point where another flow path has already gone through. So, whenever a flow path is traced, it is marked on the terrain itself to make sure that the same flow path isn't traced again and again. The resulting algorithm requires time $O(n + k)$, where k is the complexity of the drainage network.

Drainage basins, catchment areas, generalization, and spurious pits. The study of drainage on a terrain using the definition of Yu et al. [129] can be extended in various ways. Since it incorporates the notion of flow and accumulation, it becomes possible to determine the basins of the different river systems, and also the area of the terrain that drains into each river system. For any point on the terrain, we define the *catchment area* to be the part of the terrain that drains through that point, eventually. The definition above of the drainage network includes exactly the points that have a catchment area that is 2-dimensional, no matter how small the area. As a consequence, many small streams will be included in the drainage network.

We can also define a *generalized* drainage network by selecting the points of which the catchment area has at least a certain size. Since it is possible to determine the size of the catchment area for each point on the terrain—in particular, on the original drainage network—we can also compute the generalized drainage network. Other methods to compute the generalized drainage network include using stream orders to decide which streams can be omitted, or generalizing the terrain and then computing the drainage network [124].

Terrains that don't cover a large area of land usually don't have many pits on them. When a drainage network algorithm detects a number of pits, some of them will actually be the result of imprecision in the data acquisition, or errors. Other pits that occur often are not the end of a river system, but lakes may start to form that overflow into another river system, making the two connected. Avoiding pits on a terrain is also called *drainage enforcement* or *spurious pit removal* [63, 64, 82]. For every pit, there is one pass where water will overflow first when the pit is filled. The overflowing water may start a new stream that joins some other river system. Which pits should be removed can be based on the pit perimeter, the elevation difference between pit and pass, or the lake capacity. The pit perimeter is the length of the polygon that lies on the contour line through the pass and containing the pit.

8 Miscellaneous Applications

In this section we'll mention some other applications and algorithms on terrains with references. It is meant rather as an annotated bibliography than as a survey.

8.1 Paths in Terrains

Planning routes or networks in mountainous regions is of interest to civil engineers. Problems like determining the best road to connect two places on a terrain, best location of a bridge over a valley, and the like are sometimes addressed in a geographic information system. There can be several optimization criteria, like minimizing the length of a route, finding a route that stays as low as possible, with minimum cost for construction, and optimizing the resulting travel time. Algorithms for abstract versions of these problems have been given mostly in the computational geometry literature [7, 17, 86, 87, 116, 117].

8.2 Viewshed Analysis

Viewshed analysis is the general name for visibility problems on terrains. The standard viewshed analysis problem is simply the question: "What parts of the terrain are visible from a specific point?" Optimization problems of visibility are minimization of visibility (horizon pollution of planned buildings, routes not visible from enemies) and maximization of visibility (observation posts for fire detection, scenic routes). Viewshed problems have been studied in a theoretical setting, but also in more practical situations [35, 75, 109].

As a measure for the area of visible regions on a terrain, the *visibility index* can be used on grid models. The visibility index of a given pixel is the number of other pixels that are visible from the given pixel [44, 113, 119].

The computation of horizons is related to viewshed analysis. One can make a distinction between visibility above the last horizon (against the sky), visibility agains a local horizon, and visibility against a hillside, and one can also compute values for non-visible points that represent the elevation increase needed to make the point visible [32, 33].

Some other issues of visibility on terrains include moving points of view, networks of interconnected sites, and the error present in visibility analyses [2, 13, 31, 39].

8.3 Temporal Aspects of Terrains

Modelling time in a geographic information system is a topic that has received a lot of attention recently [72]. The mapping of time can be one of the themes on a static map, but it is also possible to use dynamic or animated maps for the visualization. Here a sequence of maps of the same area and the same theme is used, where each map represents the situation at a fixed moment in time. The sequence as a whole can be used to animate the changes in the mapped themes over time [24, 67, 68, 70]. An example is moving pressure fronts in weather reports. This example shows that temporal aspects and animation can be issues for elevation data as well.

Suppose that a sequence of terrains is given, each representing the terrain at a fixed moment in time. To animate the change of the terrain one could show the terrain in perspective view or by contour lines. Since the data usually is available only at a discrete set of moments, interpolation between two terrains at consecutive moments becomes necessary. If the terrains are represented by TINs, the problem comes in different forms. Firstly, it can be the case that the sequence of TINs has the same underlying set of vertices and triangulation. The only differences are the elevations of the vertices. The second—and more difficult—version of the problem has different vertex sets or triangulations for the terrains in the sequence. It can be important to use a dynamic terrain model that allows for the addition and removal of vertices and edges on the TIN [52, 61].

One more use of TINs in geographic processing is the simulation of physical processes the influence certain terrain features. For example, one can study drainage on a terrain after storms and rain showers, and analyze how long it will take before the default drainage situation is restored [107].

8.4 Statistical Analysis of Elevation Data

Statistical analysis of elevation data obtained at point samples is common in the earth sciences. The field is also called geostatistics [14, 15, 62, 123]. Important concepts include spatial interpolation, summarizing the data by mean, standard deviation, skewness, auto-correlation, and so on. One of the most important purposes of the analysis is the prediction of data at places where no measurements were made.

71

9 Conclusions

This survey has explained a number of concepts on terrains, and some algorithms for various computations. The emphasis has been on TIN algorithms, because the TIN model for terrains is more elegant than the grid and contour line models. A common argument to use grids is the simplicity of the algorithms. However, the current trends in GIS research and in the field of computational geometry have shown that algorithms on TINs need not be difficult either. More programming effort is required, but this need not outweigh the advantages that TINs have to offer. We won't repeat arguments in the raster-vector debate; a summary of algorithmic methods and specific algorithms for TINs is useful in any case. The search for efficient algorithms on terrains is an interesting area of research where the GIS developers, GIS researchers, and computational geometers can work together to develop a variety of elegant and efficient solutions to practical problems on terrains. The analysis of efficiency of these solutions should be based on realistic assumptions on terrains.

Acknowledgements

Work of the author is partially supported by the ESPRIT IV LTR Project No. 21957 (CGAL).

References

1. C.L. Bajaj, V. Pascucci, and D.R. Schikore. Fast isocontouring for improved interactivity. In *Proc. IEEE Visualization*, 1996.
2. M. Bern, D. Dobkin, D. Eppstein, and R. Grossman. Visibility with a moving point of view. *Algorithmica*, 11:360–378, 1994.
3. P. Bose and L. Devroye. Intersections with random geometric objects. Manuscript.
4. P. A. Burrough. *Principles of Geographical Information Systems for Land Resourses Assessment*. Oxford University Press, New York, 1986.
5. J.R. Carter. The effect of data precision on the calculation of slope and aspect using gridded DEMs. *Cartographica*, 29:22–34, 1992.
6. K.-T. Chang and B.-W. Tsai. The effect of DEM resolution on slope and aspect mapping. *Cartography and Geographic Information Systems*, 18:69–77, 1991.
7. J. Chen and Y. Han. Shortest paths on a polyhedron. In *Proc. 6th Annu. ACM Sympos. Comput. Geom.*, pages 360–369, 1990.
8. Z.-T. Chen and J.A. Guevara. Systematic selection of very important points (vip) from digital terrain model for constructing triangular irregular networks. In *Proc. Auto-Carto 8*, pages 50–56, 1987.
9. L. P. Chew. Constrained Delaunay triangulations. *Algorithmica*, 4:97–108, 1989.
10. A.H.J. Christensen. Fitting a triangulation to contours. In *Proc. Auto-Carto 8*, pages 57–67, 1987.
11. K. C. Clarke. *Analytical and Computer Cartography*. Prentice Hall, Englewood Cliffs, NJ, 2nd edition, 1995.
12. K. L. Clarkson and P. W. Shor. Applications of random sampling in computational geometry, II. *Discrete Comput. Geom.*, 4:387–421, 1989.

13. R. Cole and M. Sharir. Visibility problems for polyhedral terrains. *J. Symbolic Comput.*, 7:11–30, 1989.
14. N.A.C. Cressie. *Statistics for Spatial Data*. Wiley, New York, 1991.
15. J.C. Davis. *Statistics and Data Analysis in Geology*. Wiley, New York, 2nd edition, 1986.
16. M. de Berg, P. Bose, K. Dobrint, M. van Kreveld, M. Overmars, M. de Groot, T. Roos, J. Snoeyink, and S. Yu. The complexity of rivers in triangulated terrains. In *Proc. 8th Canad. Conf. Comput. Geom.*, pages 325–330, 1996.
17. M. de Berg and M. van Kreveld. Trekking in the Alps without freezing or getting tired. *Algorithmica*, 18:306–323, 1997.
18. M. de Berg, M. van Kreveld, M. Overmars, and O. Schwarzkopf. *Computational Geometry – Algorithms and Applications*. Springer-Verlag, Berlin, 1997.
19. M. de Berg, M. van Kreveld, R. van Oostrum, and M. Overmars. Simple traversal of a subdivision without extra storage. *Int. J. of GIS*, 11:359–373, 1997.
20. Mark de Berg and Katrin Dobrindt. On levels of detail in terrains. In *Proc. 11th Annu. ACM Sympos. Comput. Geom.*, pages C26–C27, 1995.
21. L. De Floriani. A pyramidal data structure for triangle-based surface description. *IEEE Comput. Graph. Appl.*, 9(2):67–78, March 1989.
22. L. De Floriani and E. Puppo. A survey of constrained Delaunay triangulation algorithms for surface representaion. In G. G. Pieroni, editor, *Issues on Machine Vision*, pages 95–104. Springer-Verlag, New York, NY, 1989.
23. Leila De Floriani, Bianca Falcidieno, George Nagy, and Caterina Pienovi. Hierarchical structure for surface approximation. *Comput. Graph. (UK)*, 8(2):183–193, 1984.
24. D. DiBiase, A.M. MacEachren, J.B. Krygier, and C. Reeves. Animation and the role of map design in scientific visualization. *Cartography and Geographic Information Systems*, 19:201–214, 1992.
25. H. Edelsbrunner. Dynamic data structures for orthogonal intersection queries. Report F59, Inst. Informationsverarb., Tech. Univ. Graz, Graz, Austria, 1980.
26. H. Edelsbrunner, L. J. Guibas, and J. Stolfi. Optimal point location in a monotone subdivision. *SIAM J. Comput.*, 15:317–340, 1986.
27. I. S. Evans. The selection of class intervals. *Trans. Inst. Br. Geogrs.*, 2:98–124, 1977.
28. B. Falcidieno and C. Pienovi. A feature-based approach to terrain surface approximation. In *Proc. 4th Int. Symp. on Spatial Data Handling*, pages 190–199, 1990.
29. B. Falcidieno and M. Spagnuolo. A new method for the characterization of topographic surfaces. *Int. J. of GIS*, 5:397–412, 1991.
30. Ulrich Finke and Klaus Hinrichs. Overlaying simply connected planar subdivisions in linear time. In *Proc. 11th Annu. ACM Sympos. Comput. Geom.*, pages 119–126, 1995.
31. P. F. Fisher. Algorithm and implementation uncertainty in viewshed analysis. *Internat. J. Geogr. Inform. Syst.*, 7:331–347, 1993.
32. P. F. Fisher. Stretching the viewshed. In *Proc. 6th Internat. Sympos. Spatial Data Handling*, pages 725–738, 1994.
33. P. F. Fisher. Reconsideration of the viewshed function in terrain modelling. *Geogr. Syst.*, 3:33–58, 1996.
34. P.-O. Fjällström. Polyhedral approximation of bivariate functions. In *Proc. 3rd Canad. Conf. Comput. Geom.*, pages 187–190, 1991.

35. L. De Floriani, B. Falcidieno, C. Pienovi, D. Allen, and G. Nagy. A visibility-based model for terrain features. In *Proc. 2nd Int. Symp. on Spatial Data Handling*, pages 235–250, 1986.

36. L. De Floriani, P. Marzano, and E. Puppo. Hierarchical terrain models: survey and formalization. In *Proc. ACM Symp. on Applied Computing*, 1994.

37. L. De Floriani, D. Mirra, and E. Puppo. Extracting contour lines from a hierarchical surface model. In *Eurographics'93*, volume 12, pages 249–260, 1993.

38. L. De Floriani and E. Puppo. A hierachical triangle-based model for terrain description. In *Theories and Methods of Spatio-Temporal Reasoning in Geographic Space, proceedings*, volume 639 of *Lecture Notes in Computer Science*, pages 236–251, Berlin, 1992. Springer-Verlag.

39. L. De Floriani, E. Puppo, and G. Nagy. Computing a line-of-sight network on a terrain model. In *Proc. 5th Int. Symp. on Spatial Data Handling*, pages 672–681, 1992.

40. J. D. Foley, A. van Dam, S. K. Feiner, and J. F. Hughes. *Computer Graphics: Principles and Practice*. Addison-Wesley, Reading, MA, 1990.

41. R. J. Fowler and J. J. Little. Automatic extraction of irregular network digital terrain models. *Comput. Graph.*, 13(2):199–207, August 1979.

42. A.U. Frank, B. Palmer, and V.B. Robinson. Formal methods for the accurate definition of some fundamental terms in physical geography. In *Proc. 2nd Int. Symp. on Spatial Data Handling*, pages 583–599, 1986.

43. Wm Randolph Franklin. Compressing elevation data. In *Advances in Spatial Databases (SSD'95)*, number 951 in Lecture Notes in Computer Science, pages 385–404, Berlin, 1995. Springer-Verlag.

44. Wm. Randolph Franklin and C. K. Ray. Higher isn't necessarily better: Visibility algorithms and experiments. In *Proc. 6th Internat. Sympos. Spatial Data Handling*, pages 751–763, 1994.

45. Wm Randolph Franklin and A. Said. Lossy compression of elevation data. In *Proc. 7th Int. Symp. on Spatial Data Handling*, pages 8B.29–8B.41, 1996.

46. H. Freeman and S.P. Morse. On searching a contour map for a given terrain elevation profile. *J. of The Franklin Institute*, 284:1–25, 1967.

47. A.B. García, C.G. Nicieza, J.B.O. Meré, and A.M. Díaz. A contour line based triangulation algorithm. In *Proc. 5th Int. Symp. on Spatial Data Handling*, pages 411–421, 1992.

48. P.K. Garg and A.R. Harrison. Quantitative representation of land-surface morphology from digital elevation models. In *Proc. 4th Int. Symp. on Spatial Data Handling*, pages 273–284, 1990.

49. M. Garland and P.S. Heckbert. Fast polygonal approximation of terrains and height fields. Technical Report CMU-CS-95-181, Carnegie Mellon University, 1995.

50. C. Gold. The practical generation and use of geograhic triangular element data. In *Harvard Papers on Geographic Information Systems*, volume 5. 1978.

51. C. Gold and S. Cormack. Spatially ordered networks and topographic reconstructions. In *Proc. 2nd Int. Sympos. Spatial Data Handling*, pages 74–85, 1986.

52. C. Gold and T. Roos. Surface modelling with guaranteed consistency – an object-based approach. In *IGIS'94 Geographic Information Systems*, number 884 in Lecture Notes in Computer Science, pages 70–87, Berlin, 1994. Springer-Verlag.

53. C. M. Gold, T. D. Charters, and J. Ramsden. Automated contour mapping using triangular element data structures and an interpolant over each irregular triangular domain. *Comput. Graph.*, 11(2):170–175, 1977.

54. C.M. Gold. Neighbours, adjacency and theft – the Voronoi process for spatial analysis. In *Proc. 1st Eur. Conf. on Geographical Information Systems*, 1990.

55. C.M. Gold and U. Maydell. Triangulation and spatial ordering in computer carthography. In *Proc. Canad. Cartographic Association Annual Meeting*, pages 69–81, 1978.

56. A. Goudie, editor. *Geomorphological Techniques*. George Allen & Unwin, London, 1981.

57. L. J. Guibas, D. E. Knuth, and M. Sharir. Randomized incremental construction of Delaunay and Voronoi diagrams. *Algorithmica*, 7:381–413, 1992.

58. L. J. Guibas and R. Seidel. Computing convolutions by reciprocal search. *Discrete Comput. Geom.*, 2:175–193, 1987.

59. L. J. Guibas and J. Stolfi. Primitives for the manipulation of general subdivisions and the computation of Voronoi diagrams. *ACM Trans. Graph.*, 4:74–123, 1985.

60. M.G. Hart. *Geomorphology – pure and applied*. George Allen & Unwin, London, 1986.

61. M. Heller. Triangulation algorithms for adaptive terrain modeling. In *Proc. 4th Int. Symp. on Spatial Data Handling*, pages 163–174, 1990.

62. E. H. Isaaks and R. M Srivastava. *An Introduction to Applied Geostatistics*. Oxford University Press, New York, 1989.

63. S.K. Jenson. Automated derivation of hydrologic basin characteristics from digital elevation model data. In *Proc. ASP/ACSM*, pages 301–310, 1985.

64. S.K. Jenson and C.M. Trautwein. Methods and applications in surface depression analysis. In *Proc. Auto-Carto 8*, pages 137–144, 1987.

65. C.B. Jones, J.M. Ware, and G.L. Bundy. Multiscale spatial modelling with triangulated surfaces. In *Proc. 5th Int. Symp. on Spatial Data Handling*, pages 612–621, 1992.

66. D. G. Kirkpatrick. Optimal search in planar subdivisions. *SIAM J. Comput.*, 12:28–35, 1983.

67. A. Kousoulakou and M.-J. Kraak. Spatio-temporal maps and cartographic communication. *Cartographic Journal*, 29:101–108, 1992.

68. M.-J. Kraak and A.M. MacEachren. Visualization of the temporal component of spatial data. In *Proc. 6th Int. Symp. on Spatial Data Handling*, pages 391–409, 1994.

69. M.-J. Kraak and F.J. Ormeling. *Cartography – visualization of spatial data*. Longman, Harlow, 1996.

70. M.-J. Kraak and E. Verbree. Tetrahedrons and animated maps in 2d and 3d space. In *Proc. 5th Int. Symp. on Spatial Data Handling*, pages 63–71, 1992.

71. I.S. Kweon and T. Kanade. Extracting topographic terrain features from elevation maps. *CVGIP: Image Understanding*, 59:171–182, 1994.

72. G. Langran. *Time in Geographic Information Systems*. Taylor & Francis, London, 1992.

73. Robert Laurini and Derek Thompson. *Fundamentals of Spatial Information Systems*. Academic Press, Boston, MA, 1992.

74. D. T. Lee and B. J. Schachter. Two algorithms for constructing a Delaunay triangulation. *Internat. J. Comput. Inform. Sci.*, 9:219–242, 1980.

75. J. Lee. Analyses of visibility sites on topographic surfaces. *Internat. J. Geogr. Inform. Syst.*, 5:413–430, 1991.

76. J. Lee, P.K. Snyder, and P.F. Fisher. Modelling the effect of data errors on feature extraction from digital elevation models. *Photogrammatic Engineering and Remote Sensing*, 58:1461–1467, 1993.

77. Jay Lee. A drop heuristic conversion method for extracting irregular network for digital elevation models. In *GIS/LIS '89 Proc.*, volume 1, pages 30–39. American Congress on Surveying and Mapping, November 1989.

78. Jay Lee. Comparison of existing methods for building triangular irregular network models of terrain from grid digital elevation models. *Internat. J. Geogr. Inform. Syst.*, 5(3):267–285, July-September 1991.

79. Y. Livnat, H.-W. Shen, and C.R. Johnson. A near optimal isosurface extraction algorithm using the span space. *IEEE Transactions on Visualization and Computer Graphics*, 2:73–84, 1996.

80. D.R. Maidment. GIS and hydologic modeling. In M.F. Goodchild, B.O. Parks, and L.T. Steyaert, editors, *Environmental Modeling with GIS*, pages 147–167. Oxford University Press, New York, 1993.

81. H. G. Mairson and J. Stolfi. Reporting and counting intersections between two sets of line segments. In R. A. Earnshaw, editor, *Theoretical Foundations of Computer Graphics and CAD*, volume 40 of *NATO ASI Series F*, pages 307–325. Springer-Verlag, Berlin, West Germany, 1988.

82. D.M. Mark. Automated detection of drainage networks from digital elevation models. *Cartographica*, 21:168–178, 1984.

83. E. M. McCreight. Priority search trees. *SIAM J. Comput.*, 14:257–276, 1985.

84. A. M. J. Meijerink, H. A. M. de Brouwer, C. M. Mannaerts, and C. R. Valenzuela. *Introduction to the Use of Geographic Information Systems for Practical Hydrology*. Number 23 in ITC Publications. ITC, Enschede, 1994.

85. C. Mitchell. *Terrain Evaluation*. Longman, Harlow, 2nd edition, 1991.

86. J. S. B. Mitchell. An algorithmic approach to some problems in terrain navigation. *Artif. Intell.*, 37:171–201, 1988.

87. J. S. B. Mitchell, D. M. Mount, and C. H. Papadimitriou. The discrete geodesic problem. *SIAM J. Comput.*, 16:647–668, 1987.

88. Ernst P. Mücke, Isaac Saias, and Binhai Zhu. Fast randomized point location without preprocessing in two- and three-dimensional Delaunay triangulations. In *Proc. 12th Annu. ACM Sympos. Comput. Geom.*, pages 274–283, 1996.

89. J. Nievergelt and F. P. Preparata. Plane-sweep algorithms for intersecting geometric figures. *Commun. ACM*, 25:739–747, 1982.

90. J.F. O'Callaghan and D.M. Mark. The extraction of drainage networks from digital elevation data. *Computer Vision, Graphics, and Image Processing*, 28:323–344, 1984.

91. J. O'Rourke. *Computational Geometry in C*. Cambridge Univ. Press, NY, 1994.

92. D.J. Pennock, B.J. Zebarth, and E. de Jong. Landform classification and soil distribution in hummocky terrain, Saskatchewan, Canada. *Geoderma*, 40:297–315, 1987.

93. T.K. Peucker. Data structures for digital terrain modules: Discussion and comparison. In *Harvard Papers on Geographic Information Systems*, volume 5. 1978.

94. T.K. Peucker and D.H. Douglas. Detection of surface-specific points by local parallel processing of discrete terrain elevation data. *Computer Vision, Graphics, and Image Processing*, 4:375–387, 1975.

95. T.K. Peucker, R.J. Fowler, J.J. Little, and D.M. Mark. The triangulated irregular network. In *Proc. DTM Symp. Am. Soc. of Photogrammetry—Am. Congress on Survey and Mapping*, pages 24–31, 1978.

96. J.L. Pfaltz. Surface networks. *Geographical Analysis*, 8:77–93, 1976.

97. F. P. Preparata and D. E. Muller. Finding the intersection of n half-spaces in time $O(n \log n)$. *Theoret. Comput. Sci.*, 8:45–55, 1979.

98. F. P. Preparata and M. I. Shamos. *Computational Geometry: An Introduction.* Springer-Verlag, New York, NY, 1985.

99. M. Sambridge, J. Braun, and H. McQueen. Geophysical parameterization and interpolation of irregular data using natural neighbours. *Geophys. J. Int.*, 122:837–857, 1995.

100. H. Samet. *The Design and Analysis of Spatial Data Structures.* Addison-Wesley, Reading, MA, 1990.

101. N. Sarnak and R. E. Tarjan. Planar point location using persistent search trees. *Commun. ACM*, 29:669–679, 1986.

102. L. Scarlatos. A compact terrain model based on critical topographic features. In *Proc. Auto-Carto 9*, pages 146–155, 1989.

103. Lori Scarlatos and Theo Pavlidis. Adaptive hierarchical triangulation. In *Proc. 10th Internat. Sympos. Comput.-Assist. Cartog.*, volume 6 of *Technical Papers 1991 ACSM-ASPRS Annual Convention*, pages 234–246, 1991.

104. A.E. Scheidegger. *Theoretical Geomorphology.* Springer-Verlag, Berlin, 3rd edition, 1991.

105. R. Seidel. A simple and fast incremental randomized algorithm for computing trapezoidal decompositions and for triangulating polygons. *Comput. Geom. Theory Appl.*, 1:51–64, 1991.

106. R. Sibson. A brief description of natural neighbour interpolation In Vic Barnot, editor, *Interpreting Multivariate Data*, pages 21–36. Wiley, Chichester, 1981.

107. A.T. Silfer, G.J. Kinn, and J.M. Hassett. A geographic information system utilizing the triangulated irregular network as a basis for hydrologic modeling. In *Auto-Carto 8*, pages 129–136, 1987.

108. C. Silva, J. S. B. Mitchell, and A. E. Kaufman. Automatic generation of triangular irregular networks using greedy cuts. In *Visualization 95*, pages 201–208, San Jose CA, 1995. IEEE Computer Society Press.

109. P. Sorensen and D. Lanter. Two algorithms for determining partial visibility and reducing data structure induced error in viewshed analysis. *Photogrammatic Engineering and Remote Sensing*, 28:1129–1132, 1993.

110. J.G. Speight. Parametric description of landform. In G.A. Stewart, editor, *Land Evaluation Papers of a CSIRO Symposium*, pages 239–250, 1968.

111. J. Star and J. Estes. *Geographic Information Systems: an Introduction.* Prentice Hall, Englewood Cliffs, 1990.

112. S. Takahashi, T. Ikeda, Y. Shinagawa, T.L. Kunii, and M. Ueda. Algorithms for extracting correct critical points and constructing topological graphs from discrete geographical elevation data. In *Eurographics'95*, volume 14, pages C–181–C–192, 1995.

113. Y. A. Teng and L. S. Davies. Visibility analysis on digital terrain models and its parallel implementation. Technical Report CAR-TR-625, Center for Automation Research, University of Maryland, 1992.

114. D.M. Theobald and M.F. Goodchild. Artifacts of TIN-based surface flow modelling. In *Proc. GIS/LIS*, pages 955–964, 1990.

115. R. van Appelen. Drainage networks on TINs. Master's thesis, Department of Computer Science, Utrecht University, 1996.

116. J. van Bemmelen, W. Quak, M. van Hekken, and P. van Oosterom. Vector vs. raster-based algorithms for cross country movement planning. In *Proc. Auto-Carto 11*, pages 304–317, 1993.

117. M. van Kreveld. On quality paths in polyhedral terrains. In *Proc. IGIS'94: Geographic Information Systems*, volume 884 of *Lecture Notes Comput. Sci.*, pages 113–122. Springer-Verlag, 1994.

118. M. van Kreveld. Efficient methods for isoline extraction from a TIN. *Int. J. of GIS*, 10:523–540, 1996.

119. M van Kreveld. Variations on sweep algorithms: efficient computation of extended viewsheds and classifications. In *Proc. 7th Int. Symp. on Spatial Data Handling*, pages 13A.15–13A.27, 1996.

120. M. van Kreveld, R. van Oostrum, C. Bajaj, V. Pascucci, and D. Schikore. Contour trees and small seed sets for isosurface traversal. In *Proc. 13th Annu. ACM Sympos. Comput. Geom.*, pages 212–220, 1997.

121. A. Voigtmann, L. Becker, and K. Hinrichs. Hierarchical surface representations using constrained Delaunay triangulations. In *Proc. 6th Int. Symp. on Spatial Data Handling*, pages 848–867, 1994.

122. W. Warntz. The topology of a socio-economic terrain and spatial flow. *Papers of the Regional Science Association*, 17:47–61, 1966.

123. R. Webster and M. A. Oliver. *Statistical Methods in Soil and Land Resource Survey*. Oxford University Press, New York, 1990.

124. R. Weibel. An adaptive methodology for automated relief generalization. In *Proc Auto-Carto 8*, pages 42–49, 1987.

125. R. Weibel and M. Heller. Digital terrain modelling. In D. J. Maguire, M. F. Goodchild, and D. W. Rhind, editors, *Geographical Information Systems – Principles and Applications*, pages 269–297. Longman, London, 1991.

126. D. Wilcox and H. Moellering. Pass location to facilitate the direct extraction of Warntz networks from grid digital elevation models. In *Proc. Auto-Carto 12*, pages 22–31, 1995.

127. G.W. Wolf. Metric surface networks. In *Proc. 4th Int. Symp. on Spatial Data Handling*, pages 844–856, 1990.

128. M.F. Worboys. *GIS: A Computing Perspective*. Taylor & Francis, London, 1995.

129. S. Yu, M. van Kreveld, and J. Snoeyink. Drainage queries in TINs: from local to global and back again. In *Proc. 7th Int. Symp. on Spatial Data Handling*, pages 13A.1–13A.14, 1996.

Chapter 4. Visualization of TINs

Mark de Berg

Dept. of Computer Science
Utrecht University
The Netherlands
markdb@cs.ruu.nl

1 Introduction

One of the fundamental tasks any GIS has to perform is the visualization of
geographic data. Often the data will be elevation data describing a terrain. How
to visualize such data depends on the representation used for the terrain, that
is, on the digital elevation model used. Three of the most popular models are
the regular square grid, the contour-line model, and the triangulated irregular
network—see Chapter 4 for an overview of these models. In this chapter we shall
concentrate on the visualization of triangulated irregular networks, or TINs for
short. In this model a triangulation of a finite set of 2-dimensional data points
is stored, together with the elevation of each data point. For our purposes we
can regard a TIN as a collection of triangles in 3-dimensional space forming
a connected surface which is z-monotone, that is, which is such that no two
triangles intersect when projected vertically onto the xy-plane. Fig. 1 shows a
TIN. (The example TIN is fairly regular—the triangles do not differ too much
in size—but this need not be the case.)

We will start in Section 2 with an overview of the basics of visualization.
This section also introduces the notational conventions we shall use throughout
the chapter.

The next two sections form the main part of this chapter. They discuss two
important topics that arise in the visualization of TINs.

The first is the *hidden-surface-removal problem*: given a TIN and a view point
(or a viewing direction), we want to compute what we see of the TIN when we
look at it from the given view point (or in the given viewing direction). Section 3
describes various approaches to the hidden-surface-removal problem.

The second topic, addressed in Section 4, concerns the following. The number
of triangles in a TIN is often so large that it becomes infeasible to visualize

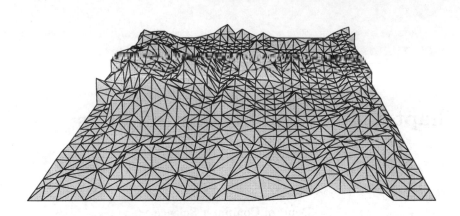

Fig. 1. A triangulated irregular network (TIN).

all triangles in a reasonable amount of time. Fortunately this is not necessary, as large parts of the TIN will be far from the view point; these parts can be visualized at a lower level of detail, using less triangles. To make this work, one should pre-compute representations of the same TIN at various *levels of detail (LOD)*.

We close the paper in Section 5 with a short discussion of some topics that arise in the visualization of TINs and that we did not consider.

2   The Basics

Below we briefly describe some of the most basic steps involved in *rendering* (that is, generating a picture of) 2- and 3-dimensional scenes. A more extensive treatment of these and many more topics in computer graphics can be found in the book by Foley et al. [15], or any other good graphics textbook.

A computer screen is composed of a large number of small dots, called *pixels*, which are arranged in a regular grid. The grid-size, or *resolution*, is typically about 1000×1000, resulting in a total of about $1,000,000$ pixels. Each pixel can be assigned a color. The colors of the pixels are stored in the *frame buffer*, an array with one entry per pixel. We denote the frame buffer by FrameBuf; thus FrameBuf$[x,y]$ stores the color of the pixel (x,y). Drawing a picture of a given scene now amounts to computing the correct colors for each pixel or, in other words, to filling in the frame buffer.

Now suppose we want to render a set of primitive objects—line segments, circles, triangles, and so on—in the plane. First we have to identify a rectangular area in the plane, the *window*, which delimits the region to be displayed on the screen. The objects are then *clipped* to the window: (the parts of) the objects that fall outside the window are discarded. Next we have to determine for each object a collection of pixels representing it, and set the entries in the frame buffer that correspond to these pixels to the color of their respective object. This process is called *scan-conversion*. Fig. 2 illustrates these processes.

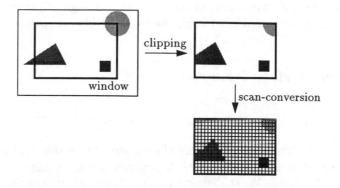

Fig. 2. Clipping and scan-converting 2-dimensional objects.

To visualize a 3-dimensional scene we need some more steps. As in the 2-dimensional case, we need to specify a region of interest. This so-called *viewing volume* is now a 3-dimensional region, which is a rectangular block for parallel projections and a truncated piramid for perspective projections. The objects are clipped to the viewing volume, and then projected onto the *viewing plane*. The projection that is used depends on the type of view: if a view point is given then a perspective projection is used with the view point as the center of projection, and if a viewing direction is given then a parallel projection in the viewing direction is used. Finally, the projected objects are scan-converted: the pixels covered by the projected objects are determined and the corresponding frame-buffer entries are assigned the color of the respective objects.

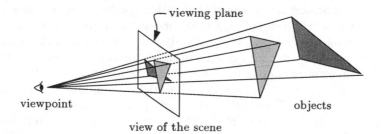

Fig. 3. Visualizing a 3D scene: projection and hidden surface removal.

There is one important step in the generation of images of 3-dimensional scenes that we have ignored in the description above: objects can completely or partly be hidden behind other objects. Somehow we have to ensure that pixels that are covered by more than one projected object are assigned the color of the right object—the one closest to the view point. This is illustrated in Fig. 3, where the light grey triangle hides part of the dark grey triangle. Computing what is visible of a given scene and what is hidden is called *hidden-surface removal*.

3 Hidden-Surface Removal

There are two major approaches to perform hidden-surface removal [31].

One is to first determine which part of each object is visible, and then project and scan-convert only the visible parts. Algorithms using this approach are called *object-space algorithms*. We shall discuss some of these algorithms in Section 3.2.

The other possibility is to first project the objects, and decide during the scan-conversion for each individual pixel which object is visible at that pixel. Algorithms following this approach are called *image-space algorithms*. The Z-buffer algorithm, which is the algorithm most commonly used to perform hidden-surface removal, uses this approach. It works as follows.

Assume for simplicity that we wish to compute a parallel view of the scene. First a transformation is applied to the scene that maps the viewing direction to the positive z-direction. The algorithm needs, besides the frame buffer which stores for each pixel its color, a z-buffer. This is a 2-dimensional array ZBuf, where $\text{ZBuf}[x, y]$ stores a z-coordinate for pixel (x, y). The objects in the scene are clipped to the viewing volume, projected, and scan-converted in arbitrary order. The z-buffer stores for each pixel the z-coordinate of the object currently visible at that pixel—the object visible among the ones processed so far—and the frame buffer stores for each pixel the color of the currently visible object. The scan-conversion process is now augmented with a visibility test, as follows. When we scan-convert a (clipped and projected) object t and we discover that a pixel (x, y) is covered by t, we do not automatically write t's color into $\text{FrameBuf}[x, y]$. Instead we first check whether t is behind one of the already processed triangles at position (x, y); this is done by comparing $z_t(x, y)$, the z-coordinate of t at (x, y), to $\text{ZBuf}[x, y]$. If $z_t(x, y) < \text{ZBuf}[x, y]$, then t is in front of the currently visible object, so we set $\text{FrameBuf}[x, y] := \text{color}_t$, where color_t denotes the color of t, and we set $\text{ZBuf}[x, y] := z_t(x, y)$. (The color of t need not be uniform, so we should actually write $\text{color}_t(x, y)$ instead of just color_t.) If $z_t(x, y) \geq \text{ZBuf}[x, y]$, we leave $\text{FrameBuf}[x, y]$ and $\text{ZBuf}[x, y]$ unchanged.

The Z-buffer algorithm is easy to implement, and any graphics workstation provides it, often in hardware. Nevertheless, there are situations where other approaches can be superior. In the next two subsections we describe two such approaches.

82

3.1 Depth-Sorting Methods

Depth-sorting methods [23] for hidden-surface removal scan-convert the objects in a back-to-front order, instead of in arbitrary order as the z-buffer algorithm does. This means that whenever an object is scan-converted, we know it is in front of all objects scan-converted thusfar. Hence, there is no need for the visibility test (the test on z-coordinate) anymore. This speeds up the algorithm, and it saves memory because the array ZBuf is no longer needed. Fig. 4 illustrates the approach. Because the algorithm resembles the way in which a painter works—

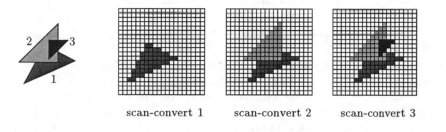

scan-convert 1   scan-convert 2   scan-convert 3

Fig. 4. The painter's algorithm.

objects in the foreground are drawn 'on top of' objects in the background—, it is often called the *painter's algorithm*.

Although the painter's algorithm avoids the visibility test in the scan-conversion phase, it adds an extra preprocessing step: a back-to-front order, or *depth order*, must be computed. This is not always easy. Even worse, it is not always possible. Fig. 5 shows three triangles that cyclically overlap each other; no ordering of the triangles will result in a correct picture of the scene. In general it is not easy to

Fig. 5. Three triangles that cyclically overlap each other.

detect and resolve such situations. (Resolving the situation is done by cutting one or several of the objects into pieces, and computing a depth order for the resulting collection of objects and object pieces.) But for TINs the problem of cyclic overlap does not occur—at least not for parallel views.

We now describe how to compute a depth order for the triangles in a TIN [2]. Let's first define more precisely what a depth order is. Assume that we want to compute a parallel view of the TIN, and let d denote the viewing direction. We say that a triangle t is *in front of* a triangle t' if there is line with direction d that first intersects t and then intersects t'. In other words, there is a point on t that hides some point on t' from the view. If t is in front of t', we write $t \prec t'$. A depth order for a collection T of triangles is an ordering $t_1, t_2, \ldots, t_n$ of the triangles in T such that $t_i \prec t_j$ implies $i > j$. This means that a triangle that is in front of another triangles should come later in the ordering.

The crucial observation that makes an efficient computation of a depth order possible for a TIN is that we can compute the order in the projection: we can project all triangles of the TIN onto the xy-plane, project the viewing direction as well, and compute a depth order for the resulting planar scene. This is possible because if a directed line in 3-dimensional space intersects two triangles, then the projected line intersects the projected triangles in the same order, provided that the projected triangles do not overlap. Fig. 6 illustrates this. Because the

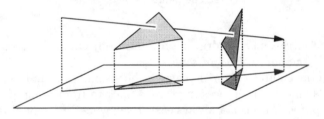

Fig. 6. The order of intersection does not change by projection.

triangles form a TIN, their projections do not overlap. From now on, we let T refer to the set of projected triangles, and d to the projected viewing direction. The in-front-of relation will also refer to the projected triangles: a projected triangle t is in front of a projected triangle t' if there is a line in the plane with direction d that first intersects t and then t'. A depth order can now be computed as follows.

Let $\mathcal{G}_T$ denote the dual graph of the TIN. This graph has a node for every triangle, and there is an arc between two nodes if the corresponding triangles are adjacent. Assuming the TIN is stored in a suitable topological structure—a doubly-connected edge list [4, 26] for instance—this graph is readily available. We turn $\mathcal{G}_T$ into a directed graph $\mathcal{G}_T(d)$ as follows: the arc connecting triangles t and t' is directed from t to t' if t is in front of t', it is directed from t' to t if t' is in front of t, and it is deleted if neither of the two is the case—see Fig. 7. (The third case can only occur if the common edge of the two triangles is parallel to the direction d.) One can prove that $\mathcal{G}_T(d)$ is acyclic.

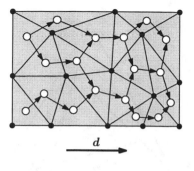

Fig. 7. The directed graph $\mathcal{G}_T(d)$.

A topological order on a directed graph is an ordering of the nodes such that if there is an arc from node ν to node ν', then ν comes before ν' in the ordering. This implies that if there is a directed path in the graph from a node ν to a node ν', then ν must come first in the topological order. Now consider two triangles t and t' in T, and assume that t is in front of t'. This means that there is a line with direction d that first intersects t and then t'. If we follow this line from t to t', it intersects a number (possible zero) adjacent triangles on the way. Hence, there is a directed path from t to t' in $\mathcal{G}_T(d)$. This implies that a topological order for $\mathcal{G}_T(d)$ corresponds to a depth order for T.

Computing a topological order on a directed acyclic graph can be done in $O(V + E)$ time, where V and E are the number of nodes and arcs of the graph, respectively [6]. In our case the number of nodes equals the number of triangles, and the number of arcs is at most three times this number, because a triangle is adjacent to at most three other triangles. We conclude that it is possible to compute a depth order for a TIN consisting of n triangles in $O(n)$ time.

So far we assumed that we are given a viewing direction. Now let's see what happens if we want to compute a perspective view from a view point p_{view}. In this case we say that a triangle t is in front of another triangle t' if there is a ray starting at p_{view} that first intersects t and then t'. A depth order is now defined as above: it is an order consistent with the in-front-of relation. We can use the same approach as before, namely project the triangles and the view point p_{view}, and compute a depth order in the projection by determining a topological ordering on the directed dual graph $\mathcal{G}_T(p_{\text{view}})$. (The graph $\mathcal{G}_T(p_{\text{view}})$ is defined in the obvious way: its nodes correspond to the projected triangles and its arcs reflect the in-front-of relation between adjacent triangles.) There is one problem with this approach: we are no longer guaranteed that $\mathcal{G}_T(p_{\text{view}})$ is acyclic, so a topological order may not exist. Fig. 8 shows this phenomenon: the directed graph obtained for the indicated view point and the shaded triangles contains a cycle. (The dotted rays show the order in which pairs of triangles are intersected, which is reflected in the direction of the arcs in the dual graph.) This problem can be solved by splitting the TIN into two parts with a line through p_{view} and

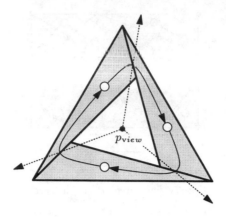

Fig. 8. Three triangles and a view point leading to a cycle in the dual graph.

treating the two halves separately. If the triangles in the TIN form a Delaunay triangulation, then one can prove that there are no cycles [8].

The approach we sketched above does not work for arbitrary 3-dimensional scenes. The problem is that the projections of the objects in an arbitrary scene are not necessarily disjoint, unlike for TINs.

An elegant way to implement the painter's algorithm for arbitrary scenes is to use BSP trees. A depth order for the scene can now be obtained by a traversal of the BSP tree [4, 17].

3.2 Object-Space Methods

Image-space methods compute the view of a scene pixel by pixel. This means that the 'structure' of the view is lost. Object-space algorithms compute a combinatorial representation of the view. Let's be more precise about this. The view of a scene is a subdivision of the viewing plane into maximal connected regions in each of which (some portion of) a single object can be seen, or no object is seen. In Fig. 3, for example, this subdivision consists of four regions; the light grey triangle is visible in one of them, the dark grey triangle is visible in two regions, and no object is seen in the fourth region. Object-space algorithms compute this subdivision—the *visibility map* of the given set of objects—as a collection of (polygonal) faces. In other words, object-space algorithms compute exactly which part of each object is visible. After that the visible parts can be projected and displayed without difficulty, because any pixel will be covered by at most one projected part.

In general, object-space hidden-surface-removal methods tend to be slower than image-space methods such as the z-buffer algorithm, because they cannot be implemented in hardware very well. However, object-space algorithms have certain advantages over image-space methods. Suppose we want to display the hidden lines in a scene dashed, instead of making them invisible. Object space algorithms compute exactly which parts of each line are visible and which parts are not, thus making it easy to display the invisible parts dashed. Image-space algorithms do not provide the information necessary to achieve this. Another weak point of image-space algorithms comes up when one wants to print the view of a scene on paper, instead of displaying it on the screen of a computer terminal. When hidden-surface removal has been done in image space, the only thing one can do is to plot every pixel separately. But this method fails to take advantage of the fact that the resolution of modern laser printers is much higher than the resolution of computer screens. If hidden-surface removal has been performed in object space then the visibility map can be processed directly, resulting in a picture of higher quality. Of course it is possible to use an image-space hidden-surface removal algorithm at the printer's resolution, but due to the very large number of pixels this tends to be slow. A third advantage of object-space algorithms is that they can be used to compute shadows in a scene, because the part of a scene lit by a light source is exactly the same as what can be seen by an observer standing at the light source. Finally, the fact that an object-space method computes exactly what is visible from a given view point means that its output is more suitable to perform *viewshed analysis* (that is, to analyze the view), which is useful in several GIS applications.

In theory, the complexity of the visibility map—the total number of vertices of the visible pieces—of a set of n triangles can be as high as $\Theta(n^2)$, even when the triangles form a TIN. This is illustrated in Fig. 9, where each of the long and 'horizontal' triangles of the hill in the back is cut into a linear number of visible pieces by the spikes in the front. This implies that any object-space hidden-

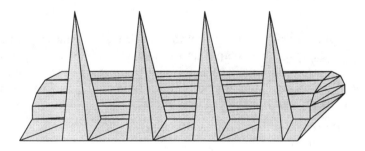

Fig. 9. A view of a TIN giving rise to a quadratic-complexity visibility map.

surface-removal algorithm must take $\Omega(n^2)$ time in the worst case. There are algorithms with an $O(n^2)$ running time [12, 22], which are thus optimal in the worst case. But obviously the worst case (a quadratic-size visibility map) usually does not occur in practice. Hence, it is useful to try and find *output-sensitive algorithms*: algorithms whose running time not only depends on n, the number of input triangles, but also on k, the complexity of the output (the visibility map in our case). Such algorithms will be faster when k is small. Dorward [13] gives an extensive overview of output-sensitive hidden-surface-removal algorithms, and de Berg's book [2] also contains an ample discussion of object-space hidden-surface removal. In the following, we shall concentrate on algorithms that are especially efficient for TINs. To simplify the description of the algorithms we assume that we want to compute a parallel view of the TIN. Let d be the viewing direction.

One approach to obtain an output-sensitive hidden-surface algorithm is the following. We have seen Section 3.1 that it is always possible to order the triangles of the TIN in a back-to-front order with respect to the viewing direction. The idea is to treat the triangles in the reverse order (that is, a front-to-back order), and to maintain the contour of the triangles—the union of their projection onto the viewing plane—processed thusfar. A triangle t is now handled as follows. First, the visible portion of t is determined by computing which part of its projection lies outside the current contour. Next, the contour is updated by computing the union of the current contour and t—see Fig. 10. Reif and Sen [27] have shown that

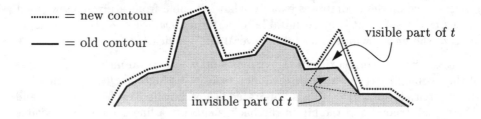

Fig. 10. Illustration of the approach of Reif and Sen.

the contour can be stored in such a way that both tasks—computing the visible portion of the triangle t and computing the new contour—can be performed in $O((k_t + 1) \log n \log \log n)$ time, where k_t is the complexity of the visible portion of t. This leads to an algorithm with a running time of $O((n + k) \log n \log \log n)$, where k is the complexity of the visibility map.

Another approach is described by Katz et al. [20]. We now describe their method in more detail.

Let T be a collection of triangles in 3-space. In our case the triangles form a TIN, but the method of Katz et al. works for any set of triangles (or other objects) provided that a depth order exists.

The first step in the algorithm is to construct a balanced binary tree $\mathcal{T}$ that stores the triangles in its leaves, where the left-to-right order of the leaves corresponds to a depth order on the triangles. For a node ν of this tree, we let $T(\nu)$ denote the set of triangles stored in the subtree rooted at ν. Thus $T(\text{root}(\mathcal{T})) = T$.

The next step is to compute for each node ν the sets $U(\nu)$ and $V(\nu)$, which are defined as follows: $U(\nu)$ is the union of the projections of the triangles of $T(\nu)$ onto the viewing plane, and $V(\nu)$ is the visible part of $U(\nu)$. More precisely,

$V(\nu)$ is defined as follows. Let $U_{\text{left}}(\nu)$ denote the union of the projections of the triangles stored in leaves to the left of $T(\nu)$. In the depth order, these triangles are the ones that come before the triangles in $\mathcal{T}(\nu)$, and so they may hide them from the view. $V(\nu)$ is now defined as $U(\nu) - U_{\text{left}}(\nu)$. Fig. 11 illustrates these definitions. If we let ν_t denote the leaf of $\mathcal{T}$ that stores triangle $t \in T$, then,

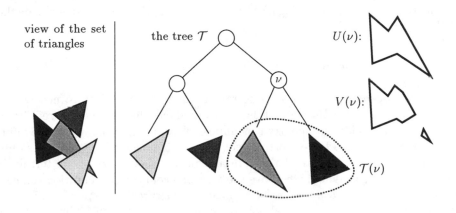

Fig. 11. The tree $\mathcal{T}$ and the sets $U(\nu)$ and $V(\nu)$.

by definition, the visible part of t is $V(\nu_t)$. Hence, after having computed the sets $V(\nu)$ for all nodes (including the leaves) in $\mathcal{T}$, we have solved the hidden-surface-removal problem. To compute the regions $V(\nu)$, we first compute the regions $U(\nu)$, using the following relation. Let lchild(ν) and rchild(ν) denote the left and right child of node ν respectively. Then we have

$$- \ U(\nu) = U(\text{lchild}(\nu)) \cup U(\text{rchild}(\nu)).$$

For a leaf ν, the region $U(\nu)$ is simply the projection of the triangle stored at that leaf. Hence, the relation above implies that the sets $U(\nu)$ can be computed by a simple procedure working in a bottom-up manner. The basic operation in this procedure is to compute the union of two polygonal sets (namely the unions of the children of a given node) in the plane. This can easily be done with a plane sweep algorithm.

Once we have computed $U(\nu)$ for each node ν we can compute the sets $V(\nu)$ in a top-down fashion by using the following relations:

$$- \ V(\text{root}(\mathcal{T})) = U(\text{root}(\mathcal{T})),$$
$$- \ V(\text{lchild}(\nu)) = V(\nu) \cap U(\text{lchild}(\nu)),$$
$$- \ V(\text{rchild}(\nu)) = V(\nu) - U(\text{lchild}(\nu)).$$

The basic operations in the top-down procedure are intersection and difference computations on polygons regions. Again, this can be done by a plane sweep.

For the efficiency of the method it is important that the regions $U(\nu)$, which are computed to aid the computation of the regions $V(\nu)$, are not too complex. If the set of triangles forms a TIN, then one can prove that this is indeed the case: the complexity of the region $U(\nu)$ is basically linear in the number of triangles in $T(\nu)$. To be precise, it is at most $O(n_\nu \alpha(n_\nu))$, where n_ν is the number of triangles in $T(\nu)$ and where $\alpha()$ denotes the extremely slowly growing functional inverse of Ackermann's function [1]. This can be used to show that the entire hidden-surface removal algorithm can be implemented so that on a TIN of n triangles it runs in $O(n\alpha(n)\log n + k\log n)$ time, where k denotes the complexity of the visibility map.

4 Levels of Detail

For a realistic representation of a terrain millions of triangles are needed. In applications such as flight simulation a terrain should be rendered at real time, but even with modern technology it is impossible to achieve this when the number of triangles is this large. Fortunately, a realistic image of the terrain is only crucial when one is close to the terrain and only a small part of the terrain is visible; when one is flying high above the terrain a coarse representation suffices. So what is needed is a *multiresolution model*: a hierarchy of representations at various levels of detail. This makes it possible for a given view point to render the terrain at an adequate level of detail. Fig. 12–14 shows an example of a hierarchy consisting of three levels. On the left perspective (and shaded) 3D views of the terrain are shown, and on the right 2D views of the underlying TIN.

One important property required from such a hierarchy is that subsequent levels should not differ too much in appearance: switching to more and more detailed representations when zooming in should not cause disturbing 'jumps' in the image. On the other hand, the reduction of the number of triangles in subsequent levels should not be too small, otherwise too many levels would be needed, resulting in an unacceptable increase in storage.

Another desirable property is the following. It is in general insufficient for rendering purposes to use only one level of detail at a time: although some part of the terrain may be close to the view point, another part—the horizon, for example—can be far away. Hence, one would like to combine parts from different levels into a single *variable-resolution representation* of the terrain, such that each part of the terrain is rendered with appropriate detail. This means that the levels cannot be completely independent, as it should be possible to glue them together smoothly.

The idea of multiresolution models is quite old, and work on it is scattered over literature in graphics, GIS, and other areas. Heckbert and Garland [18] and De Floriani et al. [10] give nice surveys of many of the existing multiresolution techniques. A related problem is the problem of simplifying general surfaces [19, 14] or terrains [16, 5, 21, 24].

The best known hierarchical data structure is probably the quadtree [29, 28]. Although quadtrees are based on regular grids, not on TINs, we discuss them briefly because they are so widely used as multiresolution model. Fig. 15 shows a three-level hierarchy based on a quadtree. The least detailed level simply consists

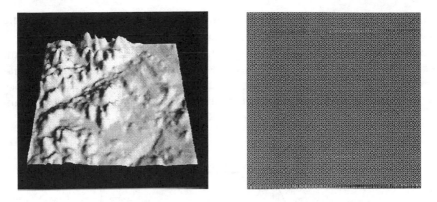

Fig. 12. The first, most detailed, level of the hierarchy.

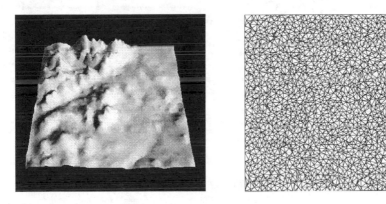

Fig. 13. The second level of the hierarchy.

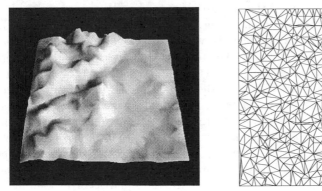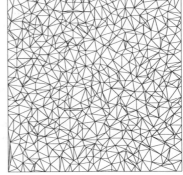

Fig. 14. The third level of the hierarchy.

of one square, namely the square corresponding to the root of the quad tree. The next level consists of the four squares corresponding to the children of the root; these squares are exactly the quadrants of the root square. In general, to obtain the next more detailed level from a given level, each square is replaced by its four quadrants. For rendering purposes, each square is split into two triangles with a diagonal. Extracting a representation where different parts of the terrain

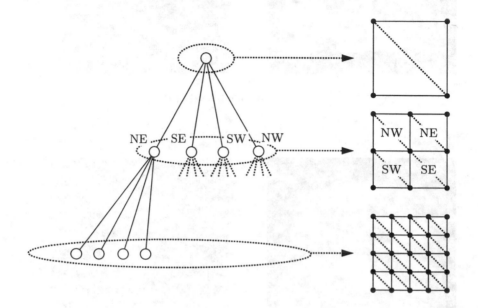

Fig. 15. A three-level hierarchy based on a quad tree.

use a different level of detail is quite easy when a quadtree is used. Consider the three-level hierarchy of Fig. 15 and suppose that we want to render the terrain for a view point located close to the SW corner of the terrain. Hence, we want a representation where the part of the terrain close to the SW corner is detailed, and parts further away are less detailed. Fig. 16 shows such a representation. In general, one simply performs a top-down traversal of the quadtree, proceeding at every node as follows: if the square corresponding to that node is detailed enough with respect to the given view point, then (the two triangles of) the square are rendered, otherwise the four children of the square are visited recursively. In the given example, the traversal starts at the root, where it is decided that the corresponding square is not detailed enough. Next, the four children of the root are visited. At three of the children it is decided that the corresponding squares are detailed enough, since they are far away from the view point. The SW square, however, is not detailed enough, since it is close to the view point. Hence, its children are visited recursively. Finally, the squares corresponding to these children are accepted, as they are detailed enough (or perhaps, in this example, because a more detailed representation is not available).

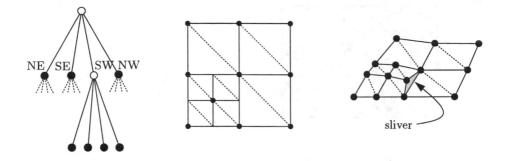

Fig. 16. Extracting a variable-resolution representation from a quad tree.

Using a quadtree as a multiresolution representation has two drawbacks. First of all, it assumes that the underlying representation is a regular grid; a quadtree it is not suitable for TINs. Second, although the parts that use different detail levels combine nicely in 2D, the terrain may no longer be continuous when the height of the sample points is taken into account. As a result, *slivers*—small 'cracks' in the terrain—may be visible, as can be seen in Fig. 16.

We now turn our attention to multiresolution models for TINs. These models can be subdivided into two categories.

In the first category one starts with a triangulation of a small subset of the data points [9, 25]. This is the coarsest representation. To obtain the next level, the triangles are refined by adding new data points inside them and retriangulating each triangle with its new interior points. Thus each triangle is replaced by a number of smaller triangles. This process is repeated until all data points have been added, or some precision criterion is met. Such a hierarchy can be modeled as a tree. The nodes in this tree correspond to the triangles in the hierarchy, and there is an arc from the node corresponding to a triangle t to the node corresponding to a triangle t' if the triangles belong to consecutive levels and t' is contained in t. There is also a root node, which is connected to all triangles of the first level. Fig. 17 shows an example of a hierarchy and the corresponding tree. Note the similarity of this model with the quad tree: in both cases an 'elementary shape' at a given level (a triangle or, for quadtrees, a square) is replaced by a number of smaller elementary shapes at the next level.

For tree-like hierarchies it is easy to combine different levels into one representation: one can use the same approach as for quadtrees, that is, perform a top-down traversal of the tree accepting the triangle of a node when it is detailed enough and visiting the children of the node otherwise.

Unfortunately, tree-like hierarchies have a serious drawback: the triangles at higher detail levels are very skinny, because the edges of the initial triangulation remain present at more detailed levels. (There are methods that try to avoid this by adding extra points on the edges of the triangles [30, 11]. The problem with this approach is that the introduction of extra vertices on the edges may cause

93

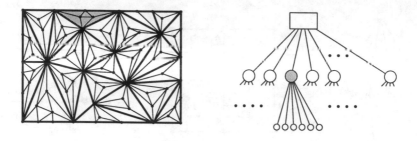

Fig. 17. A two-level hierarchy with a tree structure.

slivers.) This effect is already apparent in the two-level hierarchy of Fig. 17. Skinny triangles can cause robustness and aliasing problems.

The second category uses the Delaunay triangulation [26] of the set of data points at every level. This triangulation has the nice property that it maximizes the minimum angle of the triangles [4]. Thus robustness and aliasing problems are reduced. The hierarchies of this category can be represented by directed acyclic graphs: the nodes correspond to the triangles in the hierarchy, and there is an arc from the node corresponding to a triangle at a certain level to a triangle at the next level if these triangles intersect. An example of such a hierarchy is the *Delaunay pyramid* [7], which is obtained by always adding the data point with the maximal error and retriangulating using the Delaunay criterion.

Fig. 18 shows a two-level hierarchy of this type. Left in this figure is the Delaunay triangulation of a subset of the points, and in the middle is the Delaunay triangulation of the whole set. The corresponding graph structure is shown on the right. For these hierarchies it is not as easy to obtain a variable-resolution

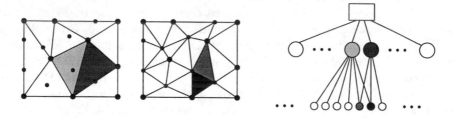

Fig. 18. A two-level hierarchy that uses the Delaunay triangulation.

representation as for tree-like hierarchies. The problem is that one cannot decide whether or not to refine a triangle indepently from the other triangles. Suppose, for instance, that in Fig. 18 we decide that the light grey triangle should be refined, but that the dark grey triangle is detailed enough (because it is slightly further from the view point). Then we have a problem, because the children of the light grey triangle overlap with the dark grey triangle.

De Berg and Dobrindt [3] have shown that if the hierarchy is constructed in a slightly different manner, then it is still possible to extract a variable-resolution representation. Their idea is to construct the hierarchy as follows. They start with the most detailed level, remove a subset of the vertices and retriangulate. The Delaunay triangulation thus obtained is the one-but-finest level of detail in the hierarchy. To obtain the next level, another subset of vertices is removed, and a triangulation of the remaining vertices is computed, and so on. The vertices to be removed in a single stage must satisfy the following property: no two should be adjacent. This means that the triangulation at a given level is obtained by replacing several groups of triangles at the previous level with groups of fewer triangles covering the same area—similar to tree-like hierarchies, where a single triangle is replaced by a group of triangles covering the same area. De Berg and Dobrindt show how to extract a variable-resolution representation, based on these groups, from the hierarchy. Fig. 19 shows an example of (a 2D view of) a variable-resolution representation obtained with their algorithm for the indicated viewpoint.

viewpoint

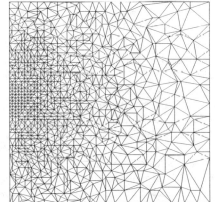

Fig. 19. A variable-resolution representation of a terrain.

As already remarked, there are many more papers dealing with multiresolution models. The examples we gave above should give a fairly good idea of the issues involved. Readers who want to know more should consult the survey papers by Heckbert and Garland[18] and by De Floriani et al. [10]. The first paper considers multiresolution models mainly from a rendering point of view and considers not only TINs but arbitrary 3D objects, whereas De Floriani et al. restrict their discussion to multiresolution models for TINs.

5 Other Issues

We have discussed two problems arising when one wants to visualize a TIN: hidden-surface removal and multiresolution models. There are many more issues that play a role, especially when one is interested in generating realistic images of a natural terrain. In that case one needs to take lighting considerations into account—which means, among other things, that one needs to compute shadows—, one needs to use so-called texture mapping, or other techniques, to make the terrain look more natural, and so on. Readers interested in these more advanced topics should consult the general graphics literature dealing with realistic rendering—Chapters 14–16 of *Computer Graphics: Principles and Practice* [15] are a good starting point.

Acknowledgements

Work of the author is supported by the ESPRIT IV LTR Project No. 21957 (CGAL).

References

1. P. K. Agarwal and M. Sharir. *Davenport-Schinzel Sequences and Their Geometric Applications.* Cambridge University Press, Cambridge, UK, 1995.
2. M. de Berg. *Ray Shooting, Depth Orders and Hidden Surface Removal.* Lecture Notes in Computer Science 703, Berlin, 1993.
3. M. de Berg and K.T. Dobrindt. On levels of detail in terrains. In *Proc. 11th Annu. ACM Sympos. Comput. Geom.*, pages C26–C27, 1995.
4. M. de Berg, M. van Kreveld, M. Overmars, and O. Schwarzkopf. *Computational Geometry: Algorithms and Applications.* Springer-Verlag, Heidelberg, 1997.
5. Z. Chen and J. A. Guevara. System selection of very important points (VIP) from digital terrain model for constructing triangular irregular networks. In *Proc. 8th Internat. Sympos. Comput.-Assist. Cartog. (Auto-Carto)*, pages 50–56, 1988.
6. T. H. Cormen, C. E. Leiserson, and R. L. Rivest. *Introduction to Algorithms.* The MIT Press, Cambridge, Mass., 1990.
7. L. De Floriani. A pyramidal data structure for triangle-based surface representation. *IEEE Comput. Graph. Appl.*, 9:67–78, March 1989.
8. L. De Floriani, B. Falcidieno, G. Nagy, and C. Pienovi. On sorting triangles in a Delaunay tessellation. *Algorithmica*, 6:522–532, 1991.
9. L. De Floriani, B. Falcidieno, G. Nagy, and C. Pienovi. Hierarchical structure for surface approximation. *Comput. Graph. (UK)*, 8(2):183–193, 1984.
10. L. De Floriani, P. Marzano, and E. Puppo. Hierarchical terrain models: Survey and formalization. In *Proc. ACM Sympos. Applied Comput.*, 1994.
11. L. De Floriani and E. Puppo. A hierarchical triangle-based model for terrain description. In *Proc. Internat. Conf. GIS: Theory and Methods of Spatio-temporal Reasoning in Geographic Space*, Lecture Notes in Computer Science, pages 236–251. Springer-Verlag, 1992.
12. F. Dévai. Quadratic bounds for hidden line elimination. In *Proc. 2nd Annu. ACM Sympos. Comput. Geom.*, pages 269–275, 1986.

13. S. E. Dorward. A survey of object-space hidden surface removal. *Internat. J. Comput. Geom. Appl.*, 4:325–362, 1994.

14. Nira Dyn, David Levin, and Samuel Rippa. Data dependent triangulations for piecewise linear interpolation. *IMA Journal of Numerical Analysis*, 10:137–154, 1990.

15. J. D. Foley, A. van Dam, S. K. Feiner, and J. F. Hughes. *Computer Graphics: Principles and Practice.* Addison-Wesley, Reading, MA, 1990.

16. R. J. Fowler and J. J. Little. Automatic extraction of irregular network digital terrain models. *Comput. Graph.*, 13(2):199–207, August 1979.

17. H. Fuchs, Z. M. Kedem, and B. Naylor. On visible surface generation by a priori tree structures. *Comput. Graph.*, 14(3):124–133, 1980. Proc. SIGGRAPH '80.

18. P. S. Heckbert and M. Garland. Multiresolution modeling for fast rendering. In *Proc. Graphics Interface '94*, pages 43–50. Canadian Inf. Proc. Soc., 1994.

19. H. Hoppe, T. DeRose, T. Duchamp, J. McDonald, and W. Stuetzle. Mesh optimization. In *Proc. SIGGRAPH '93*, pages 19–26, 1993.

20. M. J. Katz, M. H. Overmars, and M. Sharir. Efficient hidden surface removal for objects with small union size. *Comput. Geom. Theory Appl.*, 2:223–234, 1992.

21. J. Lee. A drop heuristic conversion method for extracting irregular networks for digital elevation models. In *Proc. of GIS/LIS '89*, pages 30–39, 1989.

22. M. McKenna. Worst-case optimal hidden-surface removal. *ACM Trans. Graph.*, 6:19–28, 1987.

23. M.E. Newell, R.G. Newell, and T.L. Sancha. A solution to the hidden surface problem. In *Proc. ACM Natl. Conf.*, pages 443–450, 1972.

24. Michael F. Polis and David M. McKeown, Jr. Issues in iterative TIN generation to support large scale simulations. *Proc. of 11th Intl. Symp. on Computer Assisted Cartography*, 1993.

25. J. Ponce and O. Faugeras. An object centered hierarchical representation for 3d objects: the prism tree. *Comput. Graphics and Image Proc.*, 38(1):1–28, 1987.

26. F. P. Preparata and M. I. Shamos. *Computational Geometry: An Introduction.* Springer-Verlag, New York, NY, 1985.

27. J. H. Reif and S. Sen. An efficient output-sensitive hidden-surface removal algorithms and its parallelization. In *Proc. 4th Annu. ACM Sympos. Comput. Geom.*, pages 193–200, 1988.

28. H. Samet. *Applications of Spatial Data Structures: Computer Graphics, Image Processing, and GIS.* Addison-Wesley, 1990.

29. H. Samet. *The Design and Analysis of Spatial Data Structures.* Addison-Wesley, Reading, MA, 1990.

30. L. Scarlatos and T. Pavlidis. Adaptive hierarchical triangulation. In *Proc. 10th Internat. Sympos. Comput.-Assist. Cartog. (Auto-Carto)*, pages 234–246, 1990.

31. I. E. Sutherland, R. F. Sproull, and R. A. Schumacker. A characterization of ten hidden-surface algorithms. *ACM Comput. Surv.*, 6(1):1–55, March 1974.

Chapter 5. Generalization of Spatial Data: Principles and Selected Algorithms

Robert Weibel

Dept. of Geography
University of Zürich
Switzerland

weibel@geo.unizh.ch

1 Introduction

In recent years, applications of geographic information systems (GIS) have matured, and spatial databases of considerable size have been built, and are being built and maintained continuously. Large amounts of money and time are invested into ambitious projects to build so-called spatial data infrastructures at the national level (e.g., MSC 1993, Clinton 1994, Grünreich 1992) and at the international level (e.g., EUROGI 1996). To enable the creation, maintenance, and use of such vast repositories of spatial information, a variety of methodological issues must be addressed. Besides methods for data modeling, data management and retrieval, as well as data distribution – including, for instance, standards for data documentation and exchange – techniques for automated generalization of spatial data (short: generalization) are of premier importance in this context. In GIS, generalization functions are needed for a variety of purposes, including the creation and maintenance of spatial databases at multiple scales, cartographic visualization at variable scales, and data reduction, to name just a few.

It is generally acknowledged that generalization is a complex process with ill-defined objectives, involving a good deal of subjective decisions. In order to solve the problem comprehensively, a variety of techniques including non-algorithmic solutions such knowledge-based methods and decision support systems approaches are needed. Clearly, however, data structures and algorithms form an indispensable foundation on which other approaches can build.

This survey presents an overview of principles, algorithms, and data structures for the generalization of spatial data. Sections 2 to 9 describe basic techniques, including an introduction to the principles and concepts underlying generalization, for those who are less familiar with the topic. Sections 10 , then, attempts to analyze what is wrong with basic generalization methods, and sections 11 to 13 discuss methods that extend the basic algorithms described in the first part and overcome some of their functional weaknesses. Other compilations of recent research in generalization can be found in Buttenfield and McMaster (1991), Müller et al. (1995a) and Weibel (1995a), and Molenaar (1996a).

2 What is Generalization and Why Does It Matter?

2.1 The Issue of Scale Change

Map generalization is a key element of cartography. Traditionally, spatial phenomena are cartographically portrayed on maps at different scales and for different purposes (e.g., topographic maps, geological maps, hiking maps, road maps). National topographic maps, for instance, are commonly produced at a series of scales [1], such as 1:25,000, 1:50,000, 1:100,000, 1:250,000, 1:500,000, and 1:1,000,000. The map scale is typically halved at each step in such a series (e.g., from 1:25,000 to 1:50,000). At the same time, the space available for drawing on the target map is divided by four, meaning that there is only a quarter of the space left to present the same amount of information as on the source map.

At the same time, as the *map scale* is reduced, small map objects may approach the limits of visual perceptibility. These perceptibility limits are termed *minimum dimensions* in cartography and are said to be, for instance, 0.35 mm for the length of sides of a black square (e.g., used to symbolize a building), or 0.25 mm for the distance between double lines which are often used to symbolize roads (SSC 1977). So, any map objects that would fall below these thresholds, but which the cartographer would still like to display on a map, would need to be enlarged accordingly in order to be clearly visible and discernible on the resulting map image. For example, on a map of 1:100,000, all buildings which have sidelengths smaller than 35 m – the vast majority of single family homes – would need to be enlarged to that minimum size. The same problem occurs with road objects; most roads are narrower than 25 m on the ground. So, to summarize, when reducing the scale of a map we are facing two problems which have a cumulative effect: available physical space on the map is reduced, and many objects may need to be enlarged in order to still remain visible. Both problems lead to a competition for available space among map objects. This situation is illustrated in Figure 1, which also depicts the necessary consequences. Only a subset of the original objects of the source map can be displayed on the target map and some objects may need to be displaced in order to avoid overlaps. This illustration, although schematic, also clearly shows why a mere photographic reduction would not be sufficient.

---

[1] Map *scale* is defined as the size ratio between an object (feature) in reality and its graphical representation on the map. Map scales with larger scale denominators (e.g., 1:500,000) are called 'small scales' in cartography, because they map everything to a small display area. Conversely, scales with smaller denominators (e.g., 1:10,000) are termed 'large scales'.

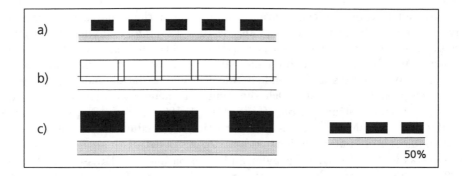

Fig. 1. Competition for space among map objects as a consequence of scale reduction. Reduction of available map space and enlargement of symbol sizes leads to overlaps and other spatial conflicts. These can be resolved by selecting only a subset of the objects of the source scale and by displacing some objects away from others (i.e., the buildings are displaced from the street).

2.2 Defining Generalization

In cartography, the process which is responsible for cartographic scale reduction is termed *generalization* (or map generalization, or cartographic generalization). It encompasses a reduction of the complexity in a map, emphasizing the essential while suppressing the unimportant, maintaining logical and unambiguous relations between map objects, and preserving aesthetic quality. The main objective then is to create maps of high graphical clarity, so that the map image is easily perceived and the message that the map intends to deliver can be readily understood. This position is expressed by the following concise definition by the International Cartographic Association:

> Selection and simplified representation of detail appropriate to the scale and/or the purpose of the map (ICA 1973: 173).

Note that map scale is not the only factor that influences generalization. Map purpose is equally important. A good map should portray the information that is essential to its intended audience. Thus, a map for bicyclists will emphasize a different selection of roads than a map targeted at car-drivers. Other factors that control traditional map generalization are the quality of the source material, the symbol specifications (e.g., the thickness and color of line symbols for roads, political boundaries, etc.), and technical reproduction capabilities (SSC 1977). The combination of these factors are called the *controls* of generalization (cf. Section 4.1).

In the context of digital cartographic systems and GIS, generalization has obtained an even wider meaning. This statement may at first sound surprising. One might expect that the transition from static paper maps to digital maps on computer screens, with the possibility to flexibly select and compose feature classes and features (objects) through queries, and the capability to interactively zoom and inspect the data at any desired scale (i.e., magnification) factor would have overcome the need for generalization. The answer clearly is negative, which is mainly due to two fundamental facts. First, spatial phenomena and processes are usually scale-dependent. Ideally, spatial data should be analyzed and viewed at the scale at which the modeled phenomena and processes are meaningful and best understood (Müller et al. 1995b). Second, generalization – which is essentially a process of abstraction and reduction of complexity – is a fundamental human intellectual activity and part of the general scientific process as well as everyday behavior and decision-making (Brassel and Weibel 1988). Without concentration on the essential aspects of a given problem we are soon lost in irrelevant details and unable to understand overriding patterns, let alone communicate them to outsiders. Thus, generalization also is of fundamental importance as a process of maximizing information content in building, maintaining, and communicating the content of spatial databases (Müller 1991). The definition of generalization in the digital context used by McMaster and Shea (1992) exhibits this close relation to the traditional process:

> Digital generalization can be defined as the process of deriving, from a data source, a symbolically or digitally-encoded cartographic data set through the application of spatial and attribute transformations. Objectives of this derivation are: to reduce in scope the amount, type, and cartographic portrayal of the mapped or encoded data consistent with the chosen map purpose and intended audience; and to maintain clarity of presentation at the target scale (McMaster and Shea 1992: 3-4).

Note that the same major generalization *controls* – scale, map purpose, intended audience – are mentioned as in the traditional view of map generalization. The 'spatial and attribute transformations' needed to realize the actual generalization process form the focus of Sections 7 to 13.

Viewed from yet another perspective, digital generalization can be understood as a process of **resolution reduction** (Ruas 1995a), affecting both the thematic and geometric domain[2]. In the thematic domain generalization implies a change of the database schema; the number of entities is reduced, attributes are eliminated, and attribute values are made less accurate (e.g., averaged). In the geometric domain generalization the resolution is reduced by eliminating objects or parts of objects, simplifying shapes, or displacing objects from one another in order to maintain good separability (cf. Fig. 1).

2.3 Motivations of Generalization

We have already alluded to some of the motivations of generalization above, but it is worthwhile making the list complete. Extending on Muller's (1991)

---

[2] Note that if temporal aspects are also modeled, generalization can be applied similarly to the temporal domain (e.g., reducing the resolution of a time series from daily to monthly averages).

discussion of requirements for generalization, we can develop a more detailed list of motivations:

1. **Develop primary database:** Build a digital model of the real world, with the resolution and content appropriate to the intended application(s), and populate it (object generalization; cf. Subsection 3.1).
   - Select objects
   - Approximate objects
2. **Use resources economically:** Minimize use of computing resources by filtering and selection within tolerable (and controllable) accuracy limits.
   - Save storage space
   - Save processing time
3. **Increase/ensure data robustness:** Build clean, lean and consistent spatial databases by reducing spurious and/or unnecessary detail.
   - Suppress unneeded high-frequency detail
   - Detect and suppress errors and random variations of data capture
   - Homogenize (standardize) resolution and accuracy of heterogeneous data for data integration
4. **Derive data and maps for multiple purposes:** From a detailed multi-purpose database, derive data and map products according to specific requirements.
   - Derive secondary scale and/or theme-specific datasets
   - Compose special-purpose maps (i.e. all new maps)
   - Avoid redundancy, increase consistency
5. **Optimize visual communication:** Develop meaningful and legible visualizations.
   - Maintain legibility of cartographic visualizations of a database
   - Convey an unambiguous message by focusing on main theme
   - Adapt to properties of varying output media

Examination of the above list reveals that classical cartographic generalization mainly relates to task 5 (visual communication) and to a lesser extent also to task 4, while tasks 1 to 3 are more specific to the digital domain (object generalization, model generalization). In task 5, an aspect of cartographic generalization germane to a GIS environment is that output may be generated for media of varying specifications, such as high-resolution plotted maps or low-resolution CRT views, requiring consideration of the resolution of the output media when composing maps for display (Spiess 1995).

3 Different Views of Digital Generalization

The reasons and motivations of *digital* generalization listed above show quite clearly how diverse the requirements are towards procedures implementing this process. Different views of the overall process are thus possible.

3.1 Generalization as a Sequence of Modeling Operations

A first view understands generalization as a process which realizes transitions between different models representing a portion of the real world at decreasing detail, while maximizing information content with respect to a given application. Figure 2 shows how transitions take place in three different areas along the database and map production workflow. The terminology used here was originally developed for the German ATKIS project (Grünreich 1992), but has since been adopted by other authors:

- as part of building a primary model of the real world (a so-called digital landscape model = DLM) – also known as *object generalization*
- as part of the derivation of special-purpose secondary models of reduced contents and/or resolution from the primary model – also known as *model generalization* (Also termed model-oriented, or statistical (database) generalization by different authors; cf. Weibel 1995b)
- as part of the derivation of cartographic visualizations (digital cartographic models = DCM) from either primary or secondary models – commonly known as *cartographic generalization*

Let us take a closer look at the scope and the objectives of these three generalization types.

Object generalization takes place at the time of defining and building the original database, called 'primary model' in Figure 2. Since databases are abstract representations of a portion of the real world, a certain degree of generalization (in the sense of abstraction, selection, and simplification) must take place, as only the subset of information relevant for the intended use(s) is represented in this database. Although seen from the perspective of generalization here, this operation is sufficiently explained by methods of semantic and geometric data modeling (define the relevant object classes and their attributes), as well as sampling methods (define the sampling strategy and its resolution), combined with human interpretation skills (e.g., if photogrammetric data capture is used). This survey will therefore not go to any further detail regarding object generalization.

While the process of object generalization had to be carried out in much the same way when preparing data for a traditional map, *model generalization* is new and specific to the digital domain. In digital systems, generalization can affect directly the map data, and not the map graphics alone. The main objective of model generalization is *controlled data reduction* for various purposes. Data reduction may be desirable for reasons of computational or storage efficiency in analysis functions, but also in light of data transfer via communication networks. It may further serve the purpose of *deriving datasets of reduced accuracy and/or resolution*. This capability is particularly useful in the integration of data sets of heterogeneous resolution and accuracy as well as in the context of multi-resolution databases. While model generalization may also be used as a *preprocessing* step to cartographic generalization, it is important to note that it is not oriented towards graphical depiction, and thus involves no artistic, intuitive components. Instead, it encompasses processes which can be modelled completely formally (Weibel 1995b); these may, however, have aestethic consequences on subsequent cartographic generalization. Model generalization is discussed further in Molenaar (1996a) and Weibel (1995b); an example of an algorithm for this class of generalization functions is presented in Section 9.

Cartographic generalization is the term commonly used to describe the generalization of spatial data for cartographic visualization. It is this the process that most people typically think of when they hear the term 'generalization'. The difference to model generalization is that it is aimed at generating visualizations, and brings about graphical symbolization of data objects. Therefore, cartographic generalization must also encompass operations to deal with problems created by symbology (cf. Fig. 1) such as object displacement, which model generalization does not. The objectives of digital cartographic generalization remain basically the same as in conventional cartography (cf. Subsection 2.1). However, technological change has also brought along new tasks with new requirements such as interactive zooming, visualization for exploratory data analysis, or progressively adapting the level of detail of 3-D perspective views to the viewing depth. The concept of cartographic generalization thus needs to be extended. On the other hand, typical maps generated in information systems are no longer complex multi-purpose maps with a multitude of feature classes involved, but rather single-purpose maps consisting of few layers. Furthermore, maps and other forms of visualizations are often presented by means of a series of different partial views in a multi-window arrangement, particularly in exploratory data analysis. Together with the capabilities of interactive direct manipulation these new forms of cartographic presentations may partially alleviate (but by no means eliminate) some of the generalization problems, or at least make them less salient for many GIS users.

3.2 Generalization Strategies

A second distinction of different views of generalization can be made with respect to the strategy used for developing digital generalization capabilities.

Process-oriented view – Deriving generalizations from a detailed database. A *process-oriented view* understands generalization as the *process* of obtaining through a series of scale and purpose-dependent transformations a database or map of reduced complexity at *arbitrary* scale or resolution, starting from a *detailed* database. As was already mentioned, generalization is a complex process, and indeed, complete solutions for all the transformation operations necessary to achieve comprehensive automated generalization largely remain to be developed.

Representation-oriented view – Multi-scale databases. A more pragmatic approach is to develop multi-scale databases in analogy to the scale series

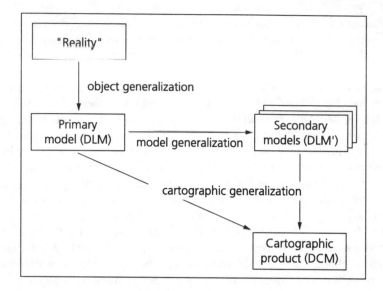

Fig. 2. Generalization as a sequence of modeling operations (modified after Grünreich 1985).

used in national topographic maps. We term this approach the *representation-oriented view*, because it attempts to develop databases that integrate single representations at different scales into a consistent multi-scale representation. Instead of devising the methods necessary to achieve the processes for transforming one level of scale into the next smaller one, scale transitions between different levels are formally coded. As one might expect, techniques of multiple representation spatial databases are needed to develop this strategy. Examples of this approach include van Oosterom and Schenkelaars (1995), Kidner and Jones (1994) and Devogele et al. (1996), to name but a few. While the representation-oriented strategy certainly overcomes the problems of missing generalization methods, it poses maintenance problems. If updates (e.g., insertions or deletions of objects) take place at a certain scale level, inconsistencies are easily introduced if they cannot be automatically propagated to all other levels – which in turn would require generalization functionality. Since both approaches have their advantages and disadvantages, it is probably safe to say that they should be exploited in conjunction during the next few years, with a gradual shift towards process-oriented generalization, since it offers more flexibility.

To summarize, this survey will mainly focus on cartographic generalization (with one exception in Section 9) and on methods to implement a process-oriented strategy towards automated generalization. *In the remainder of this survey, 'generalization' will therefore denote process-oriented cartographic generalization.*

4 Conceptual Frameworks of Generalization

In order to render a complex and holistic process such as cartographic generalization amenable to automation, conceptual frameworks need to be developed. Such theoretical models must be capable of describing the overall process and must at the same time allow to identify essential process components and steps. Of the many conceptual frameworks proposed in the literature, we briefly describe the models by Brassel and Weibel (1988) and by McMaster and Shea (1992). The latter authors discuss further models.

4.1 The Model by Brassel and Weibel

Brassel and Weibel (1988) proposed a conceptual framework of cartographic generalization which attempts to identify the major steps of the manual generalization process and transpose these concepts into the digital realm. The model departs from a view of generalization as an intellectual process which explicitly structures experienced reality into a number of individual entities, and which then selects important entities and represents them in a new form.

Figure 3 shows a schematic outline of the model. The source database is first subjected to a process termed *structure recognition* (a). This step aims at the identification of objects or aggregates, their spatial and semantic relations and the establishment of measures of relative importance (i.e., a priority order). Structure recognition is of overriding importance to the entire map generalization process, as it establishes the foundation upon which all other steps build. It is governed by the generalization controls (map purpose, source and target scale, quality of the source database, symbol specifications, minimal dimensions, etc.). Traditionally, this step is performed by visual inspection and intellectual evaluation of the map by an experienced cartographer. In the digital domain, it is possible to use methods for cartometric analysis (i.e., automated measurement in maps), but a comprehensive automation of structure recognition is non-trivial. Once the prevalent structures of the source database are known, the relevant generalization processes can be defined in *process recognition* (b). This involves the identification of both the types of data modifications (process types) and the parameters controlling these procedures (process parameters) necessary to yield the desired target map. This step too is influenced by the generalization controls. Next is *process modeling* (c), which compiles rules and procedures from a process library. Digital generalization takes place with *process execution* (d), where the rules and procedures are applied to the source database in order to create the generalized target database. As a last process, *data display* (e) converts the target database into a fully symbolized target map.

The main contribution of this model at the time of its publication was the distinction of steps leading to a characterization of the contents and structure of the source database (steps a, b, c) from operational, mechanical steps (d, e). The analysis of the shape and structure of map elements is thus made explicit. As we will see in the remainder of this survey, a weakness found in most of the basic algorithms discussed in Sections 7 to 9 is that they take the opposite approach,

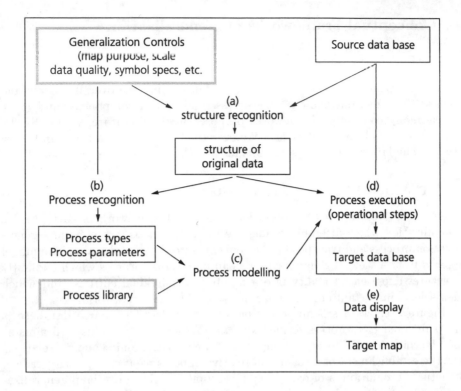

Fig. 3. The conceptual framework by Brassel and Weibel. Slightly modified after Brassel and Weibel (1988). Note that the term 'process' is equivalent to 'operator' as defined by McMaster and Shea (1992).

'hiding' shape characterization in built-in, implicit heuristics of generalization algorithms. As a result, the effectiveness and flexibility of such algorithms is limited.

4.2 The Model by McMaster and Shea

The model by Brassel and Weibel (1988) was extended by McMaster and Shea (1992), who added some missing parts and specified details for model components which were previously defined only in general terms. The resulting conceptual framework was therefore termed a comprehensive one by the authors. It decomposes the generalization process into three operational areas: (1) a consideration of the philosophical objectives of *why* to generalize; (2) a cartometric evaluation of the conditions which indicates *when* to generalize; and (3) the selection of appropriate spatial and attribute transformations which provide the techniques on *how* to generalize (McMaster and Shea 1992: 27).

Figure 4 gives a complete overview of the components of this conceptual model. McMaster and Shea (1992) first present a detailed treatment of the *philo-*

sophical objectives (why to generalize) which have been discussed here in a similar way in Subsection 2.3.

The second area of the McMaster and Shea model, *cartometric evaluation* (when to generalize) is essentially equivalent in scope to the key steps in the framework by Brassel and Weibel (1988) called structure recognition and process recognition. 'Spatial and holistic measures' are made available to characterize the source data by quantifying the density of object clustering, the spatial distribution arrangement, the length, sinuosity, and shape of map objects, and so on. These measures then serve to evaluate whether critical 'geometric conditions' are reached which trigger generalization, such as congestion (crowding) of map objects, coalescence of adjacent objects, conflicts (e.g., overlap), imperceptible objects (e.g., objects that are too small to be clearly visible), etc. Process recognition, as specified in the Brassel and Weibel model, is covered by 'transformation controls' in order to select appropriate operators, algorithms, and parameters to resolve the critical geometric conditions.

Finally in the third area, *spatial and attribute transformations* (how to generalize), a list of twelve 'generalization operators' (cf. 4.3) is proposed, sub-divided into ten operators performing spatial transformations – simplification, smoothing, aggregation, amalgamation, merging, collapse, refinement, exaggeration, enhancement and displacement – and two operators for attribute transformations – classification and symbolization. This area may be thought of as being equivalent to the 'process library' in the Brassel and Weibel model. The definition of a useful set of operators is of particular interest in the conceptual modelling of generalization, and deserves further discussion in the next paragraph.

4.3 A Closer Look at Generalization Operators

The overall process of generalization is often decomposed into individual sub-processes. Depending on the author, 'operator' may be used, or other terms such as 'operation' or 'process'. Cartographers have traditionally used terms such as 'selection', 'simplification', 'combination' or 'displacement' to describe the various facets of generalization, examples of which are the definitions given in Subsection 2.2. In the digital context, however, a functional breakdown into operators has obviously become even more important, as it clarifies identification of constituents of generalization and informs the development of specific solutions to implement these sub-problems. Naturally, given the holistic nature of the generalization process, this reductionist approach is too simple, as the whole can be expected to be more than just the sum of its parts, but it provides a useful starting point for understanding a complex of diffuse and challenging problems.

Owing to the importance of the functional decomposition of generalization various authors (e.g., Hake 1975, McMaster and Shea 1992, Ruas and Lagrange 1995, Ruas 1995a) have proposed typologies of generalization operators, each of them intended to comprehensively define the overall process. Unfortunately, no consensus has yet been reached on an all-encompassing set of operators. Even worse, authors may use different definitions for the same term or use different

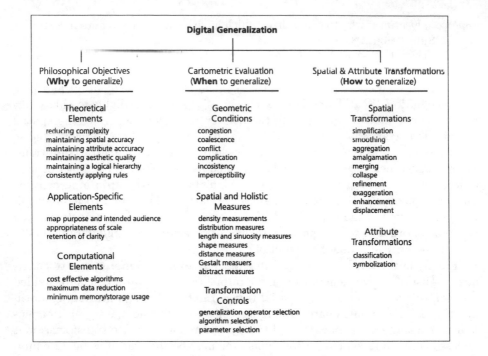

Digital Generalization

Philosophical Objectives (**Why** to generalize)	Cartometric Evaluation (**When** to generalize)	Spatial & Attribute Transformations (**How** to generalize)
Theoretical Elements	**Geometric Conditions**	**Spatial Transformations**
reducing complexity	congestion	simplification
maintaining spatial accuracy	coalescence	smoothing
maintaining attribute acccuracy	conflict	aggregation
maintaining aesthetic quality	complication	amalgamation
maintaining a logical hierarchy	incosistency	merging
consistently applying rules	imperceptibility	collaspe
		refinement
Application-Specific Elements	**Spatial and Holistic Measures**	exaggeration
		enhancement
map purpose and intended audience	density measurements	displacement
appropriateness of scale	distribution measures	
retention of clarity	length and sinuosity measures	**Attribute Transformations**
	shape measures	
Computational Elements	distance measures	classification
	Gestalt measuers	symbolization
cost effective algorithms	abstract measures	
maximum data reduction		
minimum memory/storage usage	**Transformation Controls**	
	generalization operator selection	
	algorithm selection	
	parameter selection	

Fig. 4. The conceptual framework of digital generalization by McMaster and Shea (1992).

terms for the same definition, as a recent study by Rieger and Coulson (1993) has shown.

McMaster and Shea's (1992) typology is the first detailed one which also attempts to accommodate the requirements of digital generalization, spans a variety of data types including point, line, area and volume data. Still, closer inspection of this set of operators reveals that some fundamental operators are missing (e.g. selection/elimination) and that the definitions of some operators are perhaps not sufficiently clear (e.g. refinement) or overlapping (aggregation, amalgamation, merging). This has led other authors (e.g. Ruas and Lagrange 1995, Plazanet 1996) to extend this classification by adding operators and by refining definitions of existing ones. The composition of a comprehensive set of generalization operators is still the subject of an on-going debate; it is hoped that having it would assist the development of adequate generalization algorithms as well as their integration into comprehensive workflows.

No matter what set of operators is defined, however, the relationship between generalization *operators* and generalization *algorithms* is hierarchical. An operator defines the transformation that is to be achieved; a generalization algorithm is then used to implement the particular transformation. Commonly, several algorithms are possible for each operator. In particular, a wide range of different algorithms exists for line simplification in vector mode (cf. Subsection 7.2).

5 Generalization – The Role of Algorithms

It should by now have become clear from the above discussion that generalization is a complex process. What makes generalization so particularly hard to treat is not only the complexity of geometric problems involved but also the fact that the objectives are often ill-defined, owing to subjective, intuitive elements of cartographic design. Note that this is not the case in model generalization (which is non-graphical in nature), but it certainly holds for cartographic generalization which forms the focus of this survey and also the major thrust of generalization research.

So, what is the role of algorithms if we are trying to solve problems whose objectives are so weakly defined? One consequence is that in terms of meeting the functional objectives we may not expect to develop optimal algorithms, but only plausible ones. Another effect is that algorithms are probably not the only approach that should be used to tackle the problem comprehensively.

Knowledge-based methods are often mentioned as an alternative to algorithms. Yet, a look at the history of research in cartographic generalization reveals that neither algorithmic methods (Lichtner 1979, Leberl 1986) nor knowledge-based techniques such as expert systems (Fisher and Mackaness 1987, Nickerson 1988) have been capable of solving the problem comprehensively. While the former suffered from a lack of flexibility (since they are usually designed to meet a certain task) and from weak definition of objectives, the development of the latter was impeded by the scarcity of formalized cartographic knowledge and the problems encountered in acquiring it (Weibel et al. 1995). More recent research has therefore concentrated on approaches that more closely follow the decision support system (DSS) paradigm, a strategy often used to solve ill-defined problems. A particular approach along this vein builds on the integration of algorithmic and knowledge-based techniques and has been termed *amplified intelligence* (Weibel 1991).

As visualization and generalization are essentially regarded as creative design processes, the human is kept in the loop: key decisions default explicitly to the user, who initiates and controls a range of algorithms that automatically carry out generalization tasks (Fig. 6). Algorithms are embedded in an interactive environment and complemented by various tools for structure and shape recognition giving cartometric information on object properties and clustering, spatial conflicts and overlaps, and providing decision support to the user as well as to knowledge-based components. Ideally, interactive control by the user reduces to zero for tasks which have been adequately formalized and for which automated solutions could be developed.

In such a setup, *algorithms* serve the purpose of implementing tasks for which sufficiently accurate objectives can be defined:

111

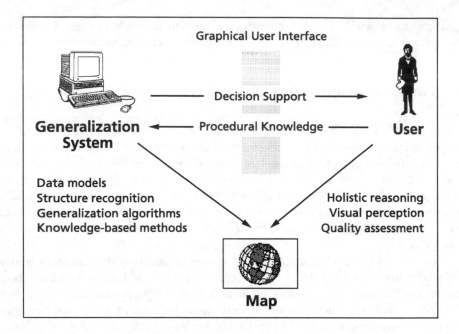

Fig. 5. The concept of amplified intelligence for map design and generalization.

- *generalization operators:* selection, simplification, aggregation, displacement, ...
- *structure recognition* (cf. Subsection 4.1): shape measures, density measures, detection of spatial conflicts, ...
- *model generalization* (cf. Section 9)

Knowledge-based methods can be used to extend the range of applicability of algorithms and code expert knowledge into the system:

- *knowledge acquisition:* machine learning may help to establish a set of parameter values that control the selection and operation of particular algorithms in a given generalization situation (Weibel et al. 1995)
- *procedural knowledge, control strategies:* once the expert knowledge is formalized, it can be used to select an appropriate set and sequence of operators and algorithms and establish a strategy to solve a particular generalization problem.

In summary, an ideal system builds on a hierarchy of control levels. The human expert takes high-level design decisions and evaluates system output. Knowledge-based methods operate at an intermediate level and are responsible for selecting appropriate operators and algorithms and for conflict resolution strategies. Finally, algorithms are the work horses of a generalization and form the foundation of everything else.

6 The Choice of Spatial Data Models – Raster or Vector?

Besides underlying theoretical principles and algorithms, spatial data models and data structures form a third component of the foundation that allows to build generalization methods. The choice of spatial data model has a great impact on the way and completeness in which properties of real world objects can be digitally represented and thus also directly governs the quality achievable by algorithms that are developed on top of them (cf. Section 13).

It is common to distinguish two major classes of spatial data models available in GIS: tessellations, of which the raster model is the most widely used, and vector models, in particular the topological vector model. The two classes of data models represent quite different concepts of representing space. The raster model, as a space-primary model, has advantages in representing continuously varying phenomena (e.g., scalar fields) or regularly sampled categorical data (e.g., landuse data derived from remote sensing imagery), and it also eases the computation of distance transformations. On the other hand, object representation is lost in raster models and severe discretization problems may be caused by the tessellation structure. The vector model, as an object-primary model, basically has reversed properties. It excels in its capabilities for object representation and accurate geometric coding, but it puts an additional burden on the computation of proximity relations and makes it almost impossible to represent continuous phenomena.

The debate over the advantages and disadvantages of raster vs. vector models has been one of the most persistent disputes in GIS research for many years. It is therefore not surprising that the debate also affected generalization research. Some authors have proposed to develop specific generalization operators for raster models different from those for vector models (McMaster and Monmonier 1989). While the differences between the two forms of representation may be considerable, it is not advisable, however, to depart from the conceptual hierarchy of tasks, operators, and algorithms. It is possible to develop solutions for all operators for both vector and raster models, although obviously some operators will be easier to implement for some data structures than others. The focus should therefore be on the generalization tasks (what are the objectives that the generalization has to meet?) and on the requirements of object representation (what data model adequately represents the structure and properties of the given real world objects?). Given these requirements, a suitable set of operators and algorithms then needs to be developed and applied, using the optimal data model. In some situations this may be a raster model, in others a vector model may be better suited. Most probably, complex problems will require a combination of different data models including auxiliary data structures, with functions to convert between them. Section 12.3 presents an example of terrain generalization which uses a combination of raster models to represent the terrain surface and 3-D vector models to represent topographic structure lines. The two models are converted into each other by object extraction and interpolation processes, respectively.

For lack of space we will focus on methods based on vector data models in this survey. Raster models are predominantly used in landuse generalization (since most landuse databases are in raster format or derived from remote sensing imagery) and as auxiliary representation to ease spatial search and distance transformations (e.g., in object displacement). A review of raster-based methods can be found in Schylberg (1993) and Jäger (1991).

7 Basic Algorithms – Context-Independent Generalization

In this section, we will describe a few basic algorithms for three simple operators: *selection/elimination, simplification and smoothing*. These operators all have in common that they are applied to individual objects independently of their spatial context. For instance, objects that are close to a line that is simplified may be affected (e.g., the new line may overlap with them), but the simplification process really only relates to the line object. This kind of generalization can be termed **context-independent generalization.** In contrast to that, **context-dependent generalization** involves operators such as aggregation or displacement which can only be triggered and controlled following an analysis of the spatial context (spatial relations of objects, object density, etc.). Context-dependent operators will only be described in Sections 11 to 13.

7.1 Object Selection/Elimination

Object selection (or defined by the antonym, object elimination) may be a simple operator, but is also an effective one as it makes space available on the map by omitting objects that are deemed irrelevant for the target map. Three questions must be addressed:

- How many objects are selected?
- Which objects are selected?
- What constraints govern the selection process?

Commonly, however, objects are not only omitted but the remaining objects are also repositioned in order to maintain the visual impression of the original arrangement of objects on the source map. Object selection thus most often only represents a first step of a series of operations.

Number of objects. The first of the above questions has been addressed in the 1960s by Töpfer and his co-workers (Töpfer 1974). Empirical rules were established in extensive studies involving the comparison of published map series. The basic empirical principle derived from these studies was termed the 'Principle of Selection' or 'Radical Law' (in German: *'Wurzelgesetz'*):

$$n_T = n_S \sqrt{\frac{s_S}{s_T}} \tag{1}$$

114

Given the number of objects at the source scale n_S and the denominators of the source scale (s_S) and target scale (s_T), this simple formula allows to compute the number of objects n_T that should be maintained on the target map.

Selecting specific objects. Töpfer's principle of selection has been extended in various ways by adding further terms to take into account special cases and specific feature classes, but no matter how detailed the equation is, it still does not give any indication which objects should be selected. The selection of an actual set of objects can only be carried out using object semantics. Assuming that each of the map features is characterized by a set of attributes, objects may be selected by a query on the attributes. If the number of attributes is large and/or attributes are at different scales of measurement (e.g., interval/ratio vs. ordinal), a ranking approach is more useful. The values of each attribute are ranked over all objects, and a total score is computed for each object, allowing to establish a rank order. Following that, the top n_T objects are selected.

Note that attributes are not limited to purely non-geometric properties of an object. Geometric properties may also contribute to object semantics. For instance, when selecting towns for a small scale map, it certainly makes sense to select the places with the largest population, but places at important highway junctions or remote settlements may also be retained regardless of their lack of population. In a desert, an oasis of just ten inhabitants (but with fresh water) is something to look forward to. Measures of proximity or remoteness can be derived from an analysis of the spatial distribution of objects, for instance, by analyzing the size of Voronoi regions of the map objects (Roos 1996).

Constraints to selection. The example of the preceding paragraph shows that objects semantics obviously exert an influence on object selection. Apart from that, selection may be governed by topological constraints. A typical example is the selection of edges in a graph. In a road network the logic of circulation must be maintained. Detached roads don't make sense; similarly, major road axes (e.g., an interstate highway) should be maintained all the way through.

River networks, which usually exhibit a tree-like structure, can only be pruned from the leaves (i.e., sources) towards the root (i.e., outlet). Quantitative geomorphology has developed a number of so-called stream ordering schemes (Horton 1945, Strahler 1957, Shreve 1966). These ordering schemes reflect the topological order of edges in the river tree from the sources to the outlet (Fig. 6) and can thus be usefully exploited for the generalization of river networks. Edges in the network can be selected at successively higher levels, ensuring the topological consistency of the resulting pruned tree. Of all the ordering schemes in use today, the Horton scheme has proven to be the most useful one for generalization (Rusak Masur and Castner 1990, Weibel 1992), because it combines topological order with metric properties (the longest branches in the tree are assigned the highest order).

115

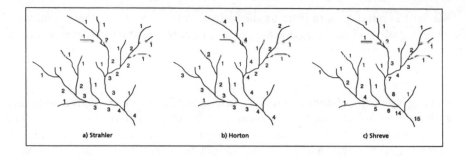

Fig. 6. Stream ordering schemes. a) Strahler order. b) Horton order. c) Shreve order.

7.2 Line Simplification

Line simplification is regarded by many as the most important generalization operator. The majority of map features are either directly represented as lines (e.g., road centerlines, streams), or form polygons which are bounded by lines (e.g., administrative regions, soil polygons, forest stands). Simplification reduces the amount of line detail and thus visibly contributes significantly to the generalization effect. If line simplification is implemented as a vertex elimination algorithm (which is the usual case), it automatically reduces data volume. Simplification algorithms are also highly useful for eliminating high frequency detail on lines digitized by continuous point sampling (stream mode digitizing) or scan-digitizing.

A seemingly countless number of line simplification algorithms has been developed over the past three decades. Commonly, simplification algorithms start with a polyline C made up of two endnodes and an arbitrary set of vertices V. C is then turned into a simplified polyline C' by reducing the number of vertices V to V', while keeping the endnodes fixed. $V\prime$ is thus a subset of V, and no further vertex locations are introduced nor vertices displaced (Fig. 7). The classical criteria which guide vertex elimination are the following: 1) minimize line distortion (e.g., no vertex of C' should be further away from C than a maximum error ε); 2) minimize V'; and 3) minimize computational complexity.

McMaster (1987a) and McMaster and Shea (1992) give overviews of some of the classical simplification algorithms used in cartography and GIS. Hershberger and Snoeyink (1992) and de Berg et al. (1995) list some algorithms from the computational geometry and the image processing literature. Lecordix et al. (1997) describe a comparative implementation of a wide range of algorithms in a research system. Based on the geometric extent of computation, simplification algorithms have been assigned to five categories by McMaster (McMaster 1987a, McMaster and Shea 1992):

1. Independent point algorithms
2. Local processing algorithms
3. Constrained extended local processing algorithms

116

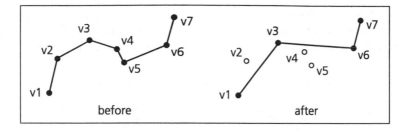

Fig. 7. Line simplification as a vertex elimination process ('weeding'). Note that although this definition is prevailing in the literature on digital generalization, it does not reflect the manual operation. In manual line drawing, simplification of the shape of a line also includes displacements along the line.

4. Unconstrained extended local processing algorithms
5. Global algorithms

Simplification algorithms may alternatively be distinguished with respect to the geometric criterion used to drive the selection of so-called critical points. Figure 8 illustrates some of these criteria, including retained length, angular change, perpendicular distance, and areal displacement.

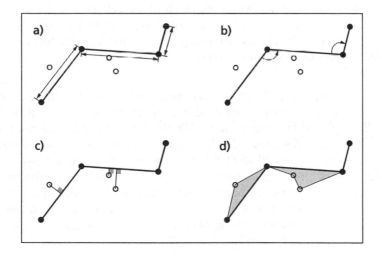

Fig. 8. Alternative geometric criteria which can be used for the selection of critical points in line simplification algorithms.

In the remainder of this subsection, we will describe a few (classical) algorithms from the area of GIS/cartography. See McMaster (1987b) and McMaster and Shea (1992) for details on related algorithms. The following discussion to some extent also reflects the historical evolution of line simplification techniques. Methods are illustrated using the same sample line shown in Figure 9.

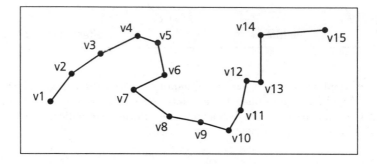

Fig. 9. The sample line used to illustrate line simplification algorithms.

Independent point algorithms. Algorithms of this class do not account for the geometric relationships with the neighboring vertices and operate independently of line topology. Examples are the nth point algorithm (every n^{th} vertex of a polyline is selected, the others eliminated) as well as random selection of vertices. Obviously, these algorithms will only very rarely pick the salient vertices along a polyline (by chance) and may thus result in major line distortions. They are therefore no longer used today.

Local processing algorithms. As the name indicates, the characteristics of immediate neighboring vertices are used in determining selection/elimination of a vertex. Examples of such local criteria are the Euclidean distance between two consecutive vertices, the perpendicular distance to a base line connecting the neighbors of a vertex, as well as angular change in a vertex (McMaster 1987a). The complexity of these algorithms is linear in the number of vertices. As an empirical study has shown (McMaster 1983, 1987b), these algorithms generate less distortion than independent point algorithms, but are inferior to algorithms described below. Nevertheless, due to their localized nature they can be usefully applied for light on-the-fly weeding in line drawing.

Constrained extended local processing algorithms. Algorithms of this class search beyond immediate vertex neighbors and evaluate sections of the polyline. The extent of the search depends on a distance, angular, or number of vertices criterion. A prominent representative of this category is the Lang algorithm, which is also one of the earliest published simplification algorithms (Lang 1969). The extent of the local search is controlled by the so-called 'look-ahead' parameter in this algorithm; the amount of filtering is governed by a perpendicular distance tolerance ε. Thus, the Lang algorithm is frequently assigned to a class of algorithms termed **tolerance band algorithms** or **bandwidth algorithms**, along with other algorithms utilizing a perpendicular distance tolerance to a base line (e.g., Douglas and Peucker 1973, Reumann and Witkam 1974, or Opheim 1982).

Figure 10 schematically depicts the working principle of the Lang algorithm for a look-ahead of 5 points. Perpendicular distances of intermediate vertices to a base line between the beginning point (1) and a floating endpoint (6 = starting point + look-ahead) are computed to evaluate if any vertices exceed the distance tolerance ε (Fig. 10 a) If so, a new floating endpoint is selected (Fig. 10 b) until all vertices fall within tolerance (Fig. 10 c), in which case they are eliminated. Subsequently, the last floating endpoint (4) becomes the new beginning point, and the algorithm continues.

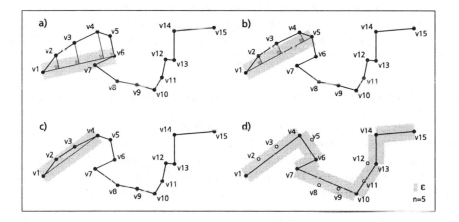

Fig. 10. The Lang algorithm. Figures **a-c** show the first iteration for a complete look-ahead. Figure **d** depicts the resulting line segments, with eliminated vertices shown in white. Note that the result of Figure **d** happens to be the same as for the Douglas-Peucker algorithm for this particular line and ε (cf. Fig. 12 d).

Unconstrained extended local processing algorithms. As with the previous class, local search extends beyond immediate neighbors of a vertex that is being tested, and sections of the line are evaluated. The extent of the search, however, is constrained by the shape complexity of the line, rather than by an arbitrary criterion. The example shown in Figure 11 depicts the algorithm by Reumann and Witkam (1974).

A search corridor made up of two parallel lines at distance ε to either side of the digitized line is extended forward until one edge intersects the digitized

119

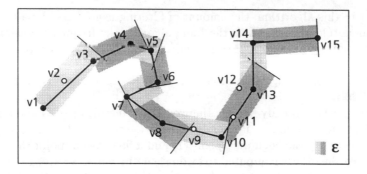

Fig. 11. The Reumann-Witkam algorithm. Vertices which were eliminated are shown in white.

line. All vertices falling within the corridor except the first and last one are eliminated. A new corridor is then extended starting with the last vertex that fell within the subsequent corridor, and the line is sequentially processed until all vertices have been tested.

Global algorithms. Global simplification algorithms consider the entire line and iteratively select critical points, while weeding out vertices within tolerance. The algorithm by Douglas and Peucker (1973) – probably one of the best known simplification algorithms – falls within this class. Just like the Lang and Reumann-Witkam algorithms, it is also a prominent representative of tolerance band algorithms. The algorithm by Visvalingam and Whyatt (1993), on the other hand, is based on an area tolerance controlling areal displacement.

Douglas-Peucker algorithm. While the Douglas-Peucker algorithm may be the line simplification method that is referenced most frequently in the literature it should be noted that nearly identical algorithms were developed independently by Ramer (1972) and Duda and Hart (1973) around nearly the same time. The method was originally developed as a weeding algorithm for removing excessive detail on digitized lines falling within the width of a source cartographic line. The algorithm is illustrated in Figure 12. It starts by connecting the two endpoints of the original line with a straight line, termed the base line or anchor line. If the perpendicular distances of all intermediate vertices are within the tolerance e from the base line, these vertices may be eliminated and the original line can be represented by the base line. If any of the intermediate vertices falls outside e, however, the line is split into two parts at the furthest vertex and the process repeated recursively on the two parts.

Several reasons may be responsible for the popularity of the Douglas-Peucker algorithm. The global tolerance band concept makes it intuitively appealing (although the related theory was only developed *post factum*; see Peucker 1975). A very practical reason for the wide-spread use of this algorithm, however, may

120

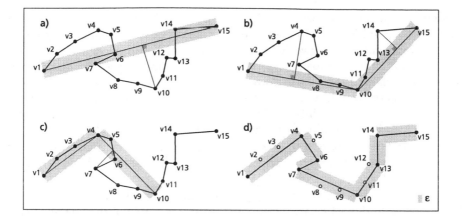

Fig. 12. The Douglas-Peucker algorithm. **a)** Initial base line with furthest vertex ($v10$).
b) First split into two parts, again with furthest vertices ($v4$, $v14$) shown. **c)** Second
split of left part. Vertices $v2$ and $v3$ are now within ε, while the second part must
be split further at vertex $v7$. **d)** Corridors that were eventually generated along the
original line. Vertices which were eliminated are shown in white. Note that the final
result happens to be the same as for the Lang algorithm for the given line and ε.

also be found in the fact that the first author made a Fortran implementation
available at an early stage, leading to the algorithm's adoption in virtually all
the GIS packages on the market.

The fact that the Douglas-Peucker algorithm recursively subdivides the original line in a hierarchical fashion has been usefully exploited by other authors.
Buttenfield (1985) has used the segmentation generated by this algorithm to
build a strip tree that represents a compact geometric description of a line. A
strip of a line segment is formed by the minimum bounding rectangle along its
base (anchor) line. For each strip, geometric measures are calculated and stored.
Van Oosterom and van den Bos (1989) and Cromley (1991) have independently
proposed to build a tree data structure for on-the-fly generalization. Using the
Douglas-Peucker algorithm, the elimination sequence and the perpendicular distance to the base line are pre-computed for all vertices except the endpoints and
stored in a binary tree called the Binary Line Generalization (BLG) tree (van
Oosterom and van den Bos 1989) or simplification tree (Cromley 1991), respectively. Once this data structure has been built, the retrieval of vertices, to the
desired tolerance, becomes a simple tree search. Only those vertices with a perpendicular distance greater than the specified tolerance are retrieved on-the-fly
from the tree for line drawing.

Visvalingam-Whyatt algorithm. McMaster (1983, 1987b) developed two classes
of measures to assess the quality of line simplification algorithms similar to
the geometric criteria shown in Figure 8. The first class relates to *attributes*
of the cartographic line such as length, total angularity and curvilinearity; the

121

second class includes measures which characterize the amount of *displacement* induced by simplification, expressed by the length of displacement vectors and the displacement area between the original and the simplified line. In an empirical study, tolerance band algorithms – in particular the Douglas-Peucker algorithm – showed superior performance relative to these measures (McMaster (1987b). Similar results were reported by perceptual studies involving subject testing (Marino 1979, White 1985), based on the concept of critical points as a psychological measure of curve similarity. However, it should be noted that the algorithms compared in these studies are either extremely simple techniques (e.g., nth point) or themselves representatives of the tolerance band approach. Results can therefore be expected to be biased.

Visvalingam and Whyatt (1993), point out a few deficiencies of the tolerance band approach. In particular, they argue that the selection of the furthest vertex outside the tolerance band as a critical point to be retained is unreliable because this point may be located on spikes (errors) and on minor features. In an attempt to preserve salient shapes and entire features rather than selecting specific points they present an algorithm which eliminates vertices on a line based on their effective area. The effective area E of a vertex is defined as the area of the triangle formed by the vertex and its immediate neighbors (Fig. 13). It represents the area by which the line would be displaced if the vertex was discarded.

The algorithm is simple. It makes multiple passes over the line. On each pass, the vertex with the smallest effective area is considered as least significant and removed. When a vertex is eliminated the effective areas of adjacent vertices need to be recalculated before the next pass. The algorithm repeats until all vertices except the endpoints are tagged with their effective area and their elimination sequence is recorded. The tagged vertices may then be filtered at runtime by interactive selection of the tolerance value for E. This approach of first pre-computing the elimination sequence is similar to the approach used for the Douglas-Peucker algorithm by van Oosterom and van den Bos (1989) and Cromley (1991). As the empirical study presented in Visvalingam and Williamson (1995) suggests, the Visvalingam-Whyatt algorithm indeed seems to perform better on the elimination of entire shapes (caricatural generalization), while the Douglas-Peucker method appears to be better at minor weeding (minimal simplification).

Simplification as an optimization problem. All of the above algorithms have in common that they exploit some kind of heuristic to determine which vertices along a line should be retained. These heuristics may produce adequate results in many cases, but it is difficult to say whether the result is better than that of another algorithm, let alone to determine whether it is optimal with respect to a particular geometric criterion. Only *a posteriori* empirical analysis (McMaster 1983, 1987b) can assess the geometric performance of such heuristic methods.

In reaction to this weakness of heuristic techniques, Cromley and Campbell (1991, 1992) re-formulated the line simplification problem as an optimization problem. Initially, they presented an algorithm that produces an optimal

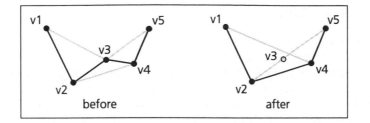

Fig. 13. The Visvalingam-Whyatt algorithm. Effective areas are computed for each vertex except the endpoints using the area of the triangle formed by each vertex and its immediate neighbors. Vertex $v3$ then is the first one to be eliminated, and the area of neighboring vertices $v2$ and $v4$ needs to be recomputed (after Visvalingam and Whyatt 1993).

simplification with respect to the tolerance band criterion using mathematical programming techniques (Cromley and Campbell 1991). This method was subsequently extended by integrating qualitative criteria such as those shown in Figure 8. Using these types of criteria, line simplification is stated as the problem of minimizing (or maximizing) a particular geometric property of a line (e.g., maximize line length, minimize areal displacement), subject to a constraint on the number of individual vertices retained in the simplified line. The maximum number of retained vertices is obtained from Töpfer's selection formula (cf. Subsection 7.11). To solve this multi-criteria problem, the digitized line is considered as a directed acyclic graph of all possible $\frac{n(n-1)}{2}$ segments which connect the n vertices (Fig. 14). Each segment is attributed a cost value c_{ij} which represents the cost of traversing the segment. Each c_{ij} corresponds to a geometric performance measure such as the line length or areal displacement associated with a line segment. Optimal line simplification is then approached as a form of shortest path problem, in which a path through the graph is to be found that minimizes the total cost.

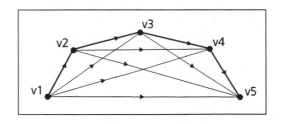

Fig. 14. Alternative paths for a simplified line connecting endpoints 1 and 5 (after Cromley and Campbell 1992).

123

7.3 Line Smoothing

Line smoothing in many ways forms a complement to line simplification. According to the definition commonly used, line smoothing techniques "shift the position of points in order to improve the appearance [of a line]. Smoothing algorithms relocate points in an attempt to plane away small perturbations and capture only the most significant trends of the line. Thus smoothing is used primarily for cosmetic modification" (McMaster and Shea 1992: 84-85). Figure 15 illustrates this process. Note, however, that this definition of line smoothing is currently under re-examination by several authors (Ruas and Lagrange 1995, Plazanet 1996, Weibel 1996).

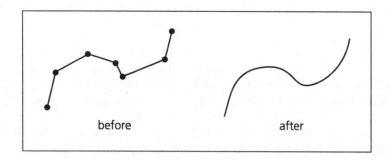

before   after

Fig. 15. Principle of line smoothing.

McMaster (1989) distinguishes three groups of smoothing algorithms:

1. Weighted averaging techniques
2. Epsilon filtering techniques
3. Mathematical approximation

Weighted averaging techniques are based on averaging of vertex coordinates. Mathematical approximation methods encompass techniques such as Gaussian filtering as well as curve fitting (Rogers and Adams 1990). We restrict the discussion to a representative of the second group. Epsilon filtering methods are based on the paradigm that generalization is a process of reducing the spatial resolution of a map such that no detail is displayed that is smaller than the smallest perceptible size ε.

Li-Openshaw algorithm. Li and Openshaw (1992) proposed a 'natural principle' of line generalization based on a concept similar to so-called Epsilon filtering (Perkal 1966), and presented three different algorithms which implement this principle: a vector-mode algorithm, a raster-mode algorithm, and a mixed raster-vector algorithm. We describe the raster-vector algorithm only, as it produces the best generalization results according to the authors.

In a first step, the size F_c of the 'smallest visible object' (SVO) at the target scale is estimated according to the formula

$$F_c = S_t D \left(1 - \frac{S_s}{S_t} \right) \qquad (2)$$

where S_t is the scale factor of the target map; D is the diameter of the SVO at the target map scale (in map units), within which all information can be neglected; and S_s is the scale factor of the source map.

A local square grid with a spacing equal to F_c is then overlaid on each cartographic line, with the origin centered at the beginning node (Fig. 16). Next, intersections with the grid are calculated along the original line. In addition to the endpoints of the line, resulting points on the output line consist of the midpoints of each pair of consecutive intersections.

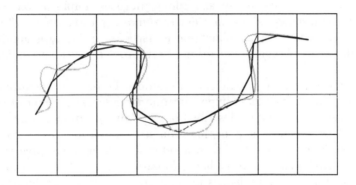

Fig. 16. The Li-Openshaw method: raster-vector algorithm. The side length of a grid cell reflects the size F_c of the smallest visible object (SVO). The heavy gray curve shows the original line. The fine line represents the segments connecting intersections of the original line with the grid. Finally, the heavy solid line represents the resulting line connecting the midpoints of the fine line segments.

8   Operator Sequencing

Subdividing the overall generalization process into individual operators alleviates the development and implementation of specific tools for generalization, allowing to address particular generalization tasks in a flexible way through a combination of different operators, algorithms, and parameter sets. On the other hand, the functional break-down into generalization operators also requires that the appropriate combination and sequence of operators (and associated algorithms) must be determined for a given generalization problem.

Unfortunately, there is no sequence of operators that is valid for all scale ranges, map purposes, and combinations of feature classes. It is possible to some extent to develop generic sequences for a particular *class* of generalization problems and product specifications (e.g., for generalization of landuse maps at medium to small scales). For each specific generalization problem, however, the operator/algorithm combination and sequence has to be fine-tuned and calibrated specifically. It is also a common fact that if two algorithms – with the same parameter values – are applied in reverse order, the result will not be the same (Monmonier and McMaster 1991, Plazanet 1996).

Research in operator and algorithm sequencing is still relatively sparse to date. Examples include Lichtner (1979) who proposed a generic sequence for large scale topographic map generalization, McMaster (1989) and Monmonier and McMaster (1991) with studies on sequential effects of line simplification and smoothing algorithms, and Lecordix et al. (1997) with empirical comparisons of line caricature algorithms. McMaster (1989) proposes a detailed procedure for generalizing linear data in which both smoothing and simplification are applied in two phases. In the first phase, smoothing precedes simplification (both using conservative parameter values) in order to remove spurious effects of digitizing. During the second phase, simplification is used for an initial generalization to target scale, with subsequent smoothing to improve the aesthetic quality of line drawing.

In today's interactive generalization systems the issue of operator and algorithm sequencing is even more important due to the large number of algorithms offered in systems of the type described by Lee (1995). Interactive systems, however, also allow to establish algorithm sequences under interactive control, with the option to fine-tune and 'train' parameter sets on representative sample data in order to subsequently apply them globally to the entire data set. Finally, some pragmatic general guidelines for operator sequences can be derived from cartographic practice:

- *Selection/elimination:* Is applied first as it eliminates insignificant details and features and increases available space.
- *Aggregation/amalgamation/merging:* These operators combine selected features and thus save space. Additionally, they induce a transition of topological type (e.g., point to area, double lines to single line, area to line) and thus must precede line processing operators (simplification, smoothing, etc.).
- *Simplification:* Reduces detail and contributes to line caricature. Should therefore be applied at an early stage.
- *Smoothing:* Contributes to aesthetical refinement. Follows simplification.
- *Displacement:* Used to resolve spatial conflicts created by previous operators.

9 Model Generalization – The Example of TIN Filtering

As was explained above model generalization functions are crucial to the development and derivation of *databases* at multiple levels of resolution. Frequently, model generalization relies on the exploitation of hierarchies which are inherent to spatial data. For instance, in the attribute (i.e., thematic) domain, the classical example of inherent hierarchies are categorical data such as land use or soil classifications. Such intrinsic relations can be formalized for storage and retrieval (Molenaar 1996b, Richardson 1994). In this section, however, we concentrate on a single example of a *geometric* model generalization process. Further examples of model generalization methods – involving aspects of thematic and temporal model generalization – can be found in the reader edited by Molenaar (1996a).

Filtering operations for data reduction are an essential component of terrain modeling systems. As more and larger datasets are being processed and new methods for high-density data collection are being put to use, the necessity for an adaptation of secondary models to the desired resolution and accuracy is becoming more urgent. For instance, it is not necessary to carry all the minute details contained in a particular model through the generation of an animated sequence if the result does not show them. TIN filtering can be desirable to eliminate redundant data points within the sampling tolerance, to detect blunders, to save storage space and processing time, to homogenize a TIN, or to convert a gridded terrain model into a TIN. It may also be used as a component of terrain generalization (Weibel 1992; cf. Subsection 12.3).

The *objective of TIN filtering* is to find an approximate representation of a field of elevations, that is, a bivariate function which nowhere deviates from the original surface by more than a specified tolerance Δz. TIN filtering thus essentially forms the 3-D equivalent of the simplification of plane curves (Subsection 7.2).

Van Kreveld (1997) gives further references to algorithms for TIN filtering. Our discussion focuses on the role of TIN filtering in model generalization and briefly presents a particular incremental algorithm developed by Heller (1990), termed 'adaptive triangular mesh (ATM) filtering'. It is based on a coherent approach of successive construction of Delaunay triangulations. However, the method can be used to reduce the data volume of both grids and TINs. A grid is just considered as a special case of a TIN, with nodes arranged in a rectangular grid. The general flow of the algorithm is as follows:

1. Start with an initial set of points: selected points on the convex hull and the *significant extremes*.
2. Triangulate these points to build an initial triangulation.
3. Determine the priority of the remaining points forming the initial priority queue. The priority of a point is calculated as the vertical distance to the current triangular mesh weighted by the inverse of the tolerance Δz.
4. Select the point with the largest priority and insert it into the triangulation, swapping edges of affected triangles to maintain the Delaunay criterion.
5. Readjust the priorities of the affected points.
6. Repeat steps 4 and 5 until no point remains whose vertical distance exceeds the user-specified tolerance Δz.

A few auxiliary data structures are used to achieve an efficient algorithm. The priority queue of points waiting to be inserted into the triangulation is organized in a heap. The points pertaining to each triangle are linked into a list. The insertion of a point requires a local retriangulation which consists of swapping all necessary triangles to maintain the Delaunay criterion, and readjusting the priorities of all affected points. It is obvious that the time required for retriangulation is proportional to the number of readjusted points and the logarithm of the number of queued points. Therefore, a heuristic is used to start the process with as many significant points as possible.

The set of initial points is formed by selected points on the convex hull and the *significant extremes*. The points which are selected on the hull include all consecutive hull points which are not collinear (i.e., not in line with respect to their planimetric location). Collinear points are handled specially, which is particularly important when the input points originate from a regular grid, since all points on the edge of the grid are collinear. A variant of the Douglas-Peucker algorithm is applied to the profile of collinear points using Δz as a distance tolerance. Local extremes form further candidates for the initial point set. The following definitions are used to select the *significant extremes*:

- A local minimum is considered as significant if it is the global minimum in a basin of depth greater or equal Δz.
- A local maximum is considered as significant if it is the global maximum on a hill of height greater or equal Δz.

These definitions lead to a straightforward approach for the determination of local minima and maxima. The local minima are sorted by their altitude by inserting them into a priority queue. Then, the following step is repeated until all minima in the queue are tested. The lowest remaining minimum z_i is selected, and the points in its neighborhood traversed radially until the lowest point along the 'wavefront' of this traversal is higher than $z_i + \Delta z$. If a local minimum is found in this process, it can be removed from the priority queue. As soon as a point is found which is lower than z_i, the traversal is aborted and the current minimum discarded. The same method is also used in an analogous way to determine significant maxima.

An example of ATM filtering is shown in Figure 17: starting from a gridded digital terrain model (68,731 points), a TIN with a tolerance of 5 m (with 11,450 points or 16.7 remaining), and a TIN with a tolerance of 10 m (5,732 points or 8.3) were obtained.

The fact that for the determination of the significance of a point the vertical distance is weighted offers the potential for useful extensions. In the normal case, the weight is set to the inverse of Δz, and therefore constant. However, the weight can also be modified individually according to the specific properties of each point. For instance, points on structure lines can be assigned higher weights than others, thus enabling the preservation of linear structural features. In a similar way, the level of detail of perspective views can be adjusted according to viewing depth. Height values of points can be weighted according to some function that is proportional to the inverse of the distance of a point to the viewpoint (Hess 1995, Misund 1996). Points near the viewer are thus assigned higher weights than distant ones, causing more points to be removed from the TIN in distant regions which are less likely to be discernible.

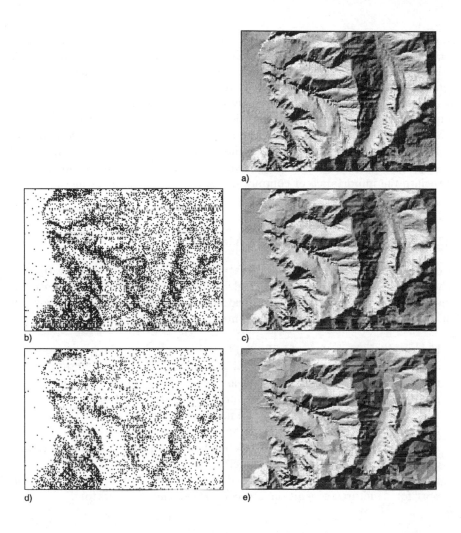

Fig. 17. Sample runs of ATM filtering for data reduction and grid-to-TIN conversion.
a) Original grid digital terrain model (311 x 221 = 68,731 points; 25 m spacing). **b)**
Remaining points (11,450 or 16.7) after ATM filtering with $\Delta z = 5$ m. **c)** Hillshading
of corresponding TIN. **d)** Remaining points (5,732 or 8.3) after ATM filtering with Δz
= 10 m. **e)** Hillshading of corresponding TIN. (DTM data courtesy of Swiss Federal
Office of Topography, DHM25 ©1997, (1263a))

10 An Assessment of Basic Algorithms

Basic generalization operators and algorithms as discussed in Sections 7 to 9
largely represent the state of the art of available generalization tools in current
commercial GIS (Schlegel and Weibel 1995) and also form the core of special-
purpose generalization systems such as Intergraph's MGE Map Generalizer (Lee
1995, Weibel and Ehrliholzer 1995). A number of deficiencies have been observed

129

and documented in the literature (Muller 1990, Beard 1991, Plazanet et al. 1995, de Berg et al. 1995, Weibel and Ehrliholzer 1995, Lecordix et al. 1997). This section presents a brief assessment of basic generalization methods, attempting to identify weaknesses as well as key areas for future research. Note that the discussion primarily focuses on *functional* deficiencies and improvements, rather than on aspects of computational efficiency. In an evaluation of the quality of today's generalization methods computational efficiency is only secondary to a functional assessment. Many methods just don't do what they are expected to do, so producing garbage fast is not really an objective. However, we certainly appreciate the importance of computationally efficient methods in the context of interactive generalization and databases of increasing size.

10.1 What's Wrong with Basic Algorithms?

Based on the study of the above literature as well as empirical investigations (Schlegel and Weibel 1995, Weibel and Ehrliholzer 1995) we have identified a number of weaknesses of basic generalization methods with respect to algorithms and data structures, which can be summarized as follows:

- Independent processing of individual features neglects spatial context.
- Structure and shape recognition for the characterization of map objects is restricted to simple heuristics (such as the tolerance band). It is not explicitly represented in terms of shape measures and spatial relations.
- Algorithms are unspecific; they are not tailored to the properties of specific feature classes (e.g., simplification of building outlines).
- Algorithms to implement context-dependent operators (displacement, amalgamation, aggregation, caricature, etc.) are largely missing.
- Feature representations and data structures commonly used offer little support for structure recognition and context-dependent operators.

10.2 What Should Be Improved?

The necessary improvements of basic algorithmic methods and the development of more advanced algorithms basically fall into three (strongly interrelated) areas:

- *Constraint-based methods:* Algorithms must observe the spatial and semantic constraints imposed by map context.
- *Methods for structure and shape recognition:* Structure recognition must be made explicit. Analysis of shape and structure of map features must precede the execution of generalization algorithms. It is necessary to select an appropriate set of operators, algorithms and parameter values.
- *Alternative data representations and data models:* Generalization requires a rich data model encompassing a combination of different data representations and auxiliary data structures.

In the remainder of this survey, we briefly discuss selected representatives of topical work for each of these areas, rather than presenting a comprehensive review of current research. Note that beyond the problems of algorithmic nature, further deficiencies can be observed with non-algorithmic issues which prompt an equally strong need for future research:

- *Knowledge-based methods:* There is a lack of procedural knowledge in generalization, and knowledge acquisition (KA) has proven to be a major bottleneck. New methods for KA must be developed, including techniques of computational intelligence (Weibel et al. 1995). Integration of knowledge-based and algorithmic techniques is also a major issue.
- *Quality assessment:* Criteria and methods (quantitative and qualitative) for the assessment of the quality of generalization methods are largely missing. Development of criteria and measures and evaluation methods to implement them are required (Weibel 1995b).
- *Human-computer interaction:* Current user interfaces are not designed specifically for generalization. Optimized user interfaces, strategies of sharing the responsibility between system and user must be developed.
- *Practical issues:* In commercial GIS, there is still a problem with the adoption of results from advanced research (the Douglas-Peucker algorithm is frequently the only method offered). Also, current systems often offer little decision support to the user, low quality graphics function (e.g., cartographic drawing), cryptic GUIs, etc.

11 Constraint-Based Methods

Context-independent generalization algorithms as outlined in Sections 7 to 9 exhibit a fundamental problem: they process each map object individually, neglecting the context which the object is embedded in. Most basic algorithms concentrate purely on metric criteria and even the simplest topological or semantic constraints are ignored. As a result, lines may intersect with themselves, with other lines nearby, or points may fall outside polygons, to name but a few of the most frequent problems (Muller 1990, Beard 1991, de Berg et al. 1995, Fritsch and Lagrange 1995).

In terms of the development of methods that can satisfy additional non-metric constraints, the simplification of polygonal subdivisions has recently attracted research interest. Polygonal subdivisions are a frequent data type in GIS applications (political boundaries, vegetation units, geological units, etc.) and present particular problems to basic line simplification algorithms (Fig. 18). Weibel (1996) has attempted to identify the constraints that govern polygonal subdivision simplification, proposing a typology of metric, topological, semantic and Gestalt constraints and reviewing relevant previous research. Two basic alternatives exist to resolve problems such as the ones illustrated in Figure 18: 1) the problems are cleaned up in a post-processing operation or 2) the simplification algorithm incorporates the corresponding constraints and thus avoids the problems in the first place. Muller (1990) has presented a post-processing method to remove self-intersections created by spurious line simplification algorithms. Intersections are detected and affected vertices displaced to eliminate the problem. Alternatively, de Berg et al. (1995) proposed an algorithm for the simplification of chains of polygonal subdivisions that extends the basic simplification techniques and satisfies four different constraints. If C is a polygonal chain and P a set of points that model special positions inside the regions of the map (e.g., cities in countries), then it is required from its simplification C':

1. No point of the chain C has a distance to its simplification C' exceeding a prespecified error tolerance.
2. C' is a chain with no self-intersections.
3. C' may not intersect other chains of the subdivision.
4. All points of P lie to the same side of C' as of C.

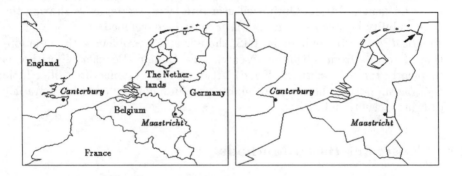

Fig. 18. An example of an inconsistent simplification of a subdivision (source: de Berg et al. 1995).

Instead of trying to satisfy all conditions at once, each condition is dealt with individually and the final result extracted from the combination of partial solutions. A polygonal chain is understood as a directed acyclic graph G, with vertices v_i forming the nodes of the graph. Each line segment, called a shortcut, that is valid relative to a particular condition is added to G. For each of the four conditions, a separate graph $G_1, \ldots, G_4$ is created. The final graph G is built from shortcuts that are allowable in all graphs G_i representing the partial solutions. In G, the resulting minimum vertex simplification of the polygonal chain C is found as the shortest path between the endpoints.

In order to build the graph G_1 that satisfies the first condition, the algorithm by Imai and Iri (1988) is used. In the version of the paper published in de Berg et al. (1995) the input chains are required to be x-monotone. The second condition is thus met automatically as no self-intersections can occur in x-monotone chains, and G_2 is equivalent to G_1. The solutions for the third and fourth condition can

132

be combined: Vertices of the chains of polygons adjacent to C are added to point set P. G_4 thus need not be established. Furthermore, only the points falling inside the convex hull of the chain C being simplified could possibly end up to the wrong side of C'. The actual number of candidate points in P can thus be reduced further. The algorithm for determining consistent shortcuts (with respect to the locations of points in P) leading to G_3 is described in detail in de Berg et al. (1995).

De Berg et al. (1995) have shown that their algorithm for simplifying a polygonal subdivision with N vertices and M extra points runs in $O(N(N+M)\log N)$ time in the worst case. Empirical studies with real data will need to establish whether this close to quadratic time behavior actually shows up.

12   Methods for Structure and Shape Recognition

12.1   Motivation and Objectives

As Section 4 discussed, structure and shape recognition (i.e., cartometric evaluation) is logically prior to the application of generalization operators (Brassel and Weibel 1988, McMaster and Shea 1992). Structure recognition allows to determine when and where generalization needs to be applied and furnishes the basis for the selection, sequencing, and parametrization of an appropriate set of generalization operators for a given problem. Cartographic data often are relatively unstructured. Entity definition in most spatial databases stops at the level of individual map features; parts of features (e.g., a hairpin bend on a road or an annex of a building) are rarely coded explicitly. Most spatial databases also contain little semantic information in terms of the relative importance of individual map objects – which is crucial in generalization since the purpose is to distinguish between important and insignificant features. Finally, little information is normally stored on shape properties of map features. The *objectives of structure recognition* are therefore the following:

- to structure the source data according to the requirements of the intended generalization
- to 'enrich' the source data
- to derive secondary metric, topologic and semantic properties:
  - metric: shape characteristics, density, distribution, object partitioning, proximity relations
  - topologic: topologic relations not represented in the source data
  - semantic: relative importance (priority) of map objects, logical relations between objects

12.2   Characterization and Segmentation of Cartographic Lines

Our first example of structure recognition is concerned with the analysis of linear data. Since such a great share of cartographic data are of linear type, this task

133

is of considerable importance. As Figure 19 shows, analysis and segmentation of cartographic lines into meaningful components is essential in order to decide how individual shapes need to be treated during generalization. Depending on shape properties such as sinuosity, but also depending on context information, different operators and algorithms may be applied.

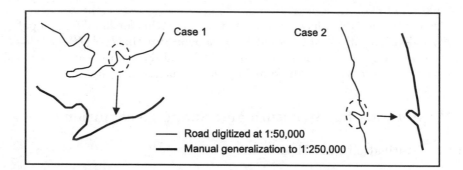

Fig. 19. Example of different generalization alternatives for the same shape (after Plazanet 1995).

First attempts at cartographic line characterization and segmentation have been made by Buttenfield (1985) who based her segmentation procedure on the partitioning scheme resulting from the Douglas-Peucker simplification algorithm (cf. Subsection 7.2). Recently, an approach which is not biased towards a particular generalization algorithm has been presented by Plazanet (Plazanet 1995, Plazanet 1996, Plazanet et al. 1995). The procedure generates a hierarchical segmentation of the line according to a *homogeneity* criterion. The resulting tree structure is called the *descriptive tree*. Figure 20 illustrates such a tree. The homogeneity of the individual sections is intuitively apparent. The homogeneity definition used to split up the line is based on the variation of the distances between consecutive inflection points. Inflection points where the sign of the difference between the mean of distances and the distance between the current and consecutive inflection point changes are considered as 'critical points' for segmentation (Fig. 21).

The objective of segmentation is mainly to obtain sections of the line that are geometrically sufficiently homogeneous to be tractable by the same generalization algorithm and parameter values. Further information may be added to the descriptive tree which characterizes the *sinuosity* (and thus the prevailing geometric character) of each line section. A variety of measures can be obtained from the deviation of the cartographic line from a trend line formed by the *base line* connecting consecutive inflection points. These measures are calculated for each individual bend (Fig. 22) and then averaged to yield the values for the corresponding line section. Using the proposed segmentation procedure and shape measures, lines can be segmented and the resulting line sections classified

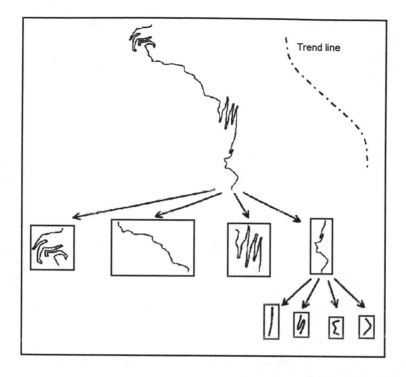

Fig. 20. An example of line segmentation (courtesy of C. Plazanet, IGN France).

according to their degree of sinuosity (e.g., using cluster analysis). Results of classification can be found in Plazanet (1995), Plazanet et al. (1996) or Plazanet (1996).

12.3 Terrain generalization

Our second example of structure recognition is intended to show the use of structural and shape information in terrain generalization. Several techniques for terrain generalization have been reported in the literature (see Weibel 1992 for a brief review). Weibel (1992) proposes three classes of generalization methods which are integrated into a common strategy whose purpose it is to select the appropriate method based on an analysis of the character of the terrain represented by the digital terrain model (DTM). This initial analysis step is termed *global structure recognition*. It is similar in nature to the approach for 2-D line characterization by Plazanet (1995, 1996) in that it segments the continuous terrain surface into patches of homogeneous terrain character based on the computation of a set of geomorphometric measures for each DTM point.

The three classes of terrain generalization methods are termed global filtering, selective filtering and heuristic generalization. *Global filtering* consists of a variety of smoothing filters (in the spatial and frequency domain), combined

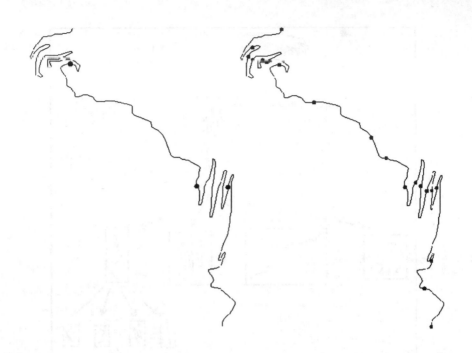

Fig. 21. Right: Detected inflection points. Left: Critical points retained automatically (courtesy of C. Plazanet, IGN France).

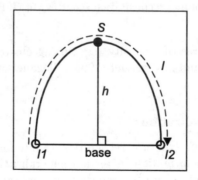

Fig. 22. Sinuosity measures (after Plazanet 1995).

with filters for enhancement. It is intended for use in smooth, rolling terrain and minor scale changes. *Selective filtering* is equivalent to ATM filtering described in Section 9. It is both aimed at cartographic generalization (minor scale change in rugged terrain) and model generalization (for the purposes outlined in Section 9). *Heuristic filtering* is potentially the most flexible method as it is the only approach which is based on explicit structure recognition. It is intended for use in complex fluvially eroded terrain.

136

Heuristic terrain generalization is based on an emulation of principles used in manual cartography. Manual cartographers use sketches of structure lines (drainage channels, ridges, and other breaklines) as a skeletal representation of the continuous terrain surface in order to guide the generalization process. Heuristic terrain generalization is thus preceded by a step called *local structure recognition*. This process yields a model called the *structure line model (SLM)*. The networks of drainage channels and ridges are extracted from the DTM, as well as the associated area features (i.e., drainage basins and hills) and additional significant points (for a review of relevant algorithms, see van Kreveld 1997). All features are tagged with descriptive geomorphometric attributes. Note again the similarity of the SLM approach to the 2-D descriptive tree of Plazanet (1995, 1996).

This rich information of the SLM is then used as a 3-D skeletal structure of the DTM and forms the basis of heuristic generalization operations. The 3-D skeleton of the SLM can be generalized by eliminating and simplifying network links, resulting in a derived version of the SLM. The reduced network still represents a 3-D, though simplified, structure. Thus, the generalized surface can be reconstructed by interpolation from the generalized structure lines. Figure 23 shows a sample terrain surface (Fig. 23 a) which was generalized using this procedure (Fig. 23 b). Compared to the original model, the derived model has been modified in many ways. Smaller landforms have been dropped or combined as a result of the elimination process. Detail has generally been reduced, the major structures have been retained, though.

Figure 23 c) shows a further processing step. By means of an interactive DTM editor, the essential landforms have been retouched and enhanced. As a consequence, the map image has become clearer and more expressive. Thus, the important structures remain clearly discernible even at small scales, as the reduced versions of Figure 23 c) demonstrate. The interactive editor used for final retouching was developed in a project that focused on dynamic modification of DTMs (Bär 1995). This editor offers a range of tools which allow to modify the surface of a grid DTM in real-time, in close analogy to tools of daily life such as spatula and iron. The individual tools are implemented as local filters in the spatial domain that can both be controlled interactively and be guided automatically along predefined lines. For any tool, its size, basic shape (footprint) and the degree of cutting and filling can be specified.

13 Alternative Data Representations and Data Models

The evaluation of geometric data structures given in Subsection 10.1 can be summarized as follows: Generalization algorithms cannot be expected to advance much beyond their current state unless certain limitations of presently used geometric data structures are overcome. That is, alternative schemes of data modeling and representation must be exploited. The problem must be addressed at two levels:

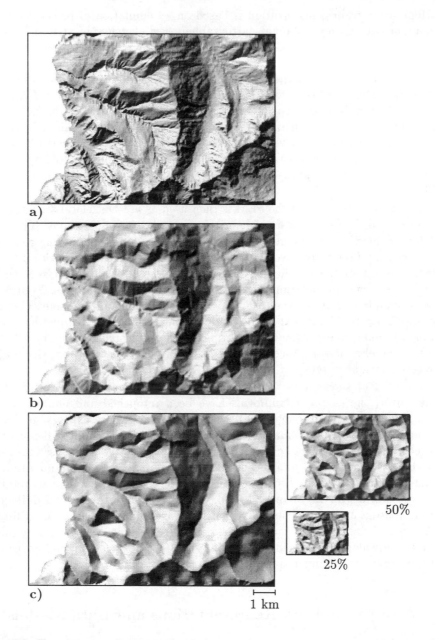

Fig. 23. Terrain generalization. **a)** Original surface. **b)** Generalized surface, resulting from the extraction and generalization of the network of topographic structure lines (channels and ridges). **c)** The automatically generalized surface has been modified further by interactive enhancement. (DTM data courtesy of Swiss Federal Office of Topography, DHM25 ⓒ1997, (1263a))

- *Representations for geometric primitives:* Basic schemes available for the representation of geometric primitives (points, lines, etc.). Examples of commonly used representations are the polygonal chain (or polyline), raster or mathematical curves.
- *Complex data models and data structures:* Complex data models allow to integrate primitives into a common model and record their spatial and semantic relations. Examples are the topological vector data model, but auxiliary data structures (uniform grid, quadtree, Delaunay triangulation, etc.) are also of use in this context.

13.1 Representations for Geometric Primitives

In vector mode generalization, polygonal chains (polylines) are by far the most commonly used scheme for representing geometric primitives. They are easy to implement and handle, intuitive to understand and they can approximate any desired shape accurately (provided the vertices are sampled sufficiently densely). On the other hand, the polyline representation also imposes severe impediments on the development of generalization algorithms (Werschlein 1996, Fritsch and Lagrange 1995). Allowable generalization operators are essentially restricted to removing points (i.e., line simplification by vertex elimination) or displacing points (i.e., line smoothing). The fact that a polyline is equivalent to a chain (i.e., sequence) of points implies that it is difficult to model entire shapes in a compact term.

The polyline representation certainly still has its merits in many generalization applications, but it should be extended by complementary representations. The work by Affholder on geometric modeling of road data (reported in Plazanet et al. 1995) is an example of fitting the representation scheme more closely to the object that needs to be represented. Road data are commonly represented as polygonal chains, neglecting the fact that these man-made features are constructed using mathematical curves rather than free-form chains of points. Affholder models roads by a series of *cubic arcs*, leading to a more compact and also more natural representation which offers potential for the development of novel algorithms. For each bend of a road between two inflection points, a pair of cubic arcs is used to approximate the left and right half of the bend, respectively.

Other representations that bear potential for complementing polylines in a useful way are parametric curve representations and wavelets. *Curvature-based curve parametrizations* can be usefully exploited for shape analysis since critical points (such as inflection points) show up as extremes (Werschlein 1996). In addition to that, the magnitude of these extremes also exhibits the size of the shape associated with a critical point and thus allows to prioritize. *Wavelets* have potential for both shape analysis (Plazanet et al. 1995, Werschlein 1996) and as a basis for novel generalization algorithms (Fritsch and Lagrange 1995, Werschlein 1996). Wavelet coefficients can be analyzed to locate critical points and shapes, and they can also be filtered yielding generalized versions of the original feature. Since wavelets are localized, it is possible to eliminate entire shapes by setting the coefficients of the wavelets supporting the shape to zero (Werschlein 1996).

13.2 Complex Data Models

While the search for alternative representations for geometric primitives is mainly guided by the requirements of shape representation and shape analysis, research for improved complex data models is driven by the need to develop adequate algorithms for the operators of *context-dependent generalization*. That is, data models used for generalization must be extended to allow improved representation of spatial and semantic relations between individual features and feature classes. The requirements for improved data models can be summarized as follows:

- Representation of relevant metric, topological and semantic relations must be possible between objects of the same feature class and across feature classes. In particular, representation of proximity relations (metric) must be improved.
- Object modeling:
  - Multiple primitives per object (e.g., a coastline is partitioned into different sections – sandy beach, estuary, rocky shore)
  - Grouping (groups of objects of the same feature class), complex objects
  - Shared primitives between objects of different feature classes
- Integration of auxiliary data structures for computing and representing proximity relations (triangulations, regular tessellations)

As a consequence of these requirements the main data model should be an object-oriented extension of the basic topological vector model (as opposed to layer-based). Data models of this kind are now beginning to appear in some commercial GIS. Integrated auxiliary data structures for proximity relations are not yet available in commercial systems, but research is under way in that direction.

Most approaches to represent proximity relations between map objects have concentrated on the use of Delaunay triangulations or Voronoi diagrams (Ruas 1995, Ruas and Plazanet 1996, Ware et al. 1995, Jones et al. 1995, Ware and Jones 1996). An example of the use of a regular triangular tessellation for line generalization has been presented by Dutton (1996a). Dutton's quaternary triangular mesh (QTM) scheme is interesting for a variety of reasons other than generalization (outlined in Dutton 1996b). It offers a method for planetary geocoding of both local and global geospatial data as an alternative to the traditional latitude/longitude coordinate notation. Starting with an octahedron inscribed to the globe, the eight faces of the octahedron are successively subdivided into four equilateral triangles (and vertices projected to the surface of the globe), yielding eight quadtree-like structures. Triangles are coded into 64-bit words. A QTM location code (QTM ID) consists of an octant number (from 1 to 8) followed by up to 30 quaternary digits (from 0 to 3) which name a leaf node in a triangular quadtree rooted in the given octant. For example, the 18-level QTM ID for the building housing the Geography Department at the University of Zurich is 1133013130312301002. This encodes the geographic location 47° 23' 48" N, 8° 33' 4" E within about 60 meters, close to the precision obtained from measuring on a 1:100,000 scale map. Adding more digits increases the locational precision. At level 30, locations are encoded to a precision of about 2 cm. This code just fills one 64-bit word, equivalent to two single precision floats used for latitude/longitude encoding. However, 32 bit latitude/longitude coordinates do not allow for the same level of precision as QTM encoding.

Not only does QTM encoding offer a very space-efficient location scheme, it also allows to adapt the level of resolution used for encoding (and therefore the locational precision that can be achieved) to the accuracy of the data. In other words, it offers a convenient way to handle locational uncertainty contained in geographical data. Since the QTM location scheme is hierarchical, it is also inherently related to scale-changing. Dutton (1996a) has proposed an algorithm for line generalization that makes use of this property. Lines whose vertices had been encoded to a certain QTM level (e.g., level 20, which relates to 1:25,000) are weeded to the locational precision of a coarser QTM level (e.g., level 18, roughly equivalent to 1:100,000). Any vertices that fall within the same 'QTM attractor' (the hexagonal region formed by the six triangles surrounding a QTM node) are replaced by the median point of the corresponding section of the line. Beyond this direct use of the hierarchical structure of QTMs, there is also potential for using it in conflict detection and spatial search.

The data models used by Ruas (1995; see also Ruas and Plazanet 1996) and by Ware et al. (1995; see also Jones et al. 1995, Ware and Jones 1996) are both based on Delaunay triangulations (cf. van Kreveld 1997). Both approaches concentrate on the support of methods for the detection and resolution of spatial conflicts between features (e.g., features that overlap, features that are too close, etc.). Additonally, both use the space subdivision scheme as a means to compute proximity relations, compute displacement vectors (in the case of conflict), and keep track of displacements. Beyond these similarities, however, the two data models are built on a different approach.

Ruas (1995) attempts to embed the use of her proposed data model in a comprehensive strategy of generalization (see also Ruas and Plazanet 1996). Generalization – in Ruas' case the generalization of built-up areas – is seen as a process of conflict detection and resolution. A conflict is defined as an infringement on a cartographic principle (such as minimum visual separability of map features, avoidance of overlaps, etc.). Conflict detection proceeds in a hierarchical fashion; it is first carried out at the global level (i.e., the entire map), then the map space is subdivided according to the hierarchy of the road network (Fig. 24). Within each of the resulting partitions, conflict detection and resolution again takes place, starting at level 1 and proceeding to finer levels.

The Delaunay triangulation is then built within each partition that is currently worked on. Note that this is an unconstrained Delaunay triangulation and that it is kept local (i.e, to the elements of the current partition only) and

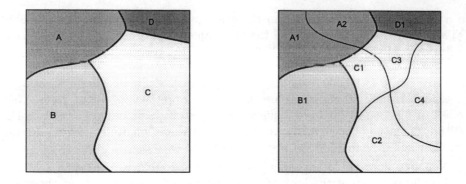

Fig. 24. Hierarchical subdivision scheme based on hierarchy of road categories (after Ruas 1995).

temporary (i.e., it is not saved). The triangulation connects the centroids of the small area objects and point objects falling within the tile, as well as projection points on the surrounding roads forming the tile boundary (Fig. 25). The edges of the triangulation are classified according to the objects they connect (Fig. 25). If the shape of a bounding road is changed or houses are enlarged or moved, the triangulation is used determine any conflicts that might have arisen. Displacement vectors are then computed from the proximity relations between objects and displacement propagation activated using decay functions (Fig. 26).

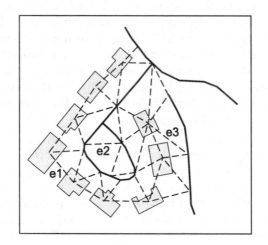

Fig. 25. Local Delaunay triangulation between buildings and adjacent roads (after Ruas 1995). Edge e_1 denotes an edge connecting two buildings; e_2 connects two vertices on a road; and e_3 connects a building and a road.

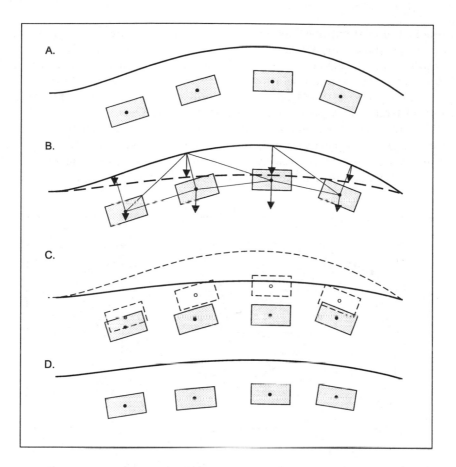

Fig. 26. Displacement of buildings after simplification of a road (after Ruas and Plazanet 1996).

There are several points in which the triangulated data model developed by researchers at the University of Glamorgan (Ware et al. 1995, Jones et al. 1995, Ware and Jones 1996) differs from the approach chosen by Ruas (1995). The triangulation forms the core of the data model. As a consequence, it is not only computed temporarily, but maintained continuously. Not only the centroids of map objects are connected, but a constrained Delaunay triangulation of all the vertices of all map objects is built (Fig. 27). The resulting simplicial data structure (SDS) is represented through a set of relations between objects, triangles, edges and vertices. These relations are stored by pointers corresponding to the entity relationships illustrated in Figure 28.

Jones et al. (1995) claim several useful properties for the SDS model: the explicit representation of all space on a 2-D map; precise representation of object boundaries from vector-structured data; ease of measurement; maintenance of topological relationships between points, lines and polygons; ease of determination of proximal polygons between objects; malleability; and a dynamic data structure. Since the SDS comprises all the geometric information of the original objects, all generalization operations are carried out directly on the SDS. The target application for the MAGE system built around the SDS is the generalization of large-scale topographic map data from the Ordnance Survey of GB. To that end, a palette of generalization operators has been developed including object exaggeration (enlargement); object collapse (constructing the centerline of road casings); operators for areal object amalgamation (direct merge, adopt merge); and building simplification using corner flipping of triangles.

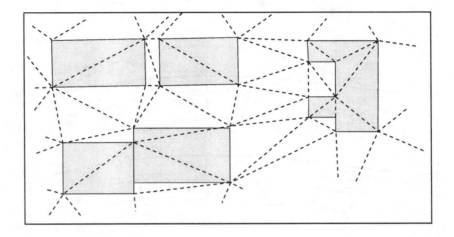

Fig. 27. Sample section of the constrained Delaunay triangulation forming the simplicial data structure (after Ware et al 1995).

The above examples all argue for an enrichment of data models used in generalization. A number of more fundamental questions, however, remain to be resolved by future research: How far does the representation of spatial and semantic relations in data models need to be extended? In what ways will this increase the cost of building spatial databases? Which relations can be determined computationally, and which ones need to be coded 'manually'? Which relations should be stored in the database, and which ones can be computed on-the-fly?

14 Conclusions

The importance of the generalization of spatial data as a function of GIS is accentuated by the current growth of the number and volume of spatial databases and by the need to produce data to specific requirements and share them among

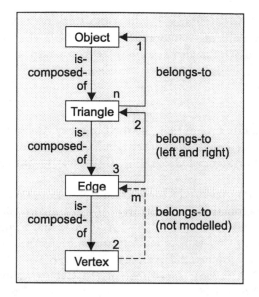

Fig. 28. Entity relationships in the simplicial data structure (after Jones et al. 1995).

different user groups. After a time of relative stagnation during the late 1970s and the 1980s, generalization has again attracted significant research interest in the GIS community and beyond. The topic is well represented at key GIS conferences and several working groups of international organizations – e.g., of the International Cartographic Association (ICA), the European Organization for Experimental Photogrammetric Research (OEEPE), and the International Society of Photogrammetry and Remote Sensing (ISPRS) – are trying to coordinate research efforts.

Although non-algorithmic methods certainly will play an increasingly significant role in automating generalization, algorithms still are of crucial importance because they form the foundation on which the other techniques must build. Computational geometry could contribute substantially to improving the functional and computational performance of current methods to address the geometrical aspects of generalization. Key areas awaiting better solutions are generalization algorithms that observe multiple constraints, robust methods for structure recognition, and the exploitation of alternative data representations and data structures to build more complex algorithms, particularly for context-dependent generalization.

Acknowledgments

I would like to thank Frank Brazile for helping with the preparation of illustrations and Geoff Dutton for reading parts of the draft manuscript. A number of people have generously provided illustrations or helped with the compilation of figures, including Corinne Plazanet and Anne Ruas of IGN France, and Chris Jones of the University of Glamorgan. Support from the Swiss NSF through project 2100-043502.95/1 is also gratefully acknowledged.

References

Bär, H.R. (1995): Interaktive Bearbeitung von Geländeoberflächen – Konzepte, Methoden, Versuche. (Ph.D. Dissertation) *Geoprocessing Series,* Department of Geography, University of Zurich, vol. **25**, 140 pgs.

Beard, K. (1991): Theory of the Cartographic Line Revisited: Implications for Automated Generalization. *Cartographica,* **28**(4): 32-58.

Brassel, K.E. and Weibel, R. (1988). A Review and Framework of Automated Map Generalization. *International Journal of Geographical Information Systems,* **2**(3): 229-44.

Buttenfield, B.P. (1985): Treatment of the Cartographic Line. *Cartographica,* **22**(2): 1-26.

Buttenfield, B.P and McMaster, R.B. (1991, eds.): *Map Generalization: Making Rules for Knowledge Representation.* London: Longman.

Clinton, W.J. (1994): Coordinating Geographic Data Acquisition and Access: The National Spatial Data Infrastructure. *Federal Register,* 13 April 1994, Executive Order 12906, 59(71): 17671-17674.

Cromley, R.G. (1991): Hierarchical Methods of Line Simplification. *Cartography and Geographic Information Systems,* **18**(2): 125-131.

Cromley, R.G., and Campbell, G.M. (1991): Noninferior Bandwidth Line Simplification: Algorithm and Structural Analysis. *Geographical Analysis,* **23**(1), 25-38.

Cromley, R.G., and Campbell, G.M. (1992): Integrating Quantitative and Qualitative Aspects of Digital Line Simplification. *The Cartographic Journal,* **29**(1), 25-30.

De Berg, M., van Kreveld, M. and Schirra, S. (1995): A New Approach to Subdivision Simplification. *ACSM/ASPRS Annual Convention and Exposition,* Vol. 4 (Proc. Auto-Carto 12): 79-88.

Devogele, T., Trevisan, J. and Raynal, L. (1996): Building a Multi-Scale Database with Scale-Transition Relationships. In: Kraak, M.J. and Molenaar, M. (eds.): *Advances in GIS research II* (Proceedings 7th International Symposium on Spatial Data Handling), London: Taylor & Francis: 6.19-6.33.

Douglas, D.H., and Peucker, Th.K. (1973): Algorithms for the Reduction of the Number of Points Required to Represent a Digitized Line or its Caricature. *The Canadian Cartographer,* **10**(2): 112-122.

Duda, R. and Hart, P. (1973): *Pattern Classification and Scene Analysis.* New York: John Wiley.

Dutton, G. (1996a): Encoding and Handling Geospatial Data with Hierarchical Triangular Meshes. In: Kraak, M.J. and Molenaar, M. (eds.):

Advances in GIS research II (Proceedings 7th International Symposium on Spatial Data Handling), London: Taylor & Francis: 8B.15-8B.28.

Dutton, G. (1996b): Improving Locational Specificity of Map Data – A Multi-Resolution, Metadata-Driven Approach and Notation. *International Journal of Geographical Information Systems,* 10(3): 253-268.

Fisher, P.F. and Mackaness, W.A. (1987): Are Cartographic Expert Systems Possible? *Proc. Auto-Carto 8* (8th Int. Symposium on Computer-Assisted Cartography), Baltimore, MD: 530-534.

EUROGI (1996): GI2000 – Towards a European Policy Framework for Geographic Information. *Discussion Paper,* European Umbrella Organisation for Geographic Information (EUROGI).

Fritsch, E. and Lagrange, J.-P. (1995): Spectral Representations of Linear Features for Generalisation. In: Frank, A.U. and Kuhn, W. (eds.): Spatial Information Theory – A Theoretical Basis for GIS (Proceedings COSIT '95). *Lecture Notes in Computer Science* **988**, Berlin, Springer-Verlag: 157-71

Grünreich, D. (1985): Computer-Assisted Generalization. *Papers CERCO Cartography Course.* Institut für Angewandte Geodäsie, Frankfurt a. M.

Grünreich, D. (1992): ATKIS – A Topographic Information System as a Basis for GIS and Digital Cartography in Germany. In: Vinken, R. (ed.): From Digital Map Series in Geosciences to Geo-Information Systems. *Geologisches Jahrbuch* Series A, Vol. **122**. Hannover: Federal Institute of Geosciences and Resources: 207-216.

Hake, G. (1975): Zum Begriffssystem der Generalisierung. *Nachrichten aus dem Karten- und Vermessungswesen,* Sonderheft zum 65. Geburtstag von Prof Knorr: 53-62.

Heller, M. (1990): Triangulation Algorithms for Adaptive Terrain Modeling. *Proceedings Fourth International Symposium on Spatial Data Handling,* Zurich, July 1990, **1**: 163-174.

Hershberger, J. and Snoeyink, J. (1992): Speeding Up the Douglas-Peucker Line-Simplification Algorithm. *Proceedings 5th International Symposium on Spatial Data Handling,* Charleston, SC, 134-143.

Hess, M. (1995): *Erweiterung von Methoden zur automatischen Erzeugung panoramischer Ansichten.* MSc Thesis, Department of Geography, University of Zurich.

Horton, R.E. (1945): Erosional Development of Streams and their Drainage Basins – Hydrophysical Approach to Quantitative Morphology. *Geological Society of America Bulletin,* **56**: 275-370.

ICA (International Cartographic Association) (1973): *Multilingual Dictionary of Technical Terms in Cartography.* Wiesbaden: Franz Steiner Verlag.

Imai, H. and Iri, M. (1988): Polygonal Approximations of a Curve – Formulations and Algorithms. In: Toussaint, G.T. (ed.): *Computational Morphology.* Elsevier Science Publishers, 71-86.

Jäger, E. (1991): Investigations on Automated Feature Displacement for Small Scale Maps in Raster Format. *Proceedings 15th International Cartographic Conference,* Bournemouth (UK): 245-256

Jones, C.B., Bundy, G.Ll. and Ware, J.M. (1995): Map Generalization with a Triangulated Data Structure. *Cartography and Geographic Information Systems,* **22**(4): 317-331.

Kidner, D.B. and Jones, C.B. (1994): A Deductive Object-Oriented GIS for Handling Multiple Representations. In: Waugh T.C. and Healey, R.G. (eds.): *Advances in GIS research* (Proceedings Sixth International Symposium on Spatial Data Handling), London: Taylor & Francis: 882-900

Lang, T. (1969): Rules for the Robot Draughtsmen. *The Geographical Magazine,* **42**(1): 50-51.

Leberl, F.W. (1986): ASTRA – A System for Automated Scale Transition. *Photogrammetric Engineering and Remote Sensing,* **52**(2): 251-258.

Lecordix, F., Plazanet, C. and Lagrange, J.-P. (1996): PlaGe: A Platform for Research in Generalization. Application to Caricature. *GeoInformatica,* **1**(1).

Lee, D. (1995): Experiment on Formalizing the Generalization Process. In: Müller, J-C., Lagrange, J.-P., and Weibel, R. (eds.): *GIS and Generalization: Methodological and Practical Issues,* London: Taylor & Francis, 219-234.

Li, Z., and Openshaw, S. (1992): Algorithms for Automated Line Generalization Based on a Natural Principle of Objective Generalization. *International Journal of Geographical Information Systems,* **6**(5): 373-389.

Lichtner, W. (1979): Computer-Assisted Processes of Cartographic Generalization in Topographic Maps. *Geo-Processing,* **1**: 183-199.

MSC Mapping Science Committee (1993): *Toward a Coordinated Spatial Data Infrastructure for the Nation.* Washington, DC: National Research Council, National Academy Press.

Marino, J.S. (1979): Identification of Characteristics along Naturally Occurring Lines: An Empirical Study. *The Canadian Cartographer,* **16**(1): 70-80.

McMaster, R.B. (1983): *A Quantitative Analysis of Mathematical Measures in Linear Simplification.* Ph.D. Thesis, Dept. of Geography and Meteorology, University of Kansas, Lawrence, Kansas.

McMaster, R.B. (1987a): Automated Line Generalization, *Cartographica,* **24**(2): 74-111.

McMaster, R.B. (1987b): The Geometric Properties of Numerical Generalization. *Geographical Analysis,* **19**(4): 330-346.

McMaster, R.B. (1989): The Integration of Simplification and Smoothing Algorithms in Line Generalization. *Cartographica,* **26**(1): 101-121.

McMaster, R.B. and Monmonier, M. (1989): A Conceptual Framework for Quantitative and Qualitative Raster-Mode Generalization. *Proceedings GIS/LIS '89,* Orlando, FL: 390-403.

McMaster, R.B., and Shea, K.S. (1992): *Generalization in Digital Cartography*. (Resource Publications in Geography). Washington, D.C.: Association of American Geographers.

Misund, G. (1996): Varioscale TIN Based Surfaces. In: Kraak, M.J. and Molenaar, M. (eds.): *Advances in GIS research II* (Proceedings 7th International Symposium on Spatial Data Handling), London: Taylor & Francis: 6.36-6.45.

Molenaar, M. (1996a, ed.): Methods for the Generalization of Geo-Databases. *Publications on Geodesy*, New Series, Delft: Netherlands Geodetic Commission, **43**.

Molenaar, M. (1996b): The role of topologic and hierarchical spatial object models in database generalization. In: Molenaar, M. (ed.): Methods for the Generalization of Geo-Databases. *Publications on Geodesy*, New Series, Delft, Netherlands Geodetic Commission **43**: 13-36.

Monmonier, M.S. and McMaster, R.B. (1991): The Sequential Effects of Geometric Operators in Cartographic Line Generalization. *International Yearbook of Cartography*.

Muller, J.-C. (1990): The Removal of Spatial Conflicts in Line Generalization. *Cartography and Geographic Information Systems*, **17**(2): 141-149.

Muller, J.-C. (1991): Generalization of Spatial Databases. In: Maguire, D.J., Goodchild, M.F., and Rhind, D.W. (eds.): *Geographical Information Systems: Principles and Applications*. London: Longman, **1**: 457-475.

Müller, J.-C., Lagrange, J.-P., and Weibel, R. (1995a, eds.): *GIS and Generalization: Methodological and Practical Issues*. London: Taylor & Francis.

Müller, J.-C., Weibel, R., Lagrange, J.-P., and Salgé, F. (1995b): Generalization: State of the Art and Issues. In: Müller, J-C., Lagrange, J.-P., and Weibel, R. (eds.): *GIS and Generalization: Methodological and Practical Issues,* London: Taylor & Francis, 3-17.

Nickerson, B.G. (1988): Automated Cartographic Generalization for Linear Features. *Cartographica*, **25**(3), 15-66.

Opheim, H. (1982): Fast Reduction of a Digitized Curve. *Geo-Processing*, **2**: 33-40.

Perkal, J. (1966): An Attempt at Objective Generalization. Michigan Inter-University Community of Mathematical Geographers, *Discussion Paper 10*, Ann Arbor: University of Michigan, Department of Geography.

Peucker, T.K. (1975): A Theory of the Cartographic Line. *International Yearbook of Cartography*, **16**: 134-143.

Plazanet, C. (1995): Measurements, Characterization, and Classification for Automated Line Feature Generalization. *ACSM/ASPRS Annual Convention and Exposition,* Vol. 4 (Proc. Auto-Carto 12): 59-68.

149

Plazanet, C., Affholder, J.-G. and Fritsch, E. (1995): The Importance of Geometric Modeling in Linear Feature Generalization. *Cartography and Geographic Information Systems*, **22**(4). 291-305.

Plazanet, C. (1996): *Analyse de la géométrie des objets linéaires pour l'enrichissement des bases de données. Intégration dans le processus de généralisation cartographique des routes.* PhD Thesis, Université Marne la Vallée.

Ramer, U. (1972): An Iterative Procedure for the Polygonal Approximation of Plane Curves. *Computer Graphics and Image Processing*, **1**: 244-256.

Reumann, K. and Witkam, A.P.M. (1974): Optimizing Curve Segmentation in Computer Graphics. *International Computing Symposium.* Amsterdam: North Holland; 467-472.

Richardson, D.E. (1994): Generalization of Spatial and Thematic Data Using Inheritance and Classification and Aggregation Hierarchies. In: Waugh T.C. and Healey, R.G. (eds.): *Advances in GIS research* (Proceedings Sixth International Symposium on Spatial Data Handling), London: Taylor & Francis: 901-20

Rieger, M. and Coulson, M. (1993): Consensus or Confusion: Cartographers' Knowledge of Generalization. *Cartographica,* **30**(1): 69-80.

Rogers, D.F. and Adams, J.A. (1990): *Mathematical Elements for Computer Graphics.* Second Edition. New York et al.: McGraw-Hill.

Roos, T. (1996): Voronoi Methods in GIS. *This volume.*

Ruas, A. (1995a): Multiple Representations and Generalization. *Lecture Notes for 1995 Nordic Cartography Seminar.* St.-Mandé: Institut Géographique National. Available as PostScript document from anonymous ftp <sturm.ign.fr>.

Ruas, A. (1995b): Multiple Paradigms for Automating Map Generalization: Geometry, Topology, Hierarchical Partitioning and Local Triangulation. *ACSM/ ASPRS Annual Convention and Exposition,* Vol. 4 (Proc. Auto-Carto 12): 69-78.

Ruas, A. and Lagrange, J.-P. (1995): Data and Knowledge Modelling for Generalization. In: Müller, J-C., Lagrange, J.-P., and Weibel, R. (eds.): *GIS and Generalization: Methodological and Practical Issues,* London: Taylor & Francis, 73-90.

Ruas, A. and Plazanet, C. (1996): Strategies for Automated Generalization. In: Kraak, M.J. and Molenaar, M. (eds.): *Advances in GIS research II* (Proceedings 7th International Symposium on Spatial Data Handling), London: Taylor & Francis: 6.1-6.18.

Rusak Mazur, E., and Castner, H.W. (1990): Horton's Ordering Scheme and the Generalisation of River Networks. *The Cartographic Journal,* **27**: 104-112.

Schlegel, A., and Weibel (1995): Extending a General-Purpose GIS for Computer-Assisted Generalization. *17th International Cartographic Congress of the ICA,* Barcelona (E), 2211-2220.

Schylberg, L. (1993): *Computational Methods for Generalization of Cartographic Data in a Raster Environment.* PhD Thesis, Department of Photogrammetry, Royal Institute of Technology, Stockholm.

Shreve, R.L. (1966): Statistical Law of Stream Number. *Journal of Geology,* **74**: 17-37.

Spiess, E. (1995): The Need for Generalization in a GIS Environment. In: Müller, J-C., Lagrange, J.-P., and Weibel, R. (eds.): *GIS and Generalization: Methodological and Practical Issues,* London: Taylor & Francis, 31-46.

Strahler, A.N. (1957): Quantitative Analysis of Watershed Geomorphology. *Transactions of the American Geophysical Union,* 8(6): 913-920.

Swiss Society of Cartography (1977): Cartographic Generalization – Topographic Maps. *Cartographic Publication Series,* vol. **2**. Zurich: Swiss Society of Cartography.

Töpfer, F. (1974): *Kartographische Generalisierung.* Gotha, Leipzig: VEB Hermann Haack.

van Kreveld, M. (1997): Digital Elevation Models: Overview and Selected TIN Algorithms. *This volume.*

van Oosterom, P. and van den Bos, J. (1989): An Object-Oriented Approach to the Design of Geographic Information Systems. *Computers & Graphics,* **13**: 409-418.

van Oosterom, P. and Schenkelaars, V. (1995): The Development of an Interactive Multiscale GIS. *International Journal of Geographical Information Systems,* **9**(5): 489-507.

Visvalingam, M., and Whyatt, J.D. (1993): Line Generalisation by Repeated Elimination of Points. *Cartographic Journal,* **30**(1): 46-51.

Visvalingam, M. and Williamson, P.J. (1995): Simplification and Generalization of Large Scale Data for Roads: A Comparison of Two Filtering Algorithms. *Cartography and Geographic Information Systems,* **22**(4): 264-275.

Ware, J.M., Jones, C.B. and Bundy, G.Ll. (1995): A Triangulated Spatial Model for Cartographic Generalisation of Areal Objects. In: Frank, A.U. and Kuhn, W. (eds.): Spatial Information Theory – A Theoretical Basis for GIS (Proceedings COSIT '95). *Lecture Notes in Computer Science* **988**, Berlin, Springer-Verlag: 173-192.

Ware, J.M. and Jones, C.B. (1996): A Spatial Model for Detecting (and Resolving) Conflict Caused by Scale Reduction. In: Kraak, M.J. and Molenaar, M. (eds.): *Advances in GIS research II* (Proceedings 7th International Symposium on Spatial Data Handling), London: Taylor & Francis: 9A.15-9A.26.

Weibel, R. (1991): Amplified Intelligence and Rule-Based Systems. In: Buttenfield, B.P., and McMaster, R.B. (eds.): *Map Generalization – Making Rules for Knowledge Representation.* London: Longman, 172-186.

Weibel, R. (1992): Models and Experiments for Adaptive Computer-Assisted Terrain Generalization. *Cartography and Geographic Information Systems,* **19**(3): 133-153.

Weibel, R. (1995a): Map Generalization. Special Issue of *Cartography and Geographic Information Systems,* **22**(4).

Weibel, R. (1995b): Three Essential Building Blocks for Automated Generalisation. In: Müller, J-C., Lagrange, J.-P., and Weibel, R. (eds.): *GIS and Generalization: Methodological and Practical Issues,* London: Taylor & Francis, 56-69.

Weibel, R. and Ehrliholzer, R. (1995): An Evaluation of MGE Map Generalizer. *Internal Report,* Department of Geography, University of Zurich, 36 + 18 pgs.

Weibel, R., Keller, St., and Reichenbacher, T. (1995): Overcoming the Knowledge Acquisition Bottleneck in Map Generalization: The Role of Interactive Systems and Computational Intelligence. In: Frank, A.U. and Kuhn, W. (eds.): Spatial Information Theory – A Theoretical Basis for GIS (Proceedings COSIT '95). *Lecture Notes in Computer Science* **988**, Berlin, Springer-Verlag: 139-156.

Weibel, R. (1996): A Typology of Constraints to Line Simplification. In: Kraak, M.J. and Molenaar, M. (eds.): *Advances in GIS research II* (Proceedings 7th International Symposium on Spatial Data Handling), London: Taylor & Francis: 9A.1-9A.14.

Werschlein, T. (1996): *Frequenzbasierte Linienrepräsentationen für die kartographische Generalisierung.* MSc Thesis, Department of Geography, University of Zurich.

White, E.R. (1985): Assessment of Line Generalization Algorithms Using Characteristic Points. *The American Cartographer,* **12**(1): 17-28.

Chapter 6.   Spatial Data Structures: Concepts and Design Choices

Jürg Nievergelt

Peter Widmayer

Dept. of Computer Science
ETH Zürich
Switzerland

jn@inf.ethz.ch

Dept. of Computer Science
ETH Zürich
Switzerland

widmayer@inf.ethz.ch

1   Goals and Structure of this Survey

The growing importance of graphic user interfaces and of applications such as computer-aided design and geo-information systems has confronted many applications programmers with a challenging new task: Processing large amounts of spatial data, off disk, correctly and efficiently. The task is daunting, as spatial data poses distinctly novel problems as compared to traditional "business data", in particular the following two:

1. Access is primarily via proximity relations to other objects that populate Euclidean space, rather than via inherent properties of objects, such as attribute values. A typical query is of the form "retrieve all objects that intersect a given region of space".
2. Guaranteed correctness in processing spatial objects has proven to be a thorny problem requiring a systematic analysis of degenerate configurations. Moreover, efficiency depends on the interplay between two techniques: one for representing an object (independently of its location in space), the other for storing the entire collection of objects in relation to their position in space.

We aim to provide a guide through the bewildering multitude of concepts, techniques and choices a programmer faces when designing a data structure for managing spatial data. We simplify this task by separating the question of how an object is represented for internal processing (an issue in computational geometry that we shall not address), from the question of how a large collection of simple objects embedded in space are managed on disk (the crucial issue of

153

spatial data management systems). Although object representation and external data structures may be intertwined, for example in image processing, they are treated separately in many important applications. In geoinformation systems, for example, objects of complex shape are routinely approximated or bounded by simple containers for retrieval purposes.

The tutorial approach we have chosen for this survey begins with the historical development of data structures in general, a trend from which spatial data in particular emerged as a separate discipline relatively late. In spite of the profound differences between spatial data on the one hand, and conventional data typical of the business applications that shaped the development of database technology on the other hand, the community of database researchers persisted in forcing "non-standard (including spatial) data" into a mold that had proven effective for other applications. It took a long time to realize that data structures inherited or adapted from single-key access had to be reconsidered, and often abandoned. Thus a brief sketch of the development of data structures in Section 2 serves the double purpose of reviewing the basic notions of data structures, and of identifying key differences between spatial and other data.

A second characteristic of our tutorial approach is to start from about a dozen concepts or features that appear to cover most of the many spatial data structures described in the literature. This approach assigns to every data structure a "profile" that facilitates assessment of its strong and weak points, and comparison with others. Thus we avoid enumerating the many spatial data structures described in the literature, which often differ only in details, and instead focus on the building blocks from which a programmer can assemble his own data structure tailored to the specific requirements of a given task.

Thus, Section 3 presents important concepts needed to discuss spatial data structures at an intuitive level. This includes a list of basic questions to be raised and considerations to be aware of before a programmer decides on data representation and storage technique. The latter have a pervasive influence on the complexity and efficiency of any application that processes spatial data.

Having assembled an arsenal of concepts and terminology, Section 4 presents a concise model that captures the essence of the majority of spatial data structures and the way they process queries. By separating three key aspects: the organization of the embedding space into a collection of cells, the organization of objects into cell populations, and the internal representation of an object, we arrive at a three-step query processing model that serves as a skeleton for understanding, assessing and comparing most spatial data structures.

So far we have merely mentioned a few examples of specific data structures, out of several dozen known today. Sections 5 and 6 continue by systematically listing the major building blocks that go into the design of a spatial data structure, and providing examples of structures for possible design choices. Section 5 treats points, the simplest kind of spatial object, whereas Section 6 discusses extended objects. By following an approach "from general concepts to specific examples", we hope to help a reader to program an appropriate data structure

154

based on half a dozen design choices, rather than by scanning a vast literature that includes hundreds of research papers.

The concluding Section 7 attempts to summarize the survey with a simple point of view: Design and choice of spatial data structures, today, is a matter of common sense more than of technical wizardry. Don't let hundreds of research papers prevent you from seeing the forest because of all the trees. With a clear understanding of a dozen fundamental concepts characteristic of spatial data, you can choose a data structure suited to your application.

2 Data Structures Old and New, and the Forces that Shaped Them

2.1 Prehistory, Logical vs. Physical Structure

The discipline of data structures, as a systematic body of knowledge, is truly a creation of computer science. The question of how to organize data was a lot simpler to answer in the days before the existence of computers: The organization had to be simple, because there was no automatic device capable of processing intricately structured data, and there is no human being with enough patience to do it. Consider two examples.

1. Manual files and catalogs, as used in business offices and libraries, exhibit several distinct organizing principles, such as sequential and hierarchical order and cross-references. From today's point of view, however, manual files are not well-defined data structures. For good reasons, people did not rigorously define those aspects that we consider essential when characterizing a data structure: what constraints are imposed on the data, both on the structure and its content; what operations the data structure must support; what constraints these operations must satisfy. As a consequence, searching and updating a manual file is not typically a process that can be automated: It requires common sense, and perhaps even expert training, as is the case for a library catalog.
2. In manual computing (with pencil and paper or a nonprogrammable calculator) the algorithm is the focus of attention, not the data structure. Most frequently, the person computing writes data (input, intermediate results, output) in any convenient place within his field of vision, hoping to find them again when he needs them. Occasionally, to facilitate highly repetitive computations (such as income tax declarations), someone designs a form to prompt the user, one operation at a time, to write each data item into a specific field. Such a form specifies both an algorithm and a data structure with considerable formality, but is necessarily special purpose.

Edge-notched cards are perhaps the most sophisticated data structures ever designed for manual use. Let us illustrate them with the example of a database of English words organized so as to help in solving crossword puzzles. We write one word per card and index it according to which vowels it contains and which

155

ones it does not contain. Across the top row of the card we punch 10 holes labeled $A, E, I, O, U,$
$\sim A, \sim E, \sim I, \sim O, \sim U$. When a word, say *ABACA*, exhibits a given vowel, such as A, we cut a notch above the hole for A; when it does not, such as E, we cut a notch above the hole for $\sim E$ (pronounced "not E"). The figure below shows the encoding of the words *BEAUTIFUL, EXETER, OMAHA, OMEGA*. For example, we search for words that contain at least one E, but no U, by sticking two needles through the pack of cards at the holes E and $\sim U$. *EXETER* and *OMEGA* will drop out. In principle it is easy to make this sample database more powerful by including additional attributes, such as "A occurs exactly once", "A occurs exactly twice", "A occurs as the first letter in the word", and so on. In practice, a few dozen attributes and thousands of cards will stretch this mechanical implementation of a multikey data structure to its limits of feasibility.

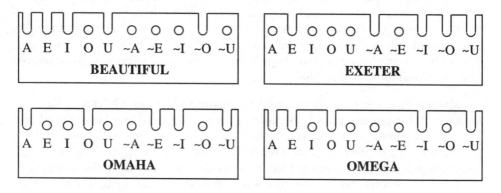

Fig. 1. Edge-notches cards as a mechanical multi-key access structure

The reader might be interested in working out the logic of evaluating queries expressed as arbitrary Boolean expressions over these attributes, observing that *AND* works in parallel with multiple needles, whereas *OR* is processed sequentially using multiple passes.

Exotic as the physical realization may appear today, the logical structure of edge-notched cards is an amazingly modern example of multi-key data structures: it organizes the search space as a hypercube, i.e. a regular grid in which every key value, whether present or not among the actual data to be stored, has its predefined place. Moreover, the physical order of the data to be stored is entirely independent of the logical order imposed on the data space, so the former can be chosen to match the physical characteristics of the hardware used (e.g. allocation of clusters of cards to boxes or, today, of records to disk blocks) without interfering with the search logic. But in order to fully appreciate these remarks we must describe the major evolutionary steps along the way from the first computerized data structures to modern multi-key data structures. We will observe that each era that focused on new application domains created new techniques to address concerns not adequately met by prior developments.

2.2 Early Scientific Computation: Static Data Sets

Numerical computation in science and engineering mostly leads to linear algebra and hence matrix computations. Matrices are static data sets: The values change, but the shape and size of a matrix rarely does - this is true even for most sparse matrices, such as band matrices, where the propagation of nonzero elements is bounded. Arrays were Goldstine and von Neumann's answer to the requirement of random access, as described in their venerable 1947 report "Planning and coding of problems for an electronic computing instrument". FORTRAN '54 supported arrays and sequential files, but no other data structures, with statements such as DIMENSION, READ TAPE, REWIND, and BACKSPACE.

Table look-up was also solved early through hashing. The software pioneers of the first decade did not look beyond address computation techniques (array indexing and hashing) because memories were so small that any structure that "wastes" space on pointers was considered a luxury. Memories containing a few K words restricted programmers to using only the very simplest of data structures, and the limited class of problems addressed let them get away with it. The discipline of data structures had yet to be created.

2.3 Commercial Data: Batch Processing of Dynamic Sets, Single Key Access

Commercial data processing led to the most prolific phase in the development of data structures. The achievements of the early days were comprehensively presented in Knuth's pioneering books on "The Art of Computer Programming" (Knuth 1968, 1973). These applications brought an entirely different set of requirements for managing data typically organized according to a single 'primary key'. When updating an ordered master file with unordered transaction files, sorting and merging algorithms determine data access patterns. The emergence of disk drives extended the challenge of data structure design to secondary storage devices. Bridging the 'memory-speed gap' became the dominant practical problem. Central memory and disk both look like random access devices, but they differ in the order of magnitude of two key parameters:

	Memory	Disk	Ratio: Disk/Memory
Access time (seconds):	$10^{-7} \ldots 10^{-6}$	$10^{-2} \ldots 10^{-1}$	$10^4 \ldots 10^5$
Size of transfer unit (bits):	$10 \ldots 10^2$	$10^4 \ldots 10^5$	$10^2 \ldots 10^3$

In recent decades technology has reduced both time parameters individually, but their ratio has remained a 'speed gap' of about 4 orders of magnitude. This fact makes the **number of disk accesses** the most relevant performance parameter of external data structures. Many data structures perform well in central memory, but disk forces us to be more selective; disks call for data structures that avoid pointer chains that indiscriminately cross disk block boundaries. The game of designing data structures suitable for disk has two main rules: the easy one is to use a small amount of central memory effectively to describe the current allocation of data on disk in a way that facilitates rapid retrieval; the hard one is to ensure that the structure adapts gracefully to the ever-changing content of the file.

Index-sequential access methods (ISAM) order records according to a single key so that a small directory, preferably kept in central memory, ideally directs any point query to the correct data bucket where the corresponding record is stored, if it is present at all. But the task of maintaining this single-disk-access performance in a dynamic file, in the presence of insertions and deletions, is far from trivial. The first widely used idea, of splitting storage into a primary area and an overflow area, suffers from now well-known defects.

Balanced trees, one of the major achievements of data structure design, provide a brilliant solution to the problem of 'maintaining large ordered indexes' without degradation. They come in many variations (e.g. Adelson-Velski et al. 1962, Bayer et al. 1972) all based on the same idea: Frequent small rebalancing operations that work in logarithmic time eliminate the need for periodic reorganization of the entire file. Trees based on comparative search derive their strength from the ease of modifying list structures in central memory. They have been so successful that we tend to apply and generalize them beyond their natural limitations. In addition to concerns about the suitability of comparative search trees for multikey access, discussed in the next section, these limitations include (Nievergelt et al. 1981):

1. The number of disk accesses grows with the height of the tree. Depending on the size of the file and the fan-out from a node, or from a page containing many nodes, the tree may well have too many levels for instantaneous retrieval.

2. Concurrency. Every node in a tree is the sole entry point to the entire subtree rooted at that node, and thus a bottleneck for concurrent processes that pass through it, even if they access different physical storage units. Early papers (e.g. Bayer et al. 1977, Kung et al. 1980) showed that concurrent access to trees implemented as lists requires elaborate protocols to insure integrity of the data.

2.4 Transaction Processing: Interactive Multikey Access to Dynamic Sets

Whereas single-key access may suffice for batch processing, transaction processing, as used in reservations or banking systems, calls for multikey access (by name, date, location, etc.). The simplest ideas were tried first. Inverted files try to salvage single-key structures by ordering data according to a 'primary key', and 'inverting' the resulting file with respect to all other keys, called 'secondary'.

158

Whereas the primary directory is compact as in ISAM, the secondary directories are voluminous: Typically, each directory has an entry for every record. Just updating the directories makes insertion and deletion time-consuming.

Comparative search trees enhanced ISAM by eliminating the need for overflow chains, so it was natural to generalize them to multikey access and improve on inverted files. This is easy enough, as first shown by k-d trees (Bentley 1975). But the resulting multi-key structures are neither as elegant nor as efficient as in the single-key case. The main hindrance is that no total order can be imposed on multidimensional space without destroying proximity relationships. As a consequence, the simple rebalancing operations that work for single-key trees fail, and rebalancing algorithms must resort to more complicated and less efficient techniques, such as general dynamization (Willard 1978, Overmars 1981).

Variations and improvements on multidimensional comparative search trees continue to appear (e.g. Lomet et al. 1989, 1990). Their main virtue, acceptable worst case bounds, comes from the fact that they partition the actual data to be stored into (nearly) equal parts. The other side of this coin is that data is partitioned *regardless of where in space it is located*. Thus the resulting space partitions exhibit no regularity, in marked contrast to radix partitions that organize space into cells of predetermined size and location.

2.5 Knowledge Representation: Associative Recall in Random Nets

There is a class of applications where data is most naturally thought of as a graph, or network, with nodes corresponding to entities and arcs to relationships among these. Library catalogs in information retrieval, hypertexts with their many links, semantic nets in artificial intelligence are examples. The characteristic access pattern is 'browsing': A probe into the net followed by a walk to adjacent nodes. Typically, a node is not accessed because of any inherent characteristic, but because it is associated with (linked to) a node currently being visited. The requirements posed by this type of problem triggered the development of list processing techniques and list processing languages.

These graphs look arbitrary, and the access patterns look like random walks - neither data nor access patterns exhibit any regular structure to be exploited. The general list structures designed for these applications have not evolved much since list processing was created, at least not when compared to the other data structures discussed. The resulting lack of sophisticated data structures for processing data collections linked as arbitrary graphs reminds us that efficient algorithms and data structures are always tailored to the presence of specific properties to be exploited, in particular to a regular structure of the data space. If the latter has no regular structure to be exploited, access degrades to exhaustive search.

Information retrieval is an example of an application where "random" structure cannot be avoided. Although a catalog entry in a library is a record with a regularly structured part, e.g. document = (author, title, publisher, year), search by content relies on index terms chosen from a large thesaurus of thousands of concepts, ranging from "Alchemy" to "Zen". It does not help to consider 1000 index terms, each with a range of perhaps only two values "relevant" and "irrelevant", as attributes of a multi-key data structures, which typically are designed to handle at most tens of access keys

2.6 Spatial Data Management: Proximity Access to Objects Embedded in Space

In typical applications that rely on spatial data, such as computer-aided design or geoinformation systems, many or all of the requirements listed so far are likely to appear: interactive transaction processing, random-looking networks of references among functionally related components, etc. In addition to such non-spatial requirements, spatial data imposes three key characteristics that sets spatial data management apart from the cases described above:

1. Data represents objects embedded in some d-dimensional Euclidean space $\mathbb{R}^d$.
2. These objects are mostly accessed through their location in space, in response to a proximity query such as intersection with some query region, or containment therein.
3. A typical spatial object has a significantly more complex structure than a 'record' in the other applications mentioned.

Although other applications share some of these characteristics to a small extent, in no other do they play a comparably important role. Let us highlight the contrast with the example of a collection of records, each with two attributes, 'social security number' (SSN) and 'year of birth'.

1. Although it may be convenient to consider such a record to be a point in a 2-d attribute space, this is not a Euclidean space; the distance between two such points, for example, or even the distance between two SSNs, is unlikely to be meaningful.
2. Partial match and orthogonal range queries are common in data processing applications, but more complex query regions are rare. In contrast, arbitrarily complex query regions are common in geometric computation (e.g. intersection of objects, or ray tracing).
3. Although a record in commercial data processing may contain a lot of data, for search purposes it is just a point. A typical spatial object, on the other hand, is a polyhedron of arbitrary complexity, and we face the additional problem of representing it using predefined primitives, such as points, edges, triangles, tetrahedra.

2.7 Concise Summary of Trends that Shaped the Development of Data Structures

Starting from the standard data processing task of the 50s and 60s: "merge the old master tape with an update file to produce a new master", we have mentioned a few of the many requirements that gradually accumulated a great variety of data handling problems. Here is a concise summary that compares yesterday's and today's requirements:

- Batch processing ⇒ interactive use:
  sequential access ⇒ random access
  static file ⇒ dynamic file
  delayed result ok ⇒ "instantaneous" response = 0.1 sec
- Simple queries ⇒ complex queries
  (e.g. access record with unique id ⇒ join in relational DB,
  proximity query in CAD)
  single key ⇒ multi-key access
  few query types ⇒ many different query types
  (e.g. point query   ⇒ region query, consistency check, ..)
  single access ⇒ multi-access transactions
- Point objects ⇒ interrelated objects of arbitrary shape
  (e.g. [name, SSN, year] ⇒ assembly of mechanical parts)

Whereas the generic data structure 'array' was able to meet most needs of numerical computation for decades, it soon became evident that no single type of data structure could be found to meet the increasing variety of data handling problems that arose when computer use expanded to many other applications. Since the sixties, the search for specialized structures designed to handle efficiently a specific set of requirements has never ceased, and the resulting zoo of data structures is impressive, perhaps frightening to the non-specialist. The next section develops concepts needed to detect some order in the wilderness of data structures.

3   Basic Concepts and Characteristics of Multi-Dimensional and Spatial Data Structures

Having presented a concise historical survey that introduced many ideas needed to understand data structures in general, we now narrow the conceptual framework to deal with multi-dimensional data, and spatial structures in particular. This section aims to be a user's guide to understand the zoo of data structures.

3.1   The Profile of a Data Structure

A relatively small number of concepts suffice to characterize any data structure according to its main features, to highlight similarities and contrasts with other structures that might be considered as alternatives, and to guide a programmer in his choice of data structure. These key concepts surface when answering the following questions:

- What type of data is to be stored? This is answered by specifying the **domain** D of key values, and all operations and relations defined on D. Well-known frequent cases include :

- D is an unordered set, such as author-defined index terms for document retrieval
- single-key: D is totally ordered w.r.t. a relation "$<$" (e.g. integers, character strings)
- multi-key: D is a Cartesian product $D_1 \times D_2 \times \ldots \times D_k$ of totally ordered domains D_i.

When a meaningful distance is defined on D, we talk about "metric data structures". The most prominent example is Euclidean space $\mathbb{R}_1 \times \mathbb{R}_2 \times \ldots \times \mathbb{R}_k$, where $\mathbb{R}$ denotes real numbers, or perhaps integers.

- How many keys (search attributes) are involved? The **dimension of the space** has a great influence on the practical complexity of a multi-key data structure. Some approaches that work well in 2 dimensions, for example, do not generalize efficiently to more dimensions. One might think that "spatial data" obviously refers to *2-d* and *3-d* Euclidean space, but this is not necessarily so. Higher-dimensional spaces arise naturally when we describe objects to be stored in terms of parameters that characterize them. The term "multi-key access", including spatial data, commonly refers to the case where we have less than 10 keys (search attributes).

- Functionality (Abstract Data Type): What operations must be supported by the data structure? The most frequently used data structures are variations of the type **dynamic table or dictionary**. They support primarily the operations **Find, Insert, Delete**, along with a host of others such as Predecessor, Successor, Min, Max, etc. Many algorithms that work on spatial data naturally use standard data structures such as stacks and queues, but there is nothing "spatial" about these.

- **How much data** is to be stored, **what storage media** are involved? The two major categories to be distinguished are internal data structures, designed for central memory, and external ones, designed for disk. Many more designs are suitable for internal data structures than for external ones.

- What type of **objects** are to be stored, how simple or complex is their description? Points are certainly the simplest case. Complex objects are often approximated or packaged in simple containers, such as a bounding box. Is the location of these objects fixed, or are they movable or subject to other transformations? In the second case it is important to separate the description of the object from its location.

- What types of **queries** occur, what do we know about their frequency and relative importance? This involves differences such as interactive or batch processing, clustered or scattered access, exact or approximate matches, and many more. Only rarely can we characterize the query population in terms of precise statistical parameters.

- How complex, how efficient are **access and update algorithms**? The efficiency of internal data structures is often well described by their asymptotic time complexity (e.g. $O(1), O(\log n), O(n), \ldots$), whereas that of external data structures is more meaningfully measured in terms of the (small) number of disk accesses needed.

162

– What **implementation techniques** are appropriate (e.g. lists, address computation)? List processing is a prime candidate for dynamic data structures in central memory, but is often less efficient for external data structures.

3.2 The Central Issue: Organizing the Embedding Space versus Organizing its Contents

The single most important issue that distinguishes spatial data structures from more traditional structures can be summarized in the phrase "organizing the embedding space versus organizing its contents". Let us illustrate this somewhat abstract idea with examples.

A record with fields such as (name, address, social security number, ...) can always be considered to be a point in some appropriate Cartesian product space. But the role and importance of this **embedding space** for query processing, whether its structure is exploited or not, depends greatly on the application and the nature of the data under consideration. Whereas the distance between two character strings, say your address and mine, has no practical relevance, the distance between two points in Euclidean space often carries information that is useful for efficient query processing. In addition, regions in Euclidean space admit relations such as "in front of", "contains", "intersects" that have no counterpart in non-spatial data. For this reason, the embedding space plays a much more important role for spatial data than for any other kind of data.

Early data structures, developed for non-spatial data, could safely ignore the embedding space and concentrate on an efficient organization of the particular data stored at any one moment. This point of view naturally favored **comparative search** techniques. These organize the data depending on the relative value of these elements to each other, regardless of the absolute location in space of any individual value. Comparative search (e.g. binary search) leads to structures that are easily balanced. Thus, they answer **statistical queries** efficiently (e.g. median, percentiles), but **not general location queries** ("who is closest to a given query point", "where are there data clusters"). Balanced trees, with their logarithmic worst-case performance for single-key data, are the most successful examples of structures that organize a specific data set.

Given the success of comparative search for non-spatial data, in particular for single-key access, it is not surprising that the first approaches to spatial data were based on them. And that the crucial role of the embedding space, independently of the data to be stored, was recognized rather late. But when comparative search is extended to multi-dimensional spatial data, some shortcomings cannot be ignored. If we generalize the idea of a balanced binary search tree to 2 dimensions, as in the following example, we generate a space partition that lacks any regularity. Such a partition does not make it easy to answer the question what cells of the partition lie within a given query region. Even the idea of dynamically balancing the tree, so as to guarantee logarithmic height and access time in the presence of insertions and deletions, does not generalize efficiently from 1 to 2 dimensions.

163

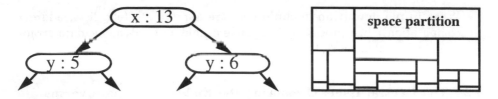

Fig. 2. By balancing data, k-d trees generate irregular space partitions of great complexity

Data structures based on regular radix partitions of the space, on the other hand, organize the domain from which data values are drawn in a systematic manner. Because they support the metric defined on this space, a prerequisite for efficient query processing, we call them metric data structures. The essential structure of these space partitions is determined before the first element is ever inserted, just as inch marks on a measuring scale are independent of what is being measured. The actual data to be stored merely determines the granularity of these regular partitions. Like the "longitude-latitude" partition of the earth, they use fixed points of reference, independent of the current contents. Thus any point on earth has a unique permanent address, regardless of its relation to any cities that may or may not be drawn on a current map.

The well-known quad-tree (Finkel et al. 1974, Hunter et al. 1979, Klinger 1971, Samet 1990 a, b) illustrates the advantages of a regular partition of the embedding space, in this example a unit square. A hierarchical partition of this square into quadrants and subquadrants, down to any desired level of granularity, provides a general-purpose scheme for organizing space, a skeleton to which any kind of spatial data can be attached for systematic access. The picture below shows a quarter circle digitized on a $16 \cdot 16$ grid, and its compact representation as a 4-level quadtree.

Most queries about spatial data involve the absolute position of objects in space, not just their relative position among each other. A typical query in computer graphics, such as a visibility computation by means of ray tracing, asks for the first object intercepted by a given ray of light. Computing the answer involves absolute position (location of the ray and objects) **and** relative order (nearest along the ray). Regular space partitions reduce search effort by providing (direct) access to any cell of the partition. Given a point with coordinates (x, y), a simple formula determines the unique index of the unique cell (at any given level of granularity) that contains (x, y). Moreover, if storage is allocated contiguously as illustrated in Fig. 4, the address of the disk block of each cell can also be computed merely on the basis of the coordinates (x, y). This allows for direct access to those objects that are near to any query point.

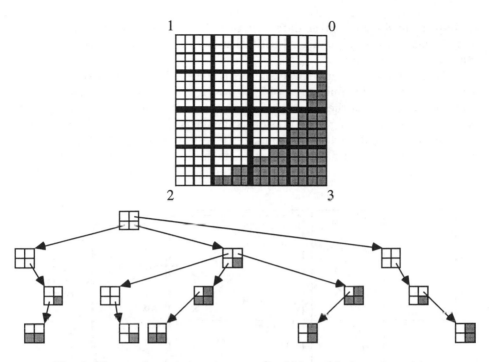

Fig. 3. The quad tree is based on a radix 4 hierarchical space partition

Breadth-first addressing: parent i <———> children 4 i + { 1, 2, 3, 4 }

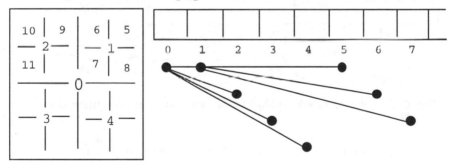

Fig. 4. Breadth-first traversal allocates quad tree cells contiguously in order of increasing depth

4 The Three Step Model of Spatial Object Retrieval

Having surveyed the main concepts the reader needs to keep in mind when exploring the space of spatial data structures, we can now compress this information into a single model that captures the essence of how most spatial data structures process the vast majority of queries. Naturally, many details remain to be filled in to define precisely how the three processing steps outlined below are implemented for any given data structure. But the point to be made is that spatial query processing is best described in terms of three steps that can be analyzed independently.

Consider the embedding space shown in Fig. 5. It is grid-partitioned into rectangular cells and populated by objects drawn as circles or ovals. A triangular query region q calls for the retrieval of all objects that intersect q. In this typical example, query processing first transforms the query q into the set of query cells surrounded by the dotted line; second, retrieves the two objects that intersect the query cells (without having to look at the horizontal oval); third, filters out the tall oval as a "false drop", and retains the circle as a "hit".

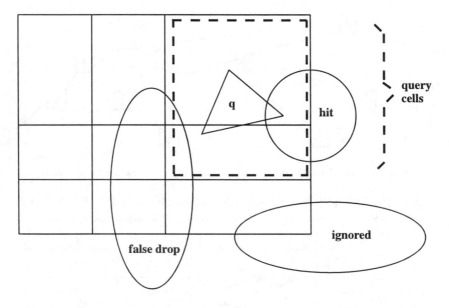

Fig. 5. A region query selects objects that populate a grid-partitioned space

A more general description of these three processing steps follows:

4.1   Cell Addressing: Query $q \to$ set $\{(i, j, \ldots)\}$ of Query Cells

Obtain the coordinates of all cells, at any desired hierarchical level, of those cells that intersect q. This address computation step depends on characteristics of the query and of the grid, but **not** directly on the population of objects. The objects affect this step only to the extent that they determine the current space partition. Thus a simple space partition, i.e. a grid of regular structure determined by a few parameters, is a pre-condition for fast cell addressing, and this is perhaps the data structure designer's major choice. He may have less control over the set of permissible queries, but fortunately query complexity is a lesser problem than partition complexity. Even if a complex query causes much computation, it need not cause any disk accesses, if the space partition has been properly designed so as to be completely described by a small amount of data that resides in central memory. And given that disk access is the efficiency bottleneck of spatial data structures, the time required by this first step is generally negligible.

4.2   Coarse Filter: Set of Query Cells Determines Candidate Objects

All the objects that populate the query cells determined in step 1 are retrieved from disk, because they might respond to the query. This coarse filter is the bottleneck of spatial data access, and the core problem of data structure design. Many issues must be resolved, e.g: what is the precise definition of "an object O populates cell C?" The picture above suggests the plausible definition "O intersects C", and if so, this coarse filter retrieves the circle and the vertical oval, a hit and a false drop, whereas the horizontal oval is ignored. More sophisticated choices are possible that avoid associating an object with all of the many cells it might intersect, but each choice requires corresponding retrieval algorithms to ensure that no objects are missed. And the main issue of data structure design revolves around the association of data buckets (disk blocks) to cells, where the aim is to allocate objects that touch neighboring cells in as few buckets as possible. However this coarse filter is designed, the disk accesses it may cause are likely to require the lion's share of query processing time.

4.3   Fine Filter: Compare Each Object to the Query

The objects selected by the coarse filter are mere candidates that need a final check to see whether they are hits, i.e. respond to the query, or false drops, i.e. passed the first but failed the second, crucial test: Does an object that appears to be close enough at first sight, really intersect the query? This intersection test is trivial or complex depending on the shape and complexity of query and object. But floating point operations are cheap compared to disk accesses, so this fine filter is unlikely to be the performance bottleneck. In any case, this step is squarely in the realm of computational geometry. Its implementation depends on the internal representation of the objects, i.e., on the data structure chosen to represent an object for processing in central memory. A complex object will have to be broken into its constituent parts, such as vertices, edges, and faces in the case of a polyhedron. This has little to do with the design of the external data structure - the topic of our survey - which considers an object as a volume of space, to be treated as an undivided unit whenever possible.

A final comment: Everything discussed above also applies to the case where the "objects", drawn as ovals in the picture, are containers, chosen for their simple shape, that hold a more complex object. In this case the circle, which was labeled a "hit", is reduced to a mere "container hit", and the fine filter must process the object hidden inside, rather than the container.

As we discuss the design choices that characterize the many spatial data structures described in subsequent sections the reader is encouraged to keep the first two query processing steps in mind: how fast is cell addressing? how is the query cell population determined, and what disk accesses are caused by the coarse filter? Such questions serve as a guide for a first assessment of any spatial data structure. Often, they suffice to eliminate from further consideration apparently plausible ideas that fail on the grounds that the first or the second step cannot be implemented efficiently.

5 A Sample of Data Structures and their Space Partitions

5.1 General Consideration

Data structures for external storage support spatial queries to a set of geometric objects by realizing a fast but inaccurate filter: The data structure returns a set of external storage blocks that together contain the requested objects (hits) and others (false drops). Thereafter, a fine filter analyzes each object retrieved to either include or exclude it from the response. The purpose of a data structure is to associate each object with a disk block in such a way that the required operations are performed efficiently.

For exact match, insert and delete operations, non-spatial data structures such as the B-tree or extendible hashing are sufficient, since a unique key can be computed from the geometric properties of each object. But whenever a query involves spatial proximity, as in range queries, a spatial data structure must take into account the shape and location of objects. Naturally, a query can be answered faster if the set of geometric objects that form the response to the query are spread over as few disk blocks as possible. This implies that for proximity queries, the objects stored in a block of an efficient data structure should be close in space. As a consequence, spatial data structures cover the data space (or a part of it) with cells and associate a storage block with each cell. For point objects, the cells partition (a part of) the space, and each point is associated with the cell in which it lies. Our illustrations and explanations of data structuring concepts always refer to 2-dimensional data, but generalization to higher dimensions is often straightforward.

Our sample of spatial data structures is limited to those suitable for external storage. We prefer simple structures over sophisticated structures that use complicated lists. We exclude trees designed to support worst-case efficient algorithms in computational geometry, such as segment trees and interval trees (they are treated elsewhere in this book), including their variants for disk storage. Segment trees and interval trees, as well as hierarchies of such trees making them multidimensional, where e.g. each node of a segment tree references an interval tree, have been studied extensively in computational geometry (Preparata et al. 1985, Edelsbrunner 1982, Iyengar et al. 1988, Overmars 1983, Samet 1988, Samet 1990a, van Kreveld 1992). Based on the segment trees designed for central memory and a worst case scenario, external storage structures have been proposed (Blankenagel 1991). They turned out to be quite complicated, and there is no evidence yet as to whether these external storage segment trees will perform well on average in practical situations. A new approach seems to be very promising (Arge 1996), but practical experience is still lacking. Among the simple structures we discuss, we emphasize address computation techniques, because they are conceptually the simplest and the easiest to implement, and they often lead to the most efficient access structures for practical situations.

Two dominant factors guide the partition of the embedding space into cells, namely the data (the objects) and the space. At one end of the spectrum, the partition is defined without attention to the location of the geometric objects; the amount of data alone determines the refinement of the partition. At the other end, the data completely determines the partition. Naturally, combinations of the two abound.

5.2 Space Driven Partitions: Multi-Dimensional Linear Hashing, Space Filling Curves

The most regular partitions of the data space are those that take into account only the amount of data to be stored (measured by the number of objects, or by the number of storage blocks needed), but disregard the objects themselves and their specific properties. The number of objects merely determines the number of cells of the partition, but not their location, size or shape. The latter are inferred from a generic partition pattern for any number of regions that is parameterized with just one parameter, namely the actual number of regions desired. As a typical example, let us look at the partitions (Fig. 6) induced by multidimensional variants of linear hashing (Enbody et al. 1988, Litwin 1980), such as multidimensional order preserving linear hashing with partial expansions (Kriegel et al. 1986) or dynamic z-hashing (Hutflesz et al. 1988a).

Linear hashing. When viewed as a spatial data structure, linear hashing (Fig. 6(a)) partitions the one-dimensional data space into intervals (one-dimensional regions) of at most two different sizes at any time, the smaller being half the larger. To the left of a separating position, all intervals are small; to the right, all intervals are large. This makes it very simple to find the interval in which a one-dimensional query point lies: Given the size of the smaller intervals and the separating position, a simple calculation returns the desired interval. With the use of an order-preserving addressing function, proximity in the 1-d data space is preserved in storage space. Dynamic modifications to the partition, induced by increasing or decreasing numbers of data points, are simple. An extra interval, for instance, is created by partitioning (splitting) the leftmost of the larger intervals into halves, distributing the data points associated so far with the split interval among the two new intervals, and adjusting the separating position. The separating position starts at the left space boundary and moves from left to right through the data space. On its move, it cuts the intervals encountered in half. After it has reached the right boundary of the data space, all intervals have been cut to the same size, thus the number of intervals has doubled (the doubling phase is complete), and the separating position is reset to the left boundary.

The simplicity of the partitioning pattern of linear hashing makes it extremely simple to keep track of the actual partition: the directory consists of only two

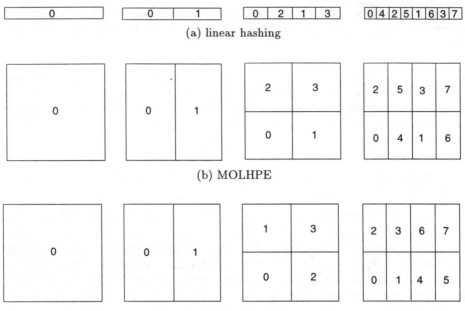

(a) linear hashing

(b) MOLHPE

(c) z-hashing

Fig. 6. Space-driven partitions and their development

values, one for the number (or size) of large (or small) intervals at the beginning of the current doubling phase, and one for the separating position. The cost for this simplicity due to the regularity of the partition is the lack of adaptivity of the partition to the actual data points. In general, it will be necessary to provide overflow blocks for intervals, since the number of data points in an interval can be larger than the block capacity. Whenever the data points are distributed evenly over the data space, the lack of adaptivity may be tolerable; otherwise, considerable inefficiency may result.

Multi-dimensional linear hashing. Multidimensional order preserving linear hashing versions are nothing but generalizations of linear hashing to higher dimensions. Therefore, they share the basic characteristics with linear hashing. They vary in the way the dimensions are involved in the addressing mechanism. The most direct extension of linear hashing is multidimensional order preserving linear hashing with partial expansions (MOLHPE, Kriegel et al. 1986), where the doubling phases cycle through the dimensions. Starting with one block, the first doubling leads to two blocks separated in the first dimension, the second doubling leads to two more blocks, separated from the first two in the second dimension, and so on (see Fig. 6(b)).

Dynamic z-hashing. Since there is some freedom in choosing the addressing function, we can even extend the scope of our considerations and ask for mappings that preserve the geometric order of the data beyond blocks. Here, we request that regions of blocks whose addresses are close tend to be close in data space. This makes sense for proximity queries whenever it is faster to read a number of consecutive blocks than to read the same number of blocks, spread out arbitrarily on the storage medium; it has been used for writing in Wang et al. (1987). Since current disks typically have much higher seek plus latency times than transfer time, they qualify as good candidates. Similar in spirit, Dröge et al. (1993), Dröge (1995) investigate space partitioning schemes for variable size storage clusters instead of fixed size blocks. An addressing function that leads to a more global preservation of order is dynamic z-hashing (Hutflesz et al. 1988a, see Fig. 6(c)). The static version of this addressing mechanism (Manola et al. 1986, Orenstein et al. 1984, Orenstein 1989, 1990) is long known to cartographers as Morton encoding (Morton 1966); it is the same as the quad code (Samet 1990a) or the locational code (Abel et al. 1983, Tropf et al. 1981). One of its nice properties is the fact that addresses can be computed easily by interleaving the bits of the coordinates, cycling through the dimensions; therefore, the technique is also known as bit interleaving.

The only reason why closeness of blocks does not match closeness of cells precisely lies in the impossiblity of embedding a higher dimensional partition in a one-dimensional one while preserving distances. That is, when applied to one-dimensional linear hashing, dynamic z-hashing fully preserves global order.

Space-filling curves. Each of the above addressing mechanisms defines a traversal of the embedding space by visiting all cells in the order of their addresses, a so-called space-filling curve. A number of space-filling curves other than the ones above have been proposed, with the goal of maintaining proximity in space also in the one-dimensional embedding the curve defines. Since the data structure based on a space-filling curve must adapt the partition pattern dynamically, space-filling curves usually have recursive definitions. Fig. 7 shows two building blocks ((a) and (c)) and the three best-known space-filling curves based on them - bit interleaving (b), the Gray code (d) (Faloutsos 1985, 1988), and Hilbert's curve (e) (Faloutsos et al. 1989, Jagadish 1990b).

Experiments comparing the efficiencies of data structures based on these curves (Abel et al. 1990, Jagadish 1990b, van Oosterom 1990) seem to indicate that for certain sets of geometric objects and sets of queries, bit interleaving and Hilbert's curve outperform the Gray code curve significantly, with Hilbert beating bit interleaving in many cases. On the other hand, an analysis of the expected behavior of space filling curves (Nievergelt et al. 1996), where all possible different queries are equally likely, indicates that all space filling curves are equally efficient, if disk seek operations are counted. This illustrates that there is a bias in the distribution of query ranges in the experiments: Queries are not chosen at random, but instead are taken to be in some sense typical of a class of applications. It is certainly useful to evaluate data structures with respect to particular query distributions; it is unfortunate that the distributions

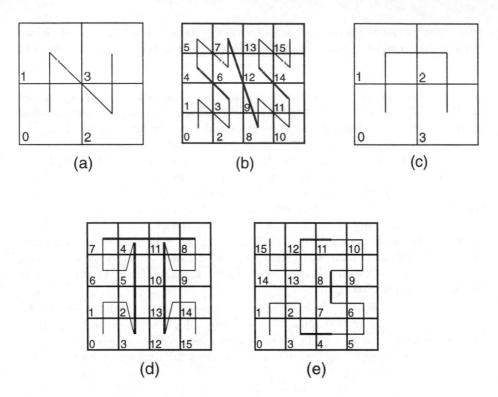

(a)   (b)   (c)

(d)   (e)

Fig. 7. Traditional one-dimensional embeddings

are not discussed explicitely. In contrast to the average case, the worst case for hierarchical space filling curves clearly depends on the curve (Asano et al. 1995).

In spite of the interesting properties of dynamic z-hashing and other proximity preserving mappings of partitions in multidimensional space to one dimension, we feel that the importance of the corresponding data structures is limited to uniformly distributed data, due to the lack of adaptivity of the partition.

5.3   Data Driven Partitions: k-d-B-Tree, hB-Tree

The most adaptive partitions of all are those defined by the set of data points. Since the partition tends to be less regular, a mechanism to keep track of the partition is needed. A natural choice for such a mechanism is a hierarchy, and hence multidimensional generalizations of one-dimensional tree structures have been proposed for that purpose. Prime examples are the k-d-B-tree (Robinson 1981), a B-tree version (Bayer et al. 1972, Comer 1979) of the k-d-tree (Bentley 1975), and a modified version of it, the hB-tree (Lomet et al. 1989, 1990).

The k-d-B-tree. A k-d-B-tree partition is created from one region for the entire data space by recursive splits of regions (see Fig. 8 (a)). The splits follow the k-d-tree structure in that they cycle through the dimensions of the data space, but do so within each node of the tree. The leaves of the tree are all on the same level, just as in B-trees, and maintain the cells of the partition. Interior nodes maintain unions of cells to direct the search through the tree. Thus, the tree is leaf-oriented and serves as a directory.

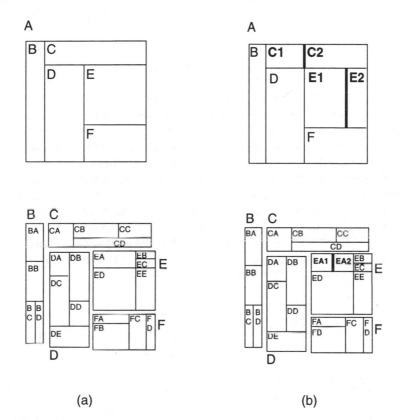

(a) (b)

Fig. 8. Data-driven partitions: (a) The *k-d*-B-tree, and (b) a split in EA forcing a split in C

Whenever a data block overflows, due to an insert operation, its region is split so as to balance the number of data points in both subregions, and the change propagates towards the root. This may necessitate a split of a directory block region, not a simple operation in a k-d-B-tree. The reason is that in order to balance the load between both directory block subregions, a split position may be chosen that cuts through a region of some child (or even several children), thereby forcing the split to propagate downwards in the tree as well. Fig. 8 (b) shows an example of a forced split for the k-d-B-tree sketch in Fig. 8 (a). For a block capacity of five entries, a split of region EA into EA1 and EA2 makes block E overfull. E splits into E1 and E2 in the most balanced way, and this makes

173

block A overfull. Splitting A in a balanced way implies cutting region C into C1 and C2. The split through C affects the subregions CB and CD of C, i.e., the split process propagates down the tree. Since the decision for the most balanced split position is made locally for a node, the forced downward split may become quite a costly operation, both in terms of runtime and of the resulting storage space utilization. As a result, no lower bound on the storage space utilization can be given for k-d-B-trees, in contrast to the 50 % guarantee for B-trees.

The hB-tree. A closer look reveals that in degenerate cases, a balanced split may be impossible in the k-d-B-tree (see Fig. 9). To remedy that situation, a variant of the k-d-B-tree has been proposed, the hB-tree (Lomet et al. 1989, 1990). It has the interesting property that the regions that form the partition of the data space are not restricted to multidimensional rectangles. Instead, subregions can be removed from a region (see Fig. 9), leaving a "holey brick". With this freedom, balanced splits are possible in degenerate situations in which a single split line fails. The hB-tree keeps track of holey brick regions by allowing directory entries to refer to the same block; that is, a holey brick region is represented in the rooted DAG directory by the union of a set of rectangular regions. Of course, holey bricks may occur on each level of the rooted DAG; changes propagate upwards, just as in B-trees. As an example, Fig. 9c shows the rooted k-d-DAG local to a directory block to be split with regions shown in Fig. 9d, and Fig. 9b shows the part of the rooted DAG that propagates upwards when the block is split as shown in Fig. 9a.

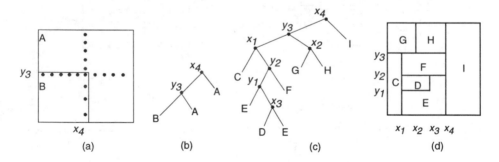

Fig. 9. Data-driven partitions: The hB-tree

Although data-driven partitions turn out to be quite complicated to maintain, they are able to cope with skew data reasonably well, while space-driven partitions fail here. For extremely skew data or whenever worst-case guarantees are more important than average behavior, data-driven partitions may be the method of choice. Due to the freedom in splitting regions, they certainly do have a great inherent flexibility that allows them to be tuned to various situations easily. For instance, the local split decision (LSD-) tree is designed to make good use of a large available main memory, while resorting to external storage as necessary (Henrich et al. 1989a, b, Henrich 1990). Except for such situations, the adaptivity of data-driven partitions will rarely compensate for the burden of their complexity and fragility. More often, combinations of space-driven and data-driven partitions will be appropriate.

5.4 Combinations of Space Driven and Data Driven Partitions: EXCELL, Grid File, Hierarchical Grid Files, BANG File

In their simplest form, these partitions follow the generic pattern of space-driven partitions, with different levels of refinement across the data space, determined by the location of the geometric objects. A typical example of a one-dimensional data structure of this type is extendible hashing (Fagin et al. 1979), where the data space is partitioned by recursively halving exactly those subspaces that contain too many data points.

EXCELL and the grid file. A direct generalization of extendible hashing to higher dimension, EXCELL (Tamminen 1981, 1982), applies the one-dimensional strategy of extendible hashing to each dimension, again running cyclically through the dimensions (see Fig. 10a). Since an extra directory is available for each dimension, plus a directory for the multidimensional product of the one-dimensional directories, and the number of blocks that can be addressed is therefore the product of the sizes of all one-dimensional directories, the sum of the sizes of the one-dimensional directories tends to be quite small in all realistic cases. This observation is used in the grid file (Nievergelt et al. 1981, 1984) to keep the directories for all single dimensions - the so-called scales - in main memory; as a result, no duplication of a directory is necessary due to a data block split, but instead a mere addition of a (multidimensional) directory entry will suffice (see Fig. 10b). Nevertheless, the grid file inherits from extendible hashing the superlinear growth of its directory (Regnier 1985, Tamminen 1985). A search operation in the grid file can always be carried out with just two external memory accesses, the first one to the directory, based on the information from the scales, and the second one to the data block referenced in the directory block. Similarly, a range query can be answered by first searching for a corner point of the query range, and then propagating to adjacent blocks as indicated by the directory. Since this entails a range query on the directory, one might organize the directory itself as a grid file.

Hierarchical grid files. This approach has been pursued in hierarchical grid files (Fig. 11) with two (Hinrichs 1985) or more directory levels (Krishnamurthy et al. 1985). The interior of each directory block is organized as a grid file directory; changes in the tree structure, due to block split or merge operations, propagate along the search path in the tree. Data structures of this type are often called hash trees. A particularly efficient example of such a structure is the buddy tree (Seeger 1989, Seeger et al. 1990). It applies a specific merge

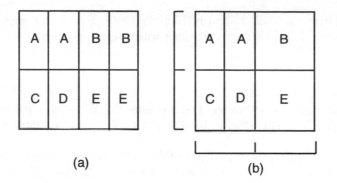

Fig. 10. EXCELL and grid file partitions: Cells with data block addresses

strategy, and it distinguishes the standard block regions, using them for insertions, from the block regions used for all search operations: The latter are the bounding boxes of the objects stored in the corresponding subtree. While the non-hierarchical grid file inherits the property of a superlinearly growing directory from extendible hashing, hash trees grow linearly, just like trees in general. This does not imply that hash trees are always the better choice in practice: It is true for many data structures that better asymptotic efficiency may come at the cost of higher conceptual complexity and therefore lead to poorer performance in practical applications.

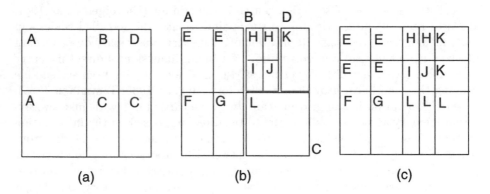

Fig. 11. Two-level (a), (b) against one-level grid file partition (c)

The BANG file. The balanced and nested grid file (BANG file, Freeston 1987) is a particularly interesting attempt at balancing the load in a regular partition pattern. It operates on a regular grid in such a way as to guarantee linear growth of the directory. Unlike the grid file, the BANG file splits a cell such that the numbers of objects in the two resulting subspaces differ as little as possible.

176

One of the two subspaces is defined by recursively halving the cell until the number of objects in the subspace is as close as possible to half the number of objects in the cell. This is achieved by repeatedly cutting the subspace that contains more than half of all objects of the cell, until a subspace S is reached for which both halves S_1 and S_2 contain at most half that number. Then, we select as the subspace resulting from the split either S, S_1, or S_2, whichever contains closest to half the number of objects. The remaining part of the cell defines the other subspace. Clearly, that other subspace will in general not be convex: It is a rectangle with one or more rectangular pieces removed. For an example, see Fig. 12a, where point numbers indicate the insertion order. As a consequence, the BANG file tends to use fewer cells than the grid file (compare Figs. 12a and c), but these cells have a more complex shape. Nevertheless, it is not a problem to maintain these cells: We simply maintain the rectangular cells without keeping track of their rectangular missing pieces (the holes), and we associate a point with the smallest of all cells containing it (Fig. 12b). In this way, the association of a point with a cell is unique, and reflects the partition of space, even though cells viewed without holes overlap. In addition, these cells can be stored in a highly compressed form: Since they are created by recursive halving that cycles through all dimensions, each cell can be stored as a pair of indices: The first index indicates the number of recursive cuts, and the second index is the relative number in some numbering scheme, for instance that of MOLHPE (Fig. 12b). Not surprisingly, the BANG file directory is organized as a tree, following the BANG file strategy recursively. This results in the necessity to propagate splits not only upwards in the hierarchy, but also downwards - the forced split phenomenon shared by quite a few hierarchical spatial access structures.

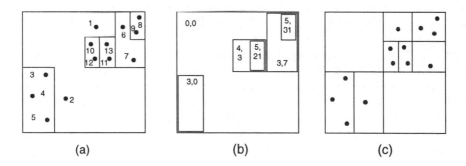

Fig. 12. A BANG file partition (a), its maintenance (b), and a grid file partition (c) for capacity 3

There is a large number of data structures whose partition is driven by a combination of space and data considerations, with quad tree based structures (Finkel et al. 1974, Samet 1990a, b) being the most prominent ones among them. Others are variants of grid file or hash tree structures (Kriegel et al. 1988, Otoo

1986, 1990, Ouksel 1985, Ozkarahan et al. 1985) or adaptive hashing schemes (Kriegel et al. 1987, 1989b). Some of them aim in particular at high storage space utilization (Hutflesz et al. 1988b, c, d), apart from the efficiency of range queries.

5.5 Redundant Data Storage

So far we have presented data structures that store every data element (point) exactly once. This natural approach is universally followed in practice, because data redundancy complicates updating and therefore is used only to enhance reliability (e.g. back-up procedures), but is not part of the access structure.

From a theoretical point of view, however, one can ask whether replicating some part of the data might speed up retrieval. This turns out to be true in particular for static files. Chazelle (1990) proves the following lower bound for a pointer machine that executes static 2-d range searches: A query time of $O(t + \log^c n)$, where t is the number of points reported and c is some constant, can only be achieved at the expense of $\Omega(n \log n / \log \log n)$ storage.

Data structures that use data redundancy to improve access time for range searching include the P-range tree (Subramanian 1995), a combination of the priority search tree and a 2-d range tree.

6 Spatial Data Structures for Extended Objects

So far we have considered the simplest of spatial objects only, points, for the good reason that any spatial data structure must be able to handle point data efficiently. Most applications, however, deal with complex spatial objects. And although complex objects are composed of simpler building blocks, we encounter a multitude of the latter: line segments, poly-lines, triangles, aligned rectangles, simple polygons, circles and ovals, and the multi-dimensional generalizations of all these. The way a spatial data structure supports extended (non-point) objects of various kinds determines whether it is generally applicable. For extended objects, each of which may intersect a number of cells, the association of an object with a cell is not as immediate as it is for points. In this case, cells usually overlap, and an object is associated most often either with a cell that contains it, or with all cells it intersects. But there are also other possibilities. Therefore, while we distinguish data structures for points merely according to the type of cells they define, we characterize data structures for extended objects also according to the way they associate objects with regions.

We consider the case where objects to be stored are relatively simple, in the sense that they have a concise description, and computations are easy and efficient. This restriction is realistic because complex objects are often approximated or bounded by a simple container, such as a bounding box, the smallest aligned (axis-parallel) multidimensional rectangle that contains the object. Such a container serves as a conservative filter for spatial proximity queries. In a range query, for instance, an object intersects the query range only if its bounding box intersects the query range; similarly, a query point can be contained in an object only if it is contained in its bounding box. Most data structures that support extended objects limit themselves to aligned rectangles (bounding boxes); exceptions include (Bruzzone et al. 1993, Günther 1988, 1989, 1992a, Günther et al. 1989, 1991, Jagadish 1990a, van Oosterom et al. 1990). Even in this restricted case, it is by no means clear how to associate a rectangle with a region in space, because a rectangle may intersect more than one of these regions. There are essentially three different extremal solutions to this problem, and a fourth one as a combination of two extremes.

6.1 Parameter Space Transformations

Since points can be maintained in any of the ways described in the previous section, there is the obvious possibility to store simple objects as points in some parameter space. Simple mappings have been proposed that transform a d-dimensional rectangle into a 2-d-dimensional point (Henrich 1990, Hinrichs 1985, Seeger 1989). The corner transformation simply takes the 2-d rectangle boundary coordinates and interprets them as the coordinates of a point in 2-d space (see Fig. 13 a for d = 1). The center transformation separates parameters for the position of the rectangle in space from parameters for the size of the rectangle: the former are the d coordinates of the rectangle center, and the latter are the extensions of the rectangles in the d dimensions - divided by two, resulting in a more evenly populated data space (see Fig. 13 b). But then, fairly small query ranges (that seem to be common in practice) may map to queries with fairly large query regions (see Fig. 13 c and d), and the distribution of points, the data space partition and the shape of the query region seem not to work well together. For the special case in which range queries are the only type of proximity queries, good transformations have actually been found (Pagel et al. 1993a). These transformations use parameters such as volume and aspect ratio, and they turn out to cluster rectangles in a way that is quite appropriate for many point data structures, especially in combination with highly adaptive space partitions such as the LSD-tree (Henrich et al. 1989b, Henrich 1990). Nevertheless, we feel that a transformation cannot be general enough to preserve the geometry of the situation entirely, and as a consequence, will not be able to support all kinds of locational queries well. In the remainder of this section, we will therefore look more closely at ways to store rectangles in the given space.

6.2 Clipping

The problem in associating a rectangle with a region in a partition is the fact that a rectangle can intersect more than one region. For the technique of answering a range query by returning all those data blocks whose regions intersect the query range, there is no other choice but to associate each rectangle with all regions in the partition that it intersects. This method is called clipping. It can be applied to any point data structure. For specific cases, in particular for clipping edges in computer graphics, explicit suggestions can be found in the literature (Nelson et

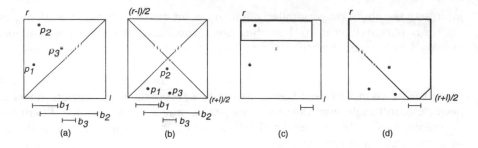

Fig. 13. Corner transformation (a) and center transformation (b) for intervals and for query regions (c), (d)

al. 1986, Samet et al. 1985, Tamminen 1981, Warnock 1969). Clipping can lead to reasonable performance in cases where the geometric object behind the bounding box can be cut at region boundaries, such as those cartographic applications in which objects are polygons with lots of corners (Schek et al. 1986, Waterfeld 1991). In these cases, the clipping technique has the advantage over many others to be conceptually simple and to preserve the geometry of the situation for any proximity query. Clipping turns out not to lead to good performance for rectangles (Six et al. 1988), though, and it can become very bad in the worst case, with a linear proportion of all rectangles even on the best possible cut line (d'Amore et al. 1993a, b, Nguyen et al. 1993, d'Amore et al. 1995).

6.3 Regions with Unbounded Overlap: R-Tree

Storing a reference point. In another straightforward way of using point data structures for storing objects, a reference point is chosen for each object - typically its center (of gravity) -, and the object is associated with the block in whose region its reference point lies. This works only if we keep track of the extensions of the objects beyond the region boundaries. In a range query, the cell to be considered for a block is not the cell defined by the partition of the data space, but instead the bounding box of all objects actually associated with the block. The latter may in general be larger than the former, and not all data structures will easily accomodate that extended region information. Hierarchical structures such as k-d-trees (Ooi 1987, Ooi et al. 1989) or the BANG file (Freeston 1989b) as well as some others (Seeger et al. 1988, Seeger 1989) can be used for that purpose. Even though this approach works well whenever only small objects are to be stored, it is inefficient in general, because no attempt is made to avoid the overlap of search regions, and therefore geometric selectivity is lost easily.

The R-tree family. With the explicit goal of high geometric selectivity, the R-tree (Guttman 1984) has been designed to maintain block cells that overlap just as much as necessary, so as to make each rectangle fall entirely within a cell. Its structure resembles the B+-tree, including restructuring operations that propagate from the leaf level towards the root. Each data cell is the bounding box of the rectangles associated with the corresponding block. Each directory block maintains a rectangular subspace of the data space; its block cell is the bounding box of the subspaces of its children. As a consequence, on each level of the tree, each stored rectangle lies entirely in at least one block cell; since block cells will overlap in general, it may actually lie in more than one cell (in Fig. 14, rectangles E and F lie in cells A and B, referenced by the root block). This fact may distort geometric selectivity: In an exact match query, we cannot restrict the search for E to either A or B, but instead we need to follow both paths (in the worst case, with no hint as to which one is more likely). Since it is essential for the R-tree to avoid excessive overlap of cells, many strategies of splitting an overflowing block into two exist, ranging from the initial suggestion (Guttman 1984) to less or more sophisticated heuristics (Beckmann et al. 1990, Greene 1989) and even to optimal splits for a number of criteria (Becker et al. 1992, Six et al. 1992). With the appropriate splitting strategy and extra restructuring operations (Beckmann et al. 1990), the R-tree seems to be one of the most efficient access structures for rectangles to date.

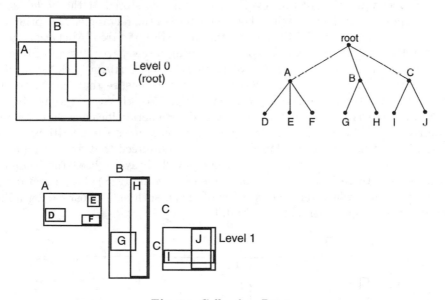

Fig. 14. Cells of an R-tree

For specific purposes, a number of variants of the R-tree have been proposed. The R+-tree avoids overlapping directory cells by clipping rectangles as necessary (Faloutsos et al. 1987, Sellis et al. 1987). It suffers, however, somewhat from the inefficiency caused by forced splits propagating downwards, similar to k-d-B-trees. For a static situation, in which almost no insertions or deletions take place, and therefore queries dominate the picture, the R-tree can be packed densely so as to increase effciency in time and space (Roussopoulos et al. 1985). Ohsawa et al. (1990) combine the R-tree and quad tree cells.

6.4 Cells with Bounded Overlap: Multilayer Structures, R-File, Guard File

Geometric selectivity increases with decreasing cell overlap. This has been the starting point for a number of attempts to maintain rectangles with cells whose overlap is under tight control

The multilayer technique. The basic idea is to cover the data space with more than one partition. Each level of a hierarchical data structure, in which the cells of a node's children are a partition of the cell of their parent, partitions the data space or a subspace. Therefore, such a hierarchy naturally offers a set of partitions with various degrees of refinement. Even though this set of partitions has not been defined with the explicit goal of covering space several times, it can be used nicely to store rectangles: Each rectangle is stored at the node as far down as possible in the hierarchy whose cell encloses the rectangle. This principle has been used in the quad-CIF tree and the MX-CIF quad tree (Abel et al. 1983, Kedem 1982). With the explicit goal of creating a few partitions according to the set of rectangles, Six et al. (1988) maintain each partition – called layer – in an extra data structure. Care has to be taken to ensure that the partitions of the different layers are actually different (Fig. 15); otherwise, the number of partitions increases more than necessary, and in general efficiency deteriorates. A fairly large number of split strategies that guarantees the partitions to be quite different have been developed. It can be guaranteed that for storing a set of small d-dimensional rectangles, $d + 1$ layers will always suffice - the technical term "small" there has a precise meaning (Six et al. 1988). Large rectangles must be clipped, whenever there are few of them, such as in most cartographic applications, clipping will not be harmful.

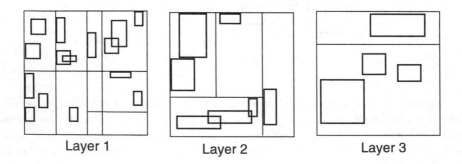

| Layer 1 | Layer 2 | Layer 3 |

Fig. 15. Cells of a three-layer grid file

In a multilayer data structure that uses a point data structure with a modified split strategy for each layer, the layers are totally ordered, from lowest to highest. A rectangle is associated with the lowest layer that has a cell of the partition containing it entirely; in that layer, the rectangle is associated with that cell, just like a point in the underlying point data structure. If there is no such layer, either a new layer is created, or the rectangle is clipped at the boundaries of the highest layer.

Note that the multilayer technique is generic in the sense that it allows any one of a number of point data structures to be used for the layers. Experiments with a multilayer grid file, for instance, show that the loss of efficiency as compared with a standard (single layer) grid file for points is tolerable, with a certain, but small, query overhead due to the fact that a directory for each layer needs to be inspected. Experiments with multilayer dynamic z-hashing (Hutflesz et al. 1992) show that the attempt to preserve global order makes range queries extremely fast, far better than any other data structure, for data and queries from cartography. In addition, the multilayer technique realizes the advantages of recursive linear hashing (Ramamohanarao et al. 1984) over linear hashing without the overhead, because there are several (recursive) layers anyway.

The R-file. Since the search parameters are the same for each layer of a multilayer structure (e.g. in a range query), it might be desirable to integrate all the directories into one directory. This is exactly what the R-file (Hutflesz et al. 1990) does: It maintains a set of overlapping cells with one directory. The cells are exactly of the same form as the rectangular regions in the BANG file, with no rectangles cut out for subspaces. Similar to the quad-CIF tree, each object is associated with the smallest cell that contains it (Fig. 16). Therefore, unlike in the BANG file, where cells are disjoint rectilinear polygons with holes, in the R-file cells overlap.

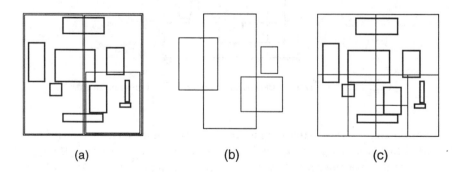

(a)   (b)   (c)

Fig. 16. R-file cells for insertion (a) and for searching (b), and grid file cells with clipping (c)

There is a limit to the association of rectangles to cells in this way: If too many rectangles intersect the split line of a cell in which they lie, they cannot be distributed to other cells. In that case, a lower dimensional R-file stores these rectangles; the one dimension in which all of these rectangles intersect the split line can be disregarded. Fortunately, no extra data structure is needed for the (d-1)-dimensional R-file; instead, a simple skip of the corresponding dimension in the cyclic turn through all dimensions will do. Experiments have shown the R-file to be a very efficient data structure for cartographic data. It suffers, however, from the disadvantage of being somewhat complicated to implement, with extra algorithmic difficulties (but not inefficiencies) such as a forced split downwards. Nevertheless, it is a good basis for a data structure that supports queries with a requested degree of detail, all within a query range (Becker et al. 1991).

The guard file. The guard file (Nievergelt et al. 1993) is designed for *fat* objects, where fatness requires that the shape of an object is sufficiently close to a sphere, in a precise, technical sense. The difficulties in maintaining arbitrary spatially extended objects, as opposed to points, comes from the fact that an object may intersect more than one cell of a space partition, that is, it may intersect split lines. For fat objects, the idea is to place guard points (not lines) in strategic positions, and then to distinguish objects according to the guards they contain. Guards are the corners of the cells of some regular space partition. Fig. 17 shows a radix tree partition, but other partitions such as triangular or hexagonal ones work just as well.

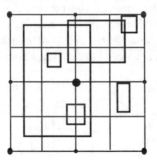

Fig. 17. A radix tree partition with guards at corner points

By imposing the radix tree as a hierarchy on the guards, range queries become efficient: The level of a guard is the level of the largest cell of which the guard is a corner point, with the root of the tree having the highest level. Any sufficiently large, fat object must contain a guard. An object containing a guard (a *guarded* object) is stored with the highest level guard that it contains. A fat object that contains no guard (an *unguarded* object) must be small; it is stored with all intersecting cells of the partition at the leaf level of the radix tree. As a consequence, a range query is carried out by inspecting the objects associated with all guards in the query range and all cells intersecting the query range. For long and skinny objects, clipping makes this data structure inefficient, but for fat objects such as those often encountered in cartography, clipping is no problem.

The guard file, however, includes an additional step to trade storage space for query time: An unguarded object is associated with exactly one cell of the partition, namely the one into which its center (of gravity) falls. Hence, each object is stored only a small, constant number of times, the number depending only on the type of regular partition being used and the precise fatness requirement. In a range query, it is no longer sufficient to inspect the cells that intersect the query range: Adjacent cells also need to be inspected. The neighborhood of adjacent cells by which a query range must be extended depends entirely on how fat the stored objects are: the fatter the objects, the smaller this neighborhood. For various types of regular partitions and various bounds on the fatness (in a precise, technical sense for arbitrary, convex geometric objects), the guard file is easy to implement and efficient at the same time.

6.5 Conclusions Drawn from the Structures Surveyed

The set of spatial data structures and paradigms has grown to a respectable size. A multitude of concepts serve as building blocks in the design of new spatial data structures for specific purposes. A number of convincing suggestions indicate that, for instance, these building blocks can be used in the design of data structures for more complex settings. Becker (1992), Brinkhoff et al. (1993), Günther (1992b), Kriegel et al. (1992a), and Shaffer et al. (1990a), among others, show how various kinds of set operations, such as spatial join, can be performed efficiently by using a spatial index. For objects and operations going beyond geometry, Ohler (1992) shows how to design access structures that involve non-geometric attributes in a query. In a closer look to the needs of cartography, a data structure of weighted geometric objects is proposed that efficiently supports queries with a varying degree of detail, as specified by the user (Becker et al. 1991). For the special case in which the objects to be stored are the polygons of a planar partition, and access to entire polygons (in containment queries with points, for instance) as well as access to the polygon boundaries (for drawing a map) is needed, there are suggestions that consider both external accesses and main memory computations as important cost factors (Kriegel et al. 1992b, Schiwietz et al. 1993); other suggestions aim at making queries efficient without storing polygon boundaries twice (Becker 1993). In addition, hierarchical spatial access structures can be designed to support queries into the past (Becker et al. 1993, Charlton et al. 1990, Xu et al. 1990). It can be seen from this list that even though a fair number of concepts of data access are readily available, it is mosten often a considerable effort to bring any two of them together without losing key features, especially efficiency.

All the arguments above concerning the efficiency of data structures are of an intuitive nature. An average case analysis is rarely given (Ang et al. 1992, Devroye 1986, Flajolet et al. 1986, Lindenbaum et al. 1995, Nelson et al. 1987, Regnier 1985, Rottke et al. 1987 are notable exceptions), because a probability model for sets of geometric objects and queries is very hard to get (Ambartzum-jan et al. 1993, Harding et al. 1974, Matheron 1975, Santalo 1976, Stoyan et al. 1987). Even experiments (Kriegel et al. 1989a, Shaffer et al. 1990b, Smith et al. 1990) that clarify some of the efficiency aspects tend to reveal not too much about the contribution of the building blocks of data structures to their over-all efficiency. Some steps towards a clarification of what is the desired average case efficiency and how to achieve it, with consequences for the data structure design, have been taken (Henrich et al. 1991, Pagel et al. 1993b, Six et al. 1992, Ang et al. 1992, Pagel 1995), but many more are needed to shed enough light on the inner workings of spatial data structures. In the meantime, the designer and programmer of spatial data structures should be aware of the existing build-ing blocks and use them with expertise and intuition, bringing them together wherever possible.

7 Summary: Points to Consider When Choosing a Spatial Data Structure

Let us conclude this bird's eye survey of the domain of spatial data structures with a few concisely stated points of view and recommendations.

Spatial data differs from all other types of data in important respects. Objects are embedded in a Euclidean space with its rich geometric structure, and most queries involve proximity rather than intrinsic properties of the objects to be retrieved. Thus data structures developed for conventional data base systems are unlikely to be efficient.

The most important feature of a spatial data structure is a systematic, regular organization of the embedding space. This serves as a skeleton that lets one address regions of space in terms of invariant quantities, rather than merely in relation to transient objects that happen to be stored. In particular, radix partitions of the space are more widely useful than data-driven space partitions based on comparative search.

The vast literature on spatial data describes a multitude of structures that differ only in technical details. Theory has not progressed to the stage where it can rank them in terms of efficiency. Performance comparisons representative of a wide range of applications and data are difficult to design, and reported results to this effect are often biased. Thus we are left to choose on the basis of common sense and conceptual simplicity.

Whereas the choice of a spatial data structure does not require a lot of detailed technical know-how, programming the computational geometry routines that implement the fine filter of Section 4.3 is a different matter. Writing robust and efficient procedures that intersect arbitrary polyhedra (say a query region and an object) is a specialist's task that requires a different set of skills.

In conclusion, this survey is neither a "how-to" recipe book that describes the multitude of data structures, nor a sales pitch for any one in particular. It intends to be a thought-provoking display of the issues to ponder and to assess.

Our message places responsibility for a competent choice where it belongs, with the programmer.

Acknowledgements

We are grateful to Martin Breunig, Thomas Ohler, Hanan Samet, Bernhard Seeger, Matti Tikkanen and an unknown referee for comments on this survey and interesting discussions.

References

Abel, D.J., J.L. Smith (1983): A data structure and algorithm based on a linear key for a rectangle retrieval problem; Computer Vision, Graphics, and Image Processing, Vol. 24, 1-13

Abel, D.J., D.M. Mark (1990): A comparative analysis of some two-dimensional orderings; International Journal of Geographical Information Systems, Vol. 4, No. 1, 21-31

Adelson-Velskii, G.M., Y.M. Landis (1962): An algorithm for the organization of information, Doklady Akademia Nauk SSSR, Vol. 146, 263-266; English translation in: Soviet Math. 3, 1259-1263

Ambartzumjan, R.V., J. Mecke, D. Stoyan (1993): Geometrische Wahrscheinlichkeiten und Stochastische Geometrie, Akademie Verlag, Berlin

Ang, C.-H., H. Samet (1992): Average storage utilization of a bucket method; Technical Report, University of Maryland

Aref, W.G., H. Samet (1994): The spatial filter revisited. In Proc. 6th Int. Symp. on Spatial Data Handling, 190-208

Arge, L. (1996): Efficient external-memory data structures and applications; PhD thesis, University of Aarhus, Denmark

Asano, T., D. Ranjan, T. Roos, P. Widmayer, E. Welzl (1995): Space filling curves and their use in the design of geometric data structures, Proc. Second Intern. Symp. of Latin American Theoretical Informatics LATIN '95, Valparaiso, Lecture Notes in Computer Science, Vol. 911, Springer-Verlag, 36-48

Bayer, R., C. McCreight (1972): Organization and maintenance of large ordered indexes; Acta Informatica, Vol. 1, No. 3, 173-189

Bayer, R., Schkolnick, M. (1977): Concurrency of operations on B-trees, Acta Informatica Vol. 1, 1-21

Becker, B. (1993): Methoden und Strukturen zur effizienten Verwaltung geometrischer Objekte in Geo-Informationssystemen, Dissertation, Mathematische Fakultät, Albert-Ludwigs-Universität, Freiburg, Germany

Becker, B., H.-W. Six, P. Widmayer (1991): Spatial priority search: An access technique for scaleless maps, Proc. ACM SIGMOD International Conference on the Management of Data, Denver, 128-137

Becker, B., P. Franciosa, S. Gschwind, T. Ohler, G. Thiemt, P. Widmayer (1992): Enclosing many boxes by an optimal pair of boxes; Proc. 9th Annual Symposium on Theoretical Aspects of Computer Science STACS, Cachan, Lecture Notes in

Computer Science, Vol. 577, 475-486

Becker, B., S. Gschwind, T. Ohler, B. Seeger, P. Widmayer (1993): On optimal multiversion access structures; 3rd International Symposium on Advances in Spatial Databases, Singapore, Lecture Notes in Computer Science, Vol. 692, Springer-Verlag, 123-141

Becker, L. (1992): A new algorithm and a cost model for join processing with grid files; Dissertation, Fachbereich Elektrotechnik und Informatik, Universität-Gesamthochschule Siegen, Germany

Beckmann, N., H.-P. Kriegel, R. Schneider, B. Seeger (1990): The R*-tree: An efficient and robust access method for points and rectangles; Proc. ACM SIGMOD International Conference on the Management of Data, Atlantic City, New Jersey, 322-331

Bentley, J.L. (1975): Multidimensional binary search used for associative searching; Communications of the ACM, Vol. 18, No. 9, 509-517

Blankenagel, G. (1991): Intervall-Indexstrukturen und externe Algorithmen für Nicht-Standard-Datenbanksysteme; Dissertation, FernUniversität Hagen

Brinkhoff, T., H.-P. Kriegel, B. Seeger (1993): Efficient processing of spatial joins using R-trees; Proc. ACM SIGMOD International Conference on the Management of Data, Washington D.C., 237-246

Brinkhoff, T. (1994): Der Spatial Join in Geo-Datenbanksystemen. Ph.D.thesis, Ludwig-Maximilians-Universität München.

Brinkhoff, T. und H.-P. Kriegel (1994): The impact of global clustering on spatial database systems. In Proc. 20th Int. Conf. on Very Large Data Bases, 168-179.

Brinkhoff, T., H.-P. Kriegel, and R. Schneider (1993): Comparison of approximations of complex objects used for approximation-based query processing in spatial database systems. In Proc. 9th IEEE Int. Conf. on Data Eng., 40-49.

Brinkhoff, T., H.-P. Kriegel, R. Schneider, and B. Seeger (1994): Multi-step processing of spatial joins. In Proc. ACM SIGMOD Int. Conf. on Management of Data, 197-208.

Bruzzone, E., L. De Floriani, M. Pellegrini (1993): A hierarchical spatial index for cell complexes; 3rd International Symposium on Advances in Spatial Databases, Singapore, Lecture Notes in Computer Science, Vol. 692, Springer-Verlag, 105-122

Charlton, M.E., S. Openshaw, C. Wymer (1990): Some experiments with an adaptive data structure in the analysis of space-time data; Proc. 4th International Symposium on Spatial Data Handling, Zurich, 1030-1039

Chazelle, B. (1990): Lower bounds for orthogonal range searching: I. The reporting case; J. ACM Vol. 37, No. 2, 200-212

Chen, L., R. Drach, M. Keating, S. Louis, D. Rotem, and A. Shoshani (1995): Access to multidimensional datasets on tertiary storage systems. Information Systems 20(2), 155-183.

Comer, D. (1979): The ubiquitous B-tree; ACM Computing Surveys, Vol. 11, No. 2, 121-138

Crain, I.K. (1990): Extremely large spatial information systems: a quantitative perspective; Proc. 4th International Symposium on Spatial Data Handling,

Zurich, 632-641

d'Amore, F., P.G. Franciosa (1993a): Separating sets of hyperrectangles, International Journal of Computational Geometry and Applications, Vol. 3, No.2, 155-165

d'Amore, F., T. Roos, P. Widmayer (1993b): An optimal algorithm for computing a best cut of a set of hyperrectangles, International Computer Graphics Conference, Bombay

d'Amore, F., V.H. Nguyen, T. Roos, P. Widmayer (1995): On optimal cuts of hyperrectangles, Computing, Vol. 55, Springer-Verlag, 191-206

Devroye, L. (1986): Lecture notes on bucket algorithms; Birkhäuser Verlag, Boston

Dröge, G., H.-J. Schek, (1993): Query-adaptive data space partitioning using variable-size storage clusters; 3rd International Symposium on Advances in Spatial Databases, Singapore, Lecture Notes in Computer Science, Vol. 692, Springer-Verlag, 337-356

Dröge, G. (1995): Eine anfrage-adaptive Partitionierungsstrategie für Raumzugriffsmethoden in Geo-Datenbanken; Dissertation ETH Zurich Nr. 11172

Edelsbrunner, H. (1982): Intersection problems in computational geometry; Dissertation, Technische Universität Graz

Enbody, R.J., H.C. Du (1988): Dynamic hashing schemes; ACM Computing Surveys, Vol. 20, No. 2, 85-113

Evangelidis, G. (1994): The hB - Tree: A Concurrent and Recoverable Multi-Attribute Index Structure. Ph.D. thesis, Northeastern University, Boston, MA.

Fagin, R., J. Nievergelt, N. Pippenger, H.R. Strong (1979): Extendible hashing - a fast access method for dynamic files; ACM Transactions on Database Systems, Vol. 4, No. 3, 315-344

Faloutsos, C. (1985): Multiattribute hashing using Gray codes; Proc. ACM SIGMOD International Conference on the Management of Data, Washington D.C., 227-238

Faloutsos, C. (1988): Gray codes for partial match and range queries; IEEE Transactions on Software Engineering, Vol. 14, 1381-1393

Faloutsos, C., S. Roseman (1989): Fractals for secondary key retrieval; Proc. 8th ACM SIGACT/SIGMOD Symposium on Principles of Database Systems, 247-252

Faloutsos, C., T. Sellis, N. Roussopoulos (1987): Analysis of object oriented spatial access methods; Proc. ACM SIGMOD International Conference on the Management of Data, San Francisco, 426-439

Finkel, R.A., J.L. Bentley (1974): Quad trees: A data structure for retrieval on composite keys; Acta Informatica, Vol. 4, No. 1, 1-9

Flajolet P., C. Puech (1986): Partial match retrieval of multidimensional data; Journal of the ACM, Vol. 33, No. 2, 371-407

Frank, A. (1981): Application of DBMS to land information systems; Proc. 7th International Conference on Very Large Data Bases, Cannes, 448-453

Freeston, M.W. (1987): The BANG file: a new kind of grid file; Proc. ACM SIGMOD International Conference on the Management of Data, San Francisco, 260-269

Freeston, M.W. (1989a): Advances in the design of the BANG file; Proc. 3rd International Conference on Foundations of Data Organization and Algorithms, Paris, Lecture Notes in Computer Science, Vol. 367, Springer-Verlag, Berlin, 322-338

Freeston, M.W. (1989b): A well-behaved file structure for the storage of spatial objects; Symposium on the Design and Implementation of Large Spatial Databases, Santa Barbara, Lecture Notes in Computer Science, Vol. 409, Springer-Verlag, Berlin, 287-300

Freeston, M. (1995): A general solution of the n-dimensional B-tree problem. In Proc. ACM SIGMOD Int. Conf. on Management of Data, 80-91.

Gaede, V. (1995): Optimal redundancy in spatial database systems. In Proc. 4th Int. Symp. on Spatial Databases (SSD'95).

Gaede, V., Günther, O. (1996): Multidimensional access methods; manuscript, Humboldt-Universität Berlin, Institut für Wirtschaftsinformatik

Goodchild, M.F. (1990): Spatial information science; Proc. 4th International Symposium on Spatial Data Handling, Zurich, 3-12

Greene, D. (1989): An implementation and performance analysis of spatial data access methods; Proc. 5th International Conference on Data Engineering, Los Angeles, 606-615

Günther, O. (1988): Efficient structures for geometric data management; Lecture Notes in Computer Science, Vol. 337, Springer-Verlag, Berlin

Günther, O. (1989): The design of the cell tree: An object oriented index structure for geometric databases; Proc. 5th International Conference on Data Engineering, Los Angeles, 598-605

Günther, O. (1992a): Evaluation of spatial access methods with oversize shelves; Geographic Data Base Management Systems, ESPRIT Basic Research Series Proceedings, Springer-Verlag, 177-193

Günther, O. (1992b): Efficient computation of spatial joins; Technical Report TR-92-029, International Computer Science Institute, Berkeley

Günther, O. (1993): Efficient computation of spatial joins. In Proc. 9th IEEE Int. Conf. on Data Eng.

Günther, O., J. Bilmes (1989): The implementation of the cell tree: design alternatives and performance evaluation; GI-Fachtagung Datenbanksysteme für Büro, Technik und Wissenschaft, Informatik-Fachberichte, Vol. 204, Springer-Verlag, Berlin, 246-265

Günther, O., J. Bilmes (1991): Tree-based access methods for spatial databases: implementation and performance evaluation; IEEE Transactions on Knowledge and Data Engineering, Vol. 3, No. 3, 342-356

Günther, O., A. Buchmann (1990): Research issues in spatial databases; IEEE CS Bulletin on Data Engineering, Vol. 13, No. 4, 35-42

Günther, o., R. Körstein, R. Müller, and P. Schmidt (1995): The MMM project: Acess to algorithms via WWW. In Proc. Third International World Wide Web Conference. URL http.//www.igd.fhg.de/www95.html.

Güting, R.H. (1989): Gral: An extensible relational database system for geometric applications; Proc. 15th International Conference on Very Large Data Bases, Amsterdam, 33-44

Guttman, A. (1984): R-trees: a dynamic index structure for spatial searching; Proc. ACM SIGMOD International Conference on the Management of Data, Boston, 47-57

Harding, E.F., D.G. Kendall (1974): Stochastic Geometry; Wiley, New York

Henrich, A. (1990): Der LSD-Baum: eine mehrdimensionale Zugriffsstruktur und ihre Einsatzmöglichkeiten in Datenbanksystemen; Dissertation, FernUniversität Hagen

Henrich, A., H.-W. Six (1991): How to split buckets in spatial data structures; Geographic Data Base Management Systems, ESPRIT Basic Research Series Proceedings, Springer-Verlag, 212-244

Henrich, A., H.-W. Six, P. Widmayer (1989a): Paging binary trees with external balancing; 15th International Workshop on Graph-Theoretic Concepts in Computer Science, Castle Rolduc, Lecture Notes in Computer Science, Vol. 411, 260-276

Henrich, A., H.-W. Six, P. Widmayer (1989b): The LSD-tree: Spatial access to multidimensional point- and non-point objects; 15th International Conference on Very Large Data Bases, Amsterdam, 45-53

Hinrichs, K.H. (1985): The grid file system: implementation and case studies of applications; Dissertation, ETH Zurich

Hutflesz, A., H.-W. Six, P. Widmayer (1988a): Globally order preserving multidimensional linear hashing; Proc. 4th International Conference on Data Engineering, Los Angeles, 572-579

Hutflesz, A., H.-W. Six, P. Widmayer (1988b): The twin grid file:A nearly space optimal index structure; Proc. International Conference on Extending Database Technology, Venice, Lecture Notes in Computer Science, Vol. 303, Springer-Verlag, Berlin, 352-363

Hutflesz, A., H.-W. Six, P. Widmayer (1988c): Twin grid files: Space optimizing access schemes; Proc. ACM SIGMOD Conference on the Management of Data, Chicago, 183-190

Hutflesz, A., H.-W. Six, P. Widmayer (1988d): Twin grid files: A performance evaluation; Proc. Workshop on Computational Geometry, CG'88, Lecture Notes in Computer Science, Vol. 333, Springer-Verlag, Berlin, 15-24

Hutflesz, A., H.-W. Six, P. Widmayer (1990): The R-file: An efficient access structure for proximity queries; Proc. 6th International Conference on Data Engineering, Los Angeles, 372-379

Hutflesz, A., P. Widmayer, C. Zimmermann (1992): Global order makes spatial access faster; Geographic Data Base Management Systems, ESPRIT Basic Research Series Proceedings, Springer-Verlag, 161-176

Hunter, G.M., K. Steiglitz (1979): Operations on images using quad trees; IEEE Transactions on Pattern Analysis and Machine Intelligence, Vol. 1, No. 2, 145-153

Iyengar, S.S., N.S.V. Rao, R.L. Kashyap, V.K. Vaishnavi (1988): Multidimensional data structures: Review and outlook; in: Advances in Computers, ed. by

Marshall C. Yovits, Academic Press, Vol. 27, 69-119

Jagadish, H.V. (1990a): Spatial search with polyhedra; Proc. 6th International Conference on Data Engineering, Los Angeles, 311 310

Jagadish, H.V. (1990b): Linear clustering of objects with multiple attributes; Proc. ACM SIGMOD International Conference on the Management of Data, Atlantic City, New Jersey, 332-342

Kamel, I. and C. Faloutsos (1994): Hilbert R-tree: An improved R-tree using fractals. In Proc. 20th Int. Conf. on Very Large Data Bases, 500-509.

Kedem, G. (1982): The quad-CIF tree: A data structure for hierarchical on-line algorithms; Proceedings of the Nineteenth Design Automation Conference, Las Vegas, 352-357

Kemper, A., M. Wallrath (1987): An analysis of geometric modelling in database systems; ACM Computing Surveys, Vol. 19, No. 1, 47-91

Klinger, A. (1971): Patterns and search statistics; In: J. S. Rustagi, ed., Optimizing methods in statistics, Academic Press, New York, 303-337

Knuth, D. E. (1968): The art of computer programming, Vol.1: Fundamental algorithms, Addison-Wesley.

Knuth, D. E. (1973): The art of computer programming, Vol. 3: Sorting and searching, Addison-Wesley.

Kriegel, H.-P., B. Seeger (1986): Multidimensional order preserving linear hashing with partial expansions; Proc. International Conference on Database Theory, Lecture Notes in Computer Science, Vol. 243, Springer-Verlag, Berlin, 203-220

Kriegel, H.-P., B. Seeger (1987): Multidimensional dynamic quantile hashing is very efficient for non-uniform record distributions; Proc. 3rd International Conference on Data Engineering, Los Angeles, 10-17

Kriegel, H.-P., B. Seeger (1988): PLOP-Hashing: a grid file without directory; Proc. 4th International Conference on Data Engineering, Los Angeles, 369-376

Kriegel, H.-P., M. Schiwietz, R. Schneider, B. Seeger (1989a): Performance comparison of point and spatial access methods; Proc. Symposium on the Design and Implementation of Large Spatial Databases, Santa Barbara, Lecture Notes in Computer Science, Vol. 409, Springer-Verlag, Berlin, 89-114

Kriegel, H.-P., B. Seeger (1989b): Multidimensional quantile hashing is very efficient for non-uniform distributions; Information Sciences, Vol. 48, 99-117

Kriegel, H.-P., T. Brinkhoff, R. Schneider (1992a): An efficient map overlay algorithm based on spatial access methods and computational geometry; Geographic Data Base Management Systems, ESPRIT Basic Research Series Proceedings, Springer-Verlag, 194-211

Kriegel, H.-P., P. Heep, S. Heep, M. Schiwietz, R. Schneider (1992b): An access method based query processor for spatial database systems; Geographic Data Base Management Systems, ESPRIT Basic Research Series Proceedings, Springer-Verlag, 273-292

Krishnamurthy, R., K.-Y. Whang (1985): Multilevel grid files; IBM T.J. Watson Research Center Report, Yorktown Heights, New York

Kung, H.T., Lehman, P.L. (1980): Concurrent manipulation of binary search trees, ACM Trans. on Database Systems, Vol. 5, No. 3, 339-353

Lindenbaum, M., H. Samet (1995): A probabilistic analysis of trie-based sorting of large collections of line segments; Technical Report, University of Maryland

Litwin, W. (1980): A new tool for file and table addressing; Proc. 6th International Conference on Very Large Data Bases, Montreal, 212-223

Lo, M. and C. Ravishankar (1994): Spatial joins using seeded trees. In Proc. ACM SIGMOD Int. Conf. on Management of Data, 209-220.

Lohmann, F., K. Neumann (1990): A geoscientific database system supporting cartography and application programming; Proc. 8th British National Conference on Databases, Pitman, London, 179-195

Lomet, D.B., B. Salzberg (1989): A robust multi-attribute search structure; Proc. 5th International Conference on Data Engineering, Los Angeles, 296-304

Lomet, D.B., B. Salzberg (1990): The hB-tree: A multiattribute indexing method with good guaranteed performance; ACM Transactions on Database Systems, Vol. 15, No. 4, 625-658

Manola, F., J.A. Orenstein (1986): Toward a general spatial data model for an object-oriented DBMS; Proc. 12th International Conference on Very Large Data Bases, Kyoto, 328-335

Matheron G. (1975): Random sets and integral geometry; Wiley, New York

Morton, G.M. (1966): A computer oriented geodetic data base and a new technique in file sequencing; IBM, Ottawa, Canada

Nelson, R.C., H. Samet (1986): A consistent hierarchical representation for vector data; Computer Graphics, Vol. 20, No. 4, 197-206

Nelson, R.C., H. Samet (1987): A population analysis for hierarchical data structures, Proc. ACM SIGMOD International Conference on the Management of Data, San Francisco, 270-277

Nguyen, V.H., T. Roos, P. Widmayer (1993): Balanced cuts of a set of hyperrectangles; Proc. 5th Canadian Conference on Computational Geometry, Waterloo, 121-126

Nievergelt, J. (1989): 7+-2 criteria for assessing and comparing spatial data structures; Symposium on the Design and Implementation of Large Spatial Databases, Santa Barbara, Lecture Notes in Computer Science, Vol. 409, Springer-Verlag, Berlin, 3-28

Nievergelt, J., H. Hinterberger and K.C. Sevcik (1981): The Grid File: An adaptable, symmetric multikey file structure. In: Trends in Information Processing Systems, Proc. 3rd ECI Conf., A. Duijvestijn and P. Lockemann (eds.), Lecture Notes in Computer Science 123, Springer Verlag, 236-251

Nievergelt, J., H. Hinterberger, K.C. Sevcik (1984): The grid file: an adaptable, symmetric multikey file structure; ACM Transactions on Database Systems, Vol. 9, No. 1, 38-71

Nievergelt, J., P. Widmayer (1993): Guard files: Stabbing and intersection queries on fat spatail objects; The Computer Journal, Vol. 36, No. 2, 107-116

Nievergelt, J., P. Widmayer (1996): All space filling curves are equally efficient. Manuscript, ETH Zurich.

Ohler, T. (1992): The multiclass grid file: An access structure for multiclass range queries; Proc. 4th International Symposium on Spatial Data Handling, 260-271

Ohler, T. (1994): On the integration of non-geometric aspects into access structures for geographic information systems; Dissertation ETH Zurich Nr. 10877

Ohler, T., P. Widmayer (1994). A brief tutorial introduction to data structures for geometric databases; in: Advances in database systems: Implementations and applications, CISM courses and lectures No. 347, Springer-Verlag Wien New York, 329-351

Ohsawa, Y., M. Sakauchi (1990): A new tree type data structure with homogeneous nodes suitable for a very large spatial database; Proc. 6th International Conference on Data Engineering, Los Angeles, 296-303

Ooi, B.C. (1987): A data structure for geographic database; GI-Fachtagung Datenbanksysteme für Büro, Technik und Wissenschaft, Informatik-Fachberichte, Vol. 136, Springer-Verlag, Berlin, 247-258

Ooi, B.C., R. Sacks-Davis, K.J. McDonell (1989): Extending a DBMS for geographic applications; Proc. 5th International Conference on Data Engineering, Los Angeles, 590-597

Orenstein, J.A. (1986): Spatial query processing in an object-oriented database system; Proc. ACM SIGMOD International Conference on the Management of Data, 326-336

Orenstein, J.A. (1989): Redundancy in spatial databases; Proc. ACM SIGMOD International Conference on the Management of Data, Portland, 294-305

Orenstein, J.A. (1990): A comparison of spatial query processing techniques for native and parameter spaces; Proc. ACM SIGMOD International Conference on the Management of Data, Atlantic City, New Jersey, 343-352

Orenstein, J.A., T.H. Merrett (1984): A class of data structures for associative searching; Proc. 3rd ACM SIGACT/SIGMOD Symposium on Principles of Database Systems, Waterloo, 181-190

Otoo, E.J. (1986): Balanced multidimensional extendible hash tree; Proc. 5th ACM SIGACT-SIGMOD International Symposium on Principles of Database Systems, Cambridge, Massachusetts, 100-113

Otoo, E.J. (1990): An adaptive symmetric multidimensional data structure for spatial searching; Proc. 4th International Symposium on Spatial Data Handling, Zurich, 1003-1015

Ouksel, M. (1985): The interpolation-based grid file; Proc. 4th ACM SIGACT-SIGMOD Symposium on Principles of Database Systems, Portland, 20-27

Overmars, M. (1981): Dynamization of order decomposable set problems, J. of Algorithms, Vol. 2, 245-260

Overmars, M. (1983): The design of dynamic data structures, Proefschrift, Rijksuniversiteit Utrecht

Ozkarahan, E.A., M. Ouksel (1985): Dynamic and order preserving data partitioning for database machines; Proc. 11th International Conference on Very Large Data Bases, Stockholm, 358-368

Pagel, B.-U. (1995): Analyse und Optimierung von Indexstrukturen in Geo-Datenbanksystemen; Dissertation, FernUniversität Hagen, Germany

Pagel, B.-U., H.-W. Six, H. Toben (1993a): The transformation technique for spatial objects revisited; 3rd International Symposium on Advances in Spatial

Databases, Singapore, Lecture Notes in Computer Science, Vol. 692, Springer-Verlag, 73-88

Pagel, B.-U., H.-W. Six, H. Toben, P. Widmayer (1993b): Towards an analysis of range query performance in spatial data structures; 12th SIGACT-SIGMOD-SIGART Symposium on Principles of Database Systems, Washington, D.C., 214-221

Preparata, F.P., M.I. Shamos (1985): Computational Geometry: An Introduction, Springer Verlag, Berlin, Heidelberg, New York

Ramamohanarao, K., R. Sacks-Davis (1984): Recursive linear hashing; ACM Transactions on Database Systems, Vol. 9, No. 3, 369-391

Regnier, M. (1985): Analysis of grid file algorithms; BIT Vol. 25, 335-357

Robinson, J.T. (1981): The K-D-B-tree: a search structure for large multidimensional dynamic indexes; Proc. ACM SIGMOD International Conference on the Management of Data, Ann Arbor, 10-18

Roman, G.-C. (1990): Formal specification of geographic data processing requirements; IEEE Transactions on Knowledge and Data Engineering, Vol. 2, No. 4, 370-380

Rottke T., H.-W. Six, P. Widmayer (1987): On the analysis of grid structures for spatial objects of non-zero size; International Workshop on Graph-Theoretic Concepts in Computer Science, Staffelstein, Lecture Notes in Computer Science, Vol. 314, Springer-Verlag, Berlin, 94-105

Roussopoulos, N., D. Leifker (1985): Direct spatial search on pictorial databases using packed R-trees; Proc. ACM SIGMOD International Conference on the Management of Data, Austin, 17 31

Sagan, H. (1994): Space-Filling Curves. Berlin/Heidelberg/New York: Springer-Verlag.

Samet, H. (1988): Hierarchical representations of collections of small rectangles; ACM Computing Surveys, Vol. 20, No. 4, 271-309

Samet, H. (1990a): The design and analysis of spatial data structures; Addison-Wesley, Reading

Samet, H. (1990b): Applications of spatial data structures; Addison-Wesley, Reading

Samet, H., R.E. Webber (1985): Storing a collection of polygons using quadtrees; ACM Transactions on Graphics, Vol. 4, No. 3, 182-222

Santalo, L.A. (1976): Integral geometry and geometric probability; Addison-Wesley, Reading

Schek, H.-J., W. Waterfeld (1986): A database kernel system for geoscientific applications; Proc. 2nd International Symposium on Spatial Data Handling, Seattle, 273-288

Schek, H.-J., A. Wolf (1993): From extensible databases to interoperability between multiple databases and GIS applications; 3rd International Symposium on Advances in Spatial Databases, Singapore, Lecture Notes in Computer Science, Vol. 692, Springer-Verlag, 207-238

Schiwietz, M., H.-P. Kriegel (1993): Query processing of spatial objects: Complexity versus redundancy, 3rd International Symposium on Advances in Spatial

Databases, Singapore, Lecture Notes in Computer Science, Vol. 692, Springer-Verlag, 377-396

Seeger, B. (1989): Entwurf und Implementierung mehrdimensionaler Zugriffsstrukturen; Dissertation, Universität Bremen

Seeger, B., H.-P. Kriegel (1988): Techniques for design and implementation of efficient spatial access methods; Proc. 14th International Conference on Very Large Data Bases, Los Angeles, 360-371

Seeger, B., H.-P. Kriegel (1990): The buddy-tree: An efficient and robust access method for spatial data base systems; Proc. 16th International Conference on Very Large Data Bases, Brisbane, 590-601

Sellis T., N. Roussopoulos, C. Faloutsos (1987): The R+-tree:A dynamic index for multi-dimensional objects; Proc. 13th International Conference on Very Large Data Bases, Brighton, 507-518

Shaffer, C.A., H. Samet (1990): Set operations for unaligned linear quadtrees; Computer Vision, Graphics, and Image Processing, Vol. 50, No. 1, 29-49

Shaffer, C.A., H. Samet, R.C. Nelson (1990b): QUILT: A geographic information system based on quadtrees; International Journal of Geographical Information Systems, Vol. 4, No. 2, 103-131

Six H.-W., P. Widmayer (1988): Spatial searching in geometric databases; Proc. 4th International Conference on Data Engineering, Los Angeles, 496-503

Six H.-W., P. Widmayer (1992): Spatial access structures for geometric databases; in: Data Structures and Efficient Algorithms, Ed. B. Monien, T. Ottmann, Final Report on the DFG Special Joint Initiative, Lecture Notes in Computer Science, Vol. 594, 214-232.

Smith, T.R., P. Gao (1990): Experimental performance evaluations on spatial access methods; Proc. 4th International Symposium on Spatial Data Handling, Zurich, 991-1002

Stoyan, D., W.S. Kendall, J. Mecke (1987): Stochastic geometry and its applications; Wiley, New York, 1987.

Subramanian, S., S. Ramaswamy (1995): The P-range tree: A new data structure for range searching in secondary memory; in Proc. Sixth Annual ACM-SIAM Symposium on Discrete Algorithms, San Francisco, CA, 378-387

Tamminen, M. (1981): The EXCELL method for efficient geometric access to data; Acta Polytechnica Scandinavica, Mathematics and Computer Science Series No. 34, Helsinki, Finland

Tamminen, M. (1982): The extendible cell method for closest point problems; BIT, Vol. 22, 27-41

Tamminen, M. (1984): Metric data structures - an overview; Helsinki University of Technology, Laboratory of Information Processing Science, Report HTKK-TKO-A25

Tamminen, M. (1985): On search by address computation; BIT Vol. 25, 135-147

Tropf, H., H. Herzog (1981): Multidimensional range search in dynamically balanced trees; Angewandte Informatik, Vol. 2, 71-77

van Oosterom, P. (1990): Reactive data structures for geographic information systems; Proefschrift, Rijksuniversiteit Leiden van Kreveld, M.J. (1992): New results on data structures in computational geometry; Proefschrift, Rijksuniversiteit Utrecht

van Oosterom, P., E. Claassen (1990): Orientation insensitive indexing methods for geometric objects; Proc. 4th International Symposium on Spatial Data Handling, Zurich, 1016-1029

van Oosterom, P. (1990): Reactive data structures for geographic information systems; Proefschrift, Rijksuniversiteit Leiden Wang, J.-H., T.-S. Yuen, D.H.-C. Du (1987): On multiple random access and physical data placement in dynamic files; IEEE Transactions on Software Engineering, Vol. 13, No. 8, 977-987

Waterfeld, W. (1991): Eine erweiterbare Speicher- und Zugriffskomponente für geowissenschaftliche Datenbanksysteme; Dissertation, Fachbereich Informatik, Technische Hochschule Darmstadt, Germany

Warnock, J.E. (1969): A hidden surface algorithm for computer generated half tone pictures; Computer Science Department TR 4-15, University of Utah, Salt Lake City

Widmayer, P. (1991): Datenstrukturen für Geodatenbanken; in: Entwicklungstendenzen bei Datenbank-Systemen, ed. G. Vossen, K.-U. Witt, Oldenbourg Verlag München Wien, 317-361

Willard, D.E., (1978): Balanced forests of k-d* trees as a dynamic data structure, TR-23-78, Aiken Computation Lab., Harvard University.

Xu, X., J. Han, W. Lu (1990): RT-tree: an improved R-tree index structure for spatiotemporal databases; Proc. 4th International Symposium on Spatial Data Handling, Zurich, 1040-1049

197

Chapter 7. Space Filling Curves versus Random Walks

Edouard Bugnion

Dept. of Computer Science
Stanford University
U.S.A.

bugnion@cs.stanford.edu

Thomas Roos

Dept. of Computer Science
ETH Zürich
Switzerland

roos@inf.ethz.ch

Roger Wattenhofer

Dept. of Computer Science
ETH Zürich
Switzerland

wattenhofer@inf.ethz.ch

Peter Widmayer

Dept. of Computer Science
ETH Zürich
Switzerland

widmayer@inf.ethz.ch

1 Introduction

Data structures for maintaining sets of multidimensional points on external storage are very useful to non-standard database systems. In contrast to the one-dimensional case, the situation in higher dimensions is far more complex, since there is no obvious total order on the points that serves all purposes.

The most frequent type of queries on these multidimensional point sets are range queries: Given the set of points and a multidimensional query interval, report all points lying in the interval. The query intervals (or aligned rectangles, as they are often called) may have arbitrary size, aspect ratio, and position. An obvious possibility for a multidimensional data structure is to map multidimensional points to one dimension, and to maintain the resulting one-dimensional points in one of the well-known data structures (such as, for instance, B^+-trees).

The mapping should be such that the query corresponding to the image of a multidimensional range in the given, original space can be supported by a data structure in image space. The ideal mapping (that does not exist) would map each multidimensional interval into a one-dimensional interval containing precisely the images of the corresponding points. With the goal of approximating this ideal in some way, several mappings have been proposed. The most prominent ones include the *z-order* [13, 15, 16, 17], the *Gray code* [6, 7], and *Hilbert's curve* [8, 12]. The *z*-order is also known as *Morton encoding* [14], *bit interleaving* [20], *quad code* [9], or *locational code* [1].

The various embeddings of multidimensional space into one dimension have mostly been studied experimentally; there are, however, theoretical studies with a different focus [18, 19]. It seems that *z*-order and Hilbert's curve are the best known numbering schemes in general [2, 12, 21].

These embeddings typically map a multidimensional interval into quite a large number of one-dimensional intervals. In order to achieve better range query efficiency, it may be worthwhile to map a multidimensional interval into *as few as possible* one-dimensional intervals. The reason is that the techniques used to index the data in the one-dimensional space try to keep neighboring objects clustered in secondary storage. This reduces the seek time and latency time of the request. In the case of a multidimensional range query, it is important to maintain the locality of access provided by the data structures. As a consequence, data structures that aim at minimizing the number of seek operations (and not the number of block accesses) have been proposed [10, 11].

Recently, Asano et al. [5] designed a new space filling curve that minimizes the number of seek operations for the image of a range query in the worst-case. They adopted a model of computation that tolerates the transfer of a linear number of blocks whose regions do not intersect the query range, when a range query is answered. Their space filling curve requires only three seek operations in the worst case, whereas other space filling curves such as the Hilbert curve require up to four seek operations in the worst case.

In this paper, we study the performance of range queries on the grid with respect to the size and aspect ratio of the query rectangle. We will limit our investigations to the average case performance of the range queries under a simple cost model that takes into account the locality of the data accesses of a range query. Thereby we model the local behavior of a space filling curve by a random walk according to the direction distributions of the space filling curve. Surprisingly this yields a model that compares quite well to the behavior of the real space filling curve.

The following section introduces the basic concepts around space filling curves. In Section 3, we define and motivate a performance model for queries. In Section 4, we prove some useful properties of SFCs in this cost model. Section 5 validates our model by comparing it with simulation results based on the average performance of real space filling curves.

2   Foundations

This section provides the necessary definitions concerning space filling curves and the model of computation. Throughout the paper, we assume a finite set S of points in the unit square $[0, 1] \times [0, 1]$ that will be stored on secondary storage according to a numbering scheme (space filling curve) of a regular grid of finite resolution. We will study the performance of range queries, modeled by a *query* *rectangle Q of size $\triangle x \times \triangle y$.*

$$Q := [x, x + \triangle x] \times [y, y + \triangle y] \subset [0,1] \times [0,1]$$

Such a query requests all the points of S that lie in Q.

2.1 Space Filling Curves

We divide, for a given number $N = 2^n$, $n \geq 0$, the unit square $[0,1] \times [0,1]$ by a two-dimensional square grid $< 1 \dots N, 1 \dots N >$ of size $N \times N$ into N^2 cells. A *numbering* $\mathcal{P}$ of this grid is a one-to-one mapping

$$\mathcal{P} : N \times N \longrightarrow \{1, \dots, N^2\}.$$

The *distance in the numbering* $\mathcal{P}$ between two cells (i,j) and (i',j') is defined to be $|\mathcal{P}(i,j) - \mathcal{P}(i',j')|$. We can display the numbering $\mathcal{P}$ by connecting any two cells (i,j) and (i',j') via an edge iff they have consecutive numbers in the numbering. This results in a graph in which each node represents a cell and the distance between nodes (u, v) and (u', v') is the same as the distance in the $\mathcal{P}$ numbering. So, one can think of these numberings as curves that cover all cells of a two-dimensional array. This motivates the term *space filling curves (SFC)* that we often use instead of *numbering*.

We call a SFC $\mathcal{P}$ for a square region with numbers $C + 1, \dots, C + N^2$, for some $C \geq 0$, *recursive* (RSFC) if $N = 1$ or $\mathcal{P}$ can be divided into four RSFCs $\mathcal{P}_0, \dots, \mathcal{P}_3$ for a partition of the region into four squares of equal size in the following way. There exists a permutation $\pi : \{0, \dots, 3\} \to \{0, \dots, 3\}$ such that $\mathcal{P}_{\pi(i)}$ is a RSFC containing the numbers

$$\left\{ C + \frac{iN^2}{4} + 1, \dots, C + \frac{(i+1)N^2}{4} \right\},$$

for $i = 0, \dots, 3$. RSFCs have several advantages over SFCs in general; e.g., they allow a compact representation of all numbers without storing the entire SFC explicitly. In fact, the number $\mathcal{P}(i,j)$ assigned to a specific cell (i,j) on the grid can be determined from i and j in $O(\log N)$ time. As a matter of fact, all well-known RSFCs, as e.g. the Hilbert curve, the z-curve, the Gray code, and the new space filling curve by Asano et al. [5] are generated by very regular permutation rules. A SFC $\mathcal{P}$ is called *continuous* if

$$\| \mathcal{P}^{-1}(i+1) - \mathcal{P}^{-1}(i) \|_\infty = 1 \quad \text{for all} \quad i = C + 1, \dots, C + N^2 - 1$$

where $\| \ \|_\infty$ denotes the maximum norm (Chebyshev distance L_∞).

The construction of RSFCs can be visualized easily by means of a grammar. A variable of the grammar represents an oriented cell; by this we mean a cell that is oriented towards some compass direction. We start with a variable that initially represents the area of the entire $N \times N$ grid as one cell. The grammar is an E0L-type (extended zero-sided Lindenmayer [23]), that is, we force rewriting to take place simultaneously at every cell of the partition, and each syntactic variable is a terminal symbol of the grammar at the same time. Each word in the language defined by the grammar defines a permutation of cells, that is, a space filling curve.

Hence, we can view the rewriting process as going through a number of iterations, where each iteration rewrites all cells. In the i-th iteration, $i = 0, \ldots, n-1$ (recall that $N = 2^n$), we get a partition of the area of the $N \times N$ grid into $2^i \cdot 2^i$ blocks, each of size $2^{n-i} \times 2^{n-i}$. In each iteration, each block is replaced by four blocks of size $2^{n-i-1} \times 2^{n-i-1}$ according to an applicable production (refinement rule) of the grammar. Our grammar allows to rotate blocks (displayed by a rotation of the corresponding symbols $\prec$, V, $\succ$).

Two dots denote the entry and exit points of the curve into a cell. In a cell that is not refined further, both dots coincide with the center of the cell; in each refinement, the dots lie in the corresponding corner cells of the refined partition. Finally, lines between dots on the right hand side of the productions represent subsequent blocks in the numbering.

Figures 1 displays the productions and the first few iterations of the Hilbert curve; notice that a single production is sufficient here.

2.2 Transformation & Query

Recall that the application of SFCs we have in mind is their use as a data structure for embedding multidimensional points into one dimension while preserving spatial proximity to a large extent. The underlying external storage and access model is the following. All points that lie in one cell of the space partition are stored on one disk block. Because the disk block is the atomic entity for disk accesses, one might say that locally, i.e. within a cell, this preserves proximity in geometric space also on disk. In this way, a SFC defines many orders of the points of point set S if all points that lie in the same cell are assigned the same number; all of them are equivalent for our purpose. The index of a cell of a point $P_i = (x_i, y_i) \in S$ in the SFC is given as

$$SFC(\lfloor x_i N \rfloor, \lfloor y_i N \rfloor)$$

A rectangular query Q is then processed in the following way (see Figure 2). We first convert the query Q into a minimum discrete query $\bar{Q}$ covering the original query. (Notice that we are not interested in the loss of performance due to discretization.) Each cell in the discrete query corresponds to a block that needs to be accessed in order to answer the query. Afterwards, the discrete query $\bar{Q}$ is transformed into a set of maximal intervals I defined by the indices of the cells of the query. For all the blocks in an interval in I, only one disk seek

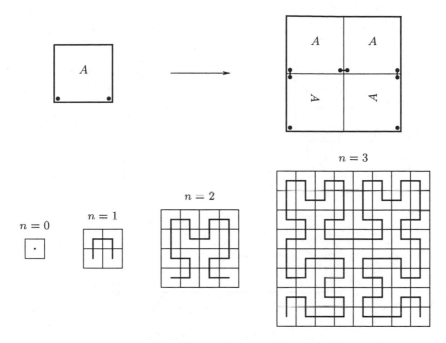

Fig. 1. The Hilbert production and the Hilbert curves for $n = 0, \ldots, 3$.

operation is necessary. That is, the intervals in I are used to access the actual objects based on their index; these intervals are called β-*intervals*. The spaces between the β-intervals are called α-*intervals*. With that, each query defines a sequence

$$\beta_0, \alpha_1, \beta_1, \ldots \alpha_I, \beta_I$$

of intervals. We note that the sum of the length of the β-intervals is always equal to the discrete query area, but that the number and length of the α-intervals is a function of the query and the space filling curve. Although the α-*intervals* are not part of the query, their presence does significantly influence the performance of range queries as they influence the locality of the data accesses. In the following section, we describe and motivate a cost model that takes this into account.

3   Model of Computation

The usage of space-filling curves to index multidimensional data allows to use well-known data structures developed for totally ordered sets. If the data set is large and resides in secondary storage, the performance of the range query will be determined by the disk accesses required to perform the range queries.

We use the following model to analyze the efficiency of the indexing scheme. We assume that the objects are physically stored in the order corresponding to the ordering based on the curve. A seek operation which moves the reading device

203

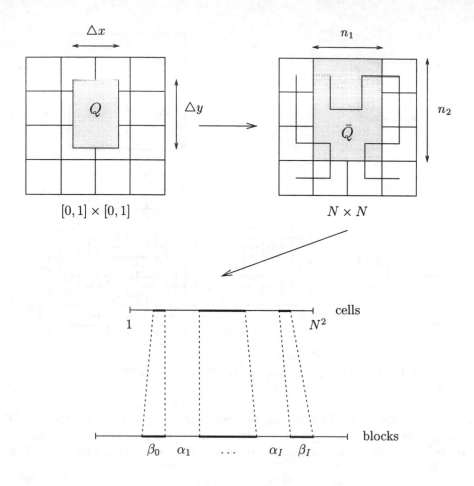

Fig. 2. Structure of the query algorithm.

to the starting block of a β-interval is assumed to have constant cost c. Reading subsequent blocks is assumed to have unit cost per block. In our model, it is imaginable to read and ignore cells if this is more efficient. Obviously, the cost of an α-interval is $min\{length(\alpha), c\}$ and the cost of a β-interval is $length(\beta)$. We assume further that the read/write head is already at the starting block of β_0 when the query is initialized.

Clearly, the model contains a number of simplifications with respect to real storage media. However, it captures one of the essential aspects associated with range queries, namely the number and length of the α-intervals generated by the query. For this model, we are going to design space filling curves that will generate only a small number of (preferably short) α-intervals.

4 Performance Analysis

4.1 Number of α-Intervals

The first problem we address is to estimate the average number of α-intervals of the discrete query $\bar{Q}$ of size $n_1 \times n_2$. For this, let $\mathcal{V}, \mathcal{H}$, and $\mathcal{D}$ denote the direction distribution of the underlying continuous SFC, i.e. the fraction of vertical, horizontal, and diagonal edges, respectively. Table 1 lists the distribution for some continuous space-filling curves (in the limit $N \to \infty$).

	$\mathcal{H}$	$\mathcal{V}$	$\mathcal{D}$
Hilbert	1/2	1/2	0
Asano et al.	7/15	7/15	1/15
Snake	1 - 1/N	1/N	0

Table 1. Distribution of $\mathcal{H}$, $\mathcal{V}$ and $\mathcal{D}$ for selected space filling curves.

Theorem 1. *For a continuous SFC, the average number I of α-intervals of a discrete query $\bar{Q}$ of size $n_1 \times n_2$ over all discrete positions of the query is*

$$I \approx \mathcal{V} \cdot n_1 + \mathcal{H} \cdot n_2 + \mathcal{D} \cdot (n_1 + n_2 - 1) - 1$$

$$\approx (1 - \mathcal{H}) \cdot n_1 + (1 - \mathcal{V}) \cdot n_2$$

Proof. We restrict ourselves to the $(N - 2n_1) \times (N - 2n_2)$ "central" cells located in the center of the grid. All central cells belong to the border of the same number of discrete query rectangles $\bar{Q}$ of size $n_1 \times n_2$. We assume each edge between two adjacent cells in the numbering to be oriented towards the cell with higher number in the SFC.

The average number of α-intervals per rectangle representing a query can be computed by considering all rectangles, and for each rectangle, the orientation of each of its border cells. In fact, swapping the summation, we can count for each central cell the number of rectangles for which it will start an α-interval:

$$\sum_{\text{queries } \bar{Q}} \text{\# cells of } \bar{Q} \text{ starting } \alpha\text{-intervals}$$

$$= \sum_{\text{cells } x} \text{\# queries with } x \text{ starting an } \alpha\text{-interval}$$

By definition, the last portion of the curve to leave the query area is not an α-interval. Notice that the above equation holds for the (comparably unimportant) boundary cells with an "$\leq$" inequality as well. This leads us directly to the first line of the formula from which, by simple arithmetic transformations, the second line can be derived. $\diamond$

4.2 Length of α-Intervals

Now, we consider the overhead generated by the α-intervals. In order to calculate the distribution of the lengths of the α-intervals, we use the following simplified model. The idea behind is an approximation of the α-intervals by random walks on the grid. Thereby the random transitions are performed according to the direction distributions $\mathcal{V}, \mathcal{H}$, and $\mathcal{D}$.

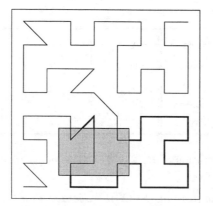

Fig. 3. The α-intervals of a query.

Figure 3 shows the α-intervals of a query in bold. We assume for simplicity that an α-interval ends on the side of the query rectangle where it has started. We now restrict ourselves to the upper side of the query rectangle and investigate an α-interval there. Notice that the probability $p_\uparrow$ of moving upwards (vertically or diagonally) is approximately equal to the probability $p_\downarrow$ of moving downwards since the SFC is continuous; to simplify the calculations, we assume equality. Thus we have

$$p_\uparrow = p_\downarrow = \frac{\mathcal{V} + \mathcal{D}}{2}.$$

Further, the probability $p_\leftrightarrow$ of staying at the same (vertical) distance from the upper side of the query rectangle is

$$p_\leftrightarrow = \mathcal{H}.$$

Clearly the first move to the outside of the query rectangle must be upward. Similarly, the last move – returning to a cell in the query rectangle – must be downward. In between, the number of down moves must be always smaller or equal to the number of upward moves (otherwise, we would re-enter the query rectangle earlier). At the end, the number of upward and downward moves must be the same. Thus, if we ignore the first and last move, we have an α-interval of length n with $m := m_\uparrow = m_\downarrow$ upward and downward moves, respectively ($m = 0, \ldots, \lfloor n/2 \rfloor$).

In order to estimate the probability of an α-interval of length n, we count the number of possibilities of making the downward moves for a fixed number of upward moves m. Before the first upward move we cannot move downward. Between the first and the second upward move, we can move downward at most once; let d_1 denote this number of downward moves. Between the second and the third upward move, the number of downward moves is bounded depending on d_1. If $d_1 = 0$, we can make at most two downward moves. If $d_1 = 1$, we can move down at most once; let d_2 denote this number of downward moves.

In general, there are d_i downward moves between the i-th and $(i + 1)$-th upward move. This gives the following restriction to d_i:

$$0 \leq d_i \leq i - \sum_{j=1}^{i-1} d_j$$

for $i = 1, \ldots, m - 1$. Since $m_\downarrow = m$, we conclude that the number of downward moves after the last upward move is determined by

$$d_m = m - \sum_{i=1}^{m-1} d_i.$$

The number of possibilities of downward placements in the sequence of upward moves is well known in combinatorics as "Catalan numbers". Notice that the m-th Catalan number can be stated explicitly by

$$\text{catalan}(m) \; := \; \frac{(2m)!}{m!(m+1)!}.$$

We can now arbitrarily insert the moves to the right or the left; their number is $n - 2m$ and we have $\binom{n}{2m}$ possibilities to place them. All in all, we have

$$\frac{(2m)!}{m!(m+1)!} \cdot \binom{n}{2m} = \frac{n!}{(n-2m)! \cdot m! \cdot (m+1)!}$$

possibilities. Now, we can combine these results to determine the probability of an α-interval of length n. We simply sum up the possibilities over all m and multiply the result with the probability of moving downward with the last move. Thus,

$$\Pr[\text{length} = n + 1] \; = \; p_\downarrow \cdot \sum_{m=0}^{\lfloor n/2 \rfloor} \frac{n!}{(n-2m)! \cdot m! \cdot (m+1)!} \cdot p_\uparrow{}^m \cdot p_\downarrow{}^m \cdot p_\leftrightarrow{}^{n-2m}$$

$$= \sum_{m=0}^{\lfloor n/2 \rfloor} \frac{n!}{(n-2m)! \cdot m! \cdot (m+1)!} \cdot \left(\frac{\mathcal{V} + \mathcal{D}}{2}\right)^{2m+1} \cdot \mathcal{H}^{n-2m}$$

If reading a cell (block) on the disk costs unit time and seeking the beginning of a new β-interval directly takes c time, it is cheaper to read and ignore all α-intervals of length at most c. Thus, the expected cost of going to the beginning of the next β-interval is

$$\gamma_c := \sum_{i=1}^{c} \Pr[\text{length} = i] \cdot i \ + \ \Pr[\text{length} > c] \cdot c$$

Notice that γ_c still depends on the parameters $\mathcal{V}$, $\mathcal{H}$, and $\mathcal{D}$. For any SFC with $\mathcal{V} = \mathcal{H} = 1/2$ and $\mathcal{D} = 0$ (e.g. the Hilbert curve), we get

$$\Pr[\text{length} = n+1] \ = \ \sum_{m=0}^{\lfloor n/2 \rfloor} \frac{n!}{(n-2m)! \cdot m! \cdot (m+1)!} \cdot \left(\frac{1}{4}\right)^{2m+1} \cdot \left(\frac{1}{2}\right)^{n-2m}$$

which can (surprisingly) be simplified to

$$\Pr[\text{length} = n] \ = \ \frac{\text{catalan}(n)}{4^n}$$

Using

$$\Pr[\text{length} > c] \ = \ \text{catalan}(c) \cdot \frac{(2c+1)}{4^c}$$

we obtain

$$\gamma_c \ = \ \sum_{i=1}^{c} \text{catalan}(i) \cdot \frac{i}{4^i} \ + \ \text{catalan}(c) \cdot \frac{(2c+1) \cdot c}{4^c}$$

$$= \ \text{catalan}(c) \cdot \frac{(2c+1) \cdot (c+2)}{4^c} \ - \ 2 \ + \ \text{catalan}(c) \cdot \frac{(2c+1) \cdot c}{4^c}$$

$$= \ \text{catalan}(c) \cdot \frac{(2c+1) \cdot (2c+2)}{4^c} \ - \ 2$$

$$= \ \binom{2c+1}{c} \cdot \frac{2c+2}{4^c} \ - \ 2$$

Notice that if we want to determine the length of an α-interval on the right or left boundary of the query, we have to exchange $\mathcal{V}$ and $\mathcal{H}$ in the formula; this leads to the value $\bar{\gamma}_c$ instead of γ_c. In general, for a specific space filling curve with direction distributions $\mathcal{V}$, $\mathcal{H}$, and $\mathcal{D}$, both γ_c and $\bar{\gamma}_c$ can (possibly) not be written explicitly, but computed very efficiently.

4.3 Final Analysis

In this section, we are going to combine the two results of the previous sections. All we have to do, is to multiply the expected number of α-intervals with the expected cost of an α-interval (depending on which side of query rectangle we are). With that, we obtain the following approximation of the expected costs generated by the α-intervals as

$$(1 - \mathcal{H}) \cdot n_1 \cdot \gamma_c + (1 - \mathcal{V}) \cdot n_2 \cdot \bar{\gamma}_c$$

This gives the following performance of our $n_1 \times n_2$ query, i.e. the ratio between the number of blocks read with respect to the number of blocks queried.

Theorem 2. *Given a continuous space filling curve with direction distribution $\mathcal{V}, \mathcal{H}$ and $\mathcal{D}$, the average performance of a $n_1 \times n_2$ query is given by*

$$per \approx 1 + \left[\frac{1 - \mathcal{H}}{n_2} \cdot \gamma_c + \frac{1 - \mathcal{V}}{n_1} \cdot \bar{\gamma}_c \right]$$

This result proves that pre-knowledge of the family of queries to be expected can influence the design decision of the space filling curve in a very effective way. For example, for very flat ($n_1 \gg n_2$) or very thin ($n_2 \gg n_1$) queries we choose $0 \ll \mathcal{H} \leq 1$ and $0 \ll \mathcal{V} \leq 1$, respectively.

Remark. One might think, that space filling curves with many diagonal elements perform better, because the probability of having short α-intervals is bigger. This means, one could overread more (and shorter) α-intervals and gain efficiency. On the other hand, the number of α-intervals generated by the query is higher if $\mathcal{D}$ is bigger. All in all, these two effects compensate more or less.

5 Simulation Results

Finally, we have checked our model by making simulations. Figure 4 displays the cumulative distribution of the α-blocks of our formula when compared with experimental results for the Hilbert curve ($N = 32768$).

For the experiments, we measured the average distribution of the lengths of the α-blocks for a given quadratic query window ($n_1 = n_2 = 1024$) over all possible positions of the query on the grid. The results indicate the surprising fact that the local behavior of space filling curves can be modeled by random walks.

The experimental results indicate that the average distribution of the lengths of the α-blocks is virtually independent of the size of the query and of the resolution of the space filling curve.

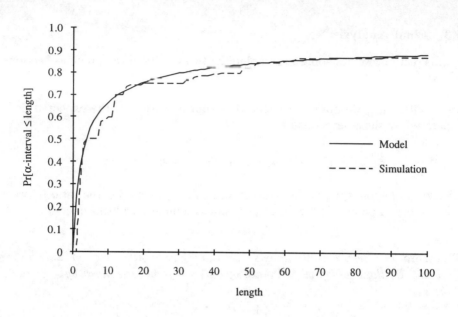

Fig. 4. Model vs. simulation for the Hilbert curve.

Acknowledgements

We are very grateful to Jürg Nievergelt for stimulating discussions. We also thank Peter Remmele and Roland Ulber for carefully reading the manuscript. The authors acknowledge the support of the Swiss National Science Foundation SNF and of the ETH Zürich. Edouard Bugnion is currently supported in part by a Graduate Research Fellowhship of the U.S. National Science Foundation.

References

1. Abel, D.J., J.L. Smith (1983): A data structure and algorithm based on a linear key for a rectangle retrieval problem; Computer Vision, Graphics, and Image Processing, Vol. 24, 1-13
2. Abel, D.J., D.M. Mark (1990): A comparative analysis of some two-dimensional orderings; Intern. J. of Geographical Information Systems, Vol. 4, No. 1, 21-31
3. Asano, T., A. Hasegawa, D. Ranjan, T. Roos (1993): Optimal and approximate digital halftoning algorithms and their experimental evaluation, Extended Abstract, Proc. Asian Conference on Computer Vision ACCV'93, Osaka
4. Asano, T., D. Ranjan, T. Roos (1996): Digital halftoning algorithms based on optimization criteria and their experimental evaluation, IEICE transactions on fundamentals of electronics, communications, and computer sciences, Vol. E79-A, No. 4, 1996, 524-532
5. Asano, T., D. Ranjan, T. Roos, E. Welzl, P. Widmayer (1995): Space filling curves and their use in the design of geometric data structures, Proc. 2nd Intern. Symp.

of Latin American Theoretical Informatics LATIN'95, Valparaiso, Chile, 1995, to appear in TCS

6. Faloutsos, C. (1985): Multiattribute hashing using Gray codes; Proc. ACM SIG-MOD International Conf. on the Management of Data, Washington D.C., 227-238

7. Faloutsos, C. (1988): Gray codes for partial match and range queries; IEEE Transactions on Software Engineering, Vol. 14, 1381-1393

8. Faloutsos, C., S. Roseman (1989): Fractals for secondary key retrieval; Proc. 8th ACM SIGACT/SIGMOD Symposium on Principles of Database Systems, 247-252

9. Finkel, R.A., J.L. Bentley (1974): Quad trees: A data structure for retrieval on composite keys; Acta Informatica, Vol. 4, No. 1, 1-9

10. Hutflesz, A., H.-W. Six, P. Widmayer (1988): Globally order preserving multidimensional linear hashing; Proc. 4th International Conference on Data Engineering, Los Angeles, 572-579

11. Hutflesz, A., P. Widmayer, C. Zimmermann (1992): Global order makes spatial access faster, International Workshop on Database Management Systems for Geographical Applications, ESPRIT Basic Research Series Proc., Springer, 161-176

12. Jagadish, H.V. (1990): Linear clustering of objects with multiple attributes; Proc. ACM SIGMOD International Conference on the Management of Data, Atlantic City, New Jersey, 332-342

13. Manola, F., J.A. Orenstein (1986): Toward a general spatial data model for an object-oriented DBMS; Proc. 12th International Conference on Very Large Data Bases, Kyoto, 328-335

14. Morton, G.M. (1966): A computer oriented geodetic data base and a new technique in file sequencing; IBM, Ottawa, Canada

15. Orenstein, J.A., T.H. Merrett (1984): A class of data structures for associative searching; Proc. 3rd ACM SIGACT/SIGMOD Symposium on Principles of Database Systems, Waterloo, 181-190

16. Orenstein, J.A. (1989): Redundancy in spatial databases; Proc. ACM SIGMOD International Conference on the Management of Data, Portland, 294-305

17. Orenstein, J.A. (1990): A comparison of spatial query processing techniques for native and parameter spaces; Proc. ACM SIGMOD International Conference on the Management of Data, Atlantic City, New Jersey, 343-352

18. Samet, H. (1996): Notes on data structures, private communications

19. Sagan, H. (1994): Space filling curves, Springer

20. Tropf, H., H. Herzog (1981): Multidimensional range search in dynamically balanced trees; Angewandte Informatik, Vol. 2, 71-77

21. van Oosterom, P. (1990): Reactive data structures for geographic information systems; Dissertation (Proefschrift), Rijksuniversiteit Leiden

22. Wattenhofer, R. (1995): Raumfüllende Kurven für den Entwurf von Räumlichen Zugriffsstrukturen, Diploma thesis, ETH Zürich, Switzerland

23. Wood, D. (1987): Theory of Computation, Harper & Row

Chapter 8. External-Memory Algorithms with Applications in GIS

Lars Arge

Center for Geometric Computing
Dept. of Computer Science
Duke University
U.S.A.

large@cs.duke.edu

1 Introduction

Traditionally when designing computer programs people have focused on the minimization of the internal computation time and ignored the time spent on Input/Output (I/O). Theoretically one of the most commonly used machine models when designing algorithms is the Random Access Machine (RAM) (see e.g. [7, 87]). One main feature of the RAM model is that its memory consists of an (infinite) array, and that any entry in the array can be accessed at the same (constant) cost. Also in practice most programmers conceptually write programs on a machine model like the RAM. In an UNIX environment for example the programmer thinks of the machine as consisting of a processor and a huge ("infinite") memory where the contents of each memory cell can be accessed at the same cost (Figure 1). The task of moving data in and out of the limited main memory is then entrusted to the operating system. However, in practice there is a huge difference in access time of fast internal memory and slower external memory such as disks. While typical access time of main memory is measured in nanoseconds, a typical access time of a disk is on the order of milliseconds [36]. So roughly speaking there is a factor of a million in difference in the access time of internal and external memory, and therefore the assumption that every memory cell can be accessed at the same cost is questionable, to say the least!

In many modern large-scale applications the communication between internal and external memory, rather than the internal computation time, is actually the bottleneck in the computation. As geographic information systems (GIS) frequently store, manipulate, and search through enormous amounts of spatial data [42, 55, 66, 79, 86] they are good examples of such large-scale applications.

213

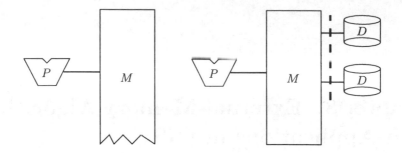

Fig. 1. A RAM-like model.　　**Fig. 2.** A more realistic model.

The amount of data manipulated in such systems is often too large to fit in main memory and must reside on disk, hence the I/O communication can become a very severe bottleneck. An especially good example is NASA's EOS project GIS system [42], which is expected to manipulate petabytes (thousands of terabytes, or millions of gigabytes) of data!

The effect of the I/O bottleneck is getting more pronounced as internal computation gets faster, and especially as parallel computing gains popularity [75]. Currently, technological advances are increasing CPU speeds at an annual rate of 40–60% while disk transfer rates are only increasing by 7–10% annually [78]. Internal memory sizes are also increasing, but not nearly fast enough to meet the needs of important large-scale applications.

Modern operating systems try to minimize the effect of the I/O bottleneck by using sophisticated paging and prefetching strategies in order to assure that data is present in internal memory when it is accessed. However, these strategies are general purpose in nature and therefore they cannot take full advantage of the properties of a specific problem. Instead we could hope to design more efficient algorithms by explicitly considering the I/O communication when designing algorithms for specific problems. This could e.g. be done by designing algorithms for a model where the memory system consists of a main memory of limited size and a number of external memory devices (Figure 2), where the memory access time depends on the type of memory accessed. Algorithms designed for such a model are often called *external-memory (or I/O) algorithms*.

1.1 Outline of Chapter

In this chapter we survey the basic paradigms for designing efficient external-memory algorithms and especially for designing external-memory algorithms for computational geometry problems with applications in GIS. As the area of external-memory algorithms is relatively young the chapter focuses on fundamental external-memory design techniques more than on algorithms for specific GIS problems. The presentation is survey-like with a more detailed discussion of the most important techniques and algorithms.

In Section 2 we first present the theoretical external memory model we will be considering (*the parallel disk model* [6, 96]). In Section 3 we then illustrate why normal internal-memory algorithms for even very simple problems can perform terribly when the problem instances get just moderately large. We also discuss the theoretical I/O lower bounds on fundamental problems like sorting. In Section 4 we discuss the fundamental paradigms for designing I/O-efficient algorithms. We do so by using the different paradigms to design theoretically optimal sorting algorithms. Many problems in computational geometry are abstractions of important GIS operations, and in Section 5 we survey techniques and algorithms in external-memory computational geometry. We also discuss some experimental results. Finally, we in Section 6 shortly describe a Transparent Parallel I/O Environment (TPIE) designed by Vengroff [88, 93] to allow programmers to write I/O-efficient programs.

We assume that the reader has some basic knowledge about e.g. fundamental sorting algorithms and data structures like balanced search trees (especially B-trees) and priority queues. We also assume that the reader is familiar with asymptotic notation ($O(\cdot)$, $\Omega(\cdot)$, $\Theta(\cdot)$). One excellent textbook covering these subjects is [37].

2 The Parallel Disk Model

Accurately modeling memory and disk systems is a complex task [78]. The primary feature of disks that we want to model is their extremely long access time relative to that of internal memory. In order to amortize the access time over a large amount of data, typical disks read or write large blocks of contiguous data at once. Therefore we use a theoretical model with the following parameters [6]:

N = number of elements in the problem instance;

M = number of elements that can fit into internal memory;

B = number of elements per disk block;

where $M < N$ and $1 \leq B \leq M/2$.

In order to study the performance of external-memory algorithms, we use the standard notion of I/O complexity [6]. We define an I/O operation to be the process of simultaneously reading or writing a block of B contiguous data elements to or from the disk. As I/O communication is our primary concern, we define the I/O complexity of an algorithm simply to be the number of I/Os it performs. Internal computation is free in the model. Thus the I/O complexity of reading all of the input data is equal to N/B. Depending on the size of the data elements, typical values for workstations and file servers in production today are on the order of $M = 10^6$ or 10^7 and $B = 10^3$. Large-scale problem instances can be in the range $N = 10^{10}$ to $N = 10^{12}$.

An increasingly popular approach to further increase the throughput of the I/O system is to use a number of disks in parallel [50, 51, 96]. Several authors have considered an extension of the above model with a parameter D denoting the number of disks in the system [21, 72, 70, 71, 96]. In *the parallel disk model* [96] one can read or write one block from each of the D disks simultaneously in one I/O. The number of disks D range up to 10^2 in current disk arrays.

The parallel disk model corresponds to the one shown in Figure 2, where we only count the number of blocks of B elements moved across the dashed line. Of course the model is designed for theoretical considerations and is thus very simple in comparison with a real system. For example one cannot always ignore internal computation time and one could try to model more accurately the fact that (in single user systems at least) reading a block from disk in most cases decreases the cost of reading the block succeeding it. Also today the memory of a real machine is typically made up of not only two but several levels of memory (e.g. on-chip data and instruction cache, secondary cache, main memory and disk) between which data are moved in blocks (Figure 3). The memory in such a hierarchy gets larger and slower the further away from the processor one gets, but as the access time of the disk is extremely large compared to that of all the other levels of memory we can in most practical situations restrict our attention to the two level case. Thus theoretical results obtained in the parallel disk model can help to gain valuable insight. This is supported by experimental results which show that implementing algorithms designed for the model can lead to significant runtime improvements in practice [32, 34, 88, 91]. We will discuss some of these experiments in later sections.

Finally, it should be mentioned that several authors have considered extended theoretical models that try to model the hierarchical nature of the memory of real machines [2, 3, 4, 5, 8, 59, 80, 94, 95, 97], but such models quickly become theoretically very complicated due to the large number of parameters. Therefore only very basic problems like sorting have been considered in these models.

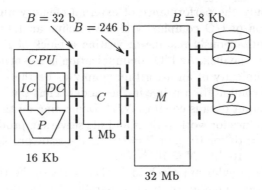

Fig. 3. A "real" machine with typical memory and block sizes [36].

3 RAM-Complexity and I/O-Complexity

In order to illustrate the difference in complexity of a problem in the RAM model and the parallel disk model, consider the following simple problem: We are given an N-vertex linked list stored as an (unsorted) sequence of vertices, each with a pointer to the successor vertex in the list (Figure 4). Our goal is to determine for each vertex the number of links to the end of the list. This problem is normally referred to as the *list ranking* problem.

Fig. 4. List ranking problem.

In internal memory the list ranking problem is easily solved in $O(N)$ time. We simply traverse the list by following the pointers, and rank the vertices $N-1, N-2$ and so on in the order we meet them. In external memory, however, this simple algorithm could perform terribly. To illustrate this imagine that we run the algorithm on a (rather memory limited) machine where the internal memory is capable of holding two blocks with two data elements each. Assume furthermore that the operating system controlling the I/O uses a least recently used (LRU) paging strategy, that is, when a data item in a block not present in internal memory is accessed the block containing the least recently used data items is flushed from internal memory in order to make room for the new block. Finally, assume that the list is blocked as indicated in Figure 5. First we load block 1 and give A rank $N-1$. Then we follow the pointer to E, that is, we load block 3 and rank E. Then we follow the pointer to D, loading block 2 while removing block 1 from internal memory. Until now we have made an I/O every time we follow a pointer. Now we follow the pointer to B, which means that we have to load block 1 again. To do so we have to remove block 3 from internal memory. Next we remove block 2 to get room for block 4. This process continues and we see that we do not utilize the blocked disk access, but do an I/O every time we access a vertex. Put another way we use $O(N)$ I/Os instead of $O(N/B)$ I/Os which would correspond to the $O(N)$ internal-memory RAM

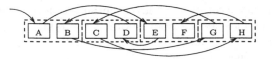

Fig. 5. List ranking in external memory ($B = 2$).

bound because $O(N/B)$ is the number of I/Os we need to read all N vertices. As typical practical values of B are measured in thousands this difference can be extremely significant in practice.

In general the above type of behavior is characteristic for internal-memory algorithms when analyzed in an I/O-environment, and the main reason why many applications experience a dramatic decrease in performance when the problem instances get larger than the available internal memory. The lesson one should learn from this is that one should be very careful about following pointers, and be careful to ensure a high degree of locality in the access to data (what is normally referred to as *locality of reference*).

3.1 Fundamental External-Memory Bounds

After illustrating how simple internal-memory algorithms can have a terrible performance in an I/O environment, let us review the fundamental theoretical bounds in the parallel disk model. For simplicity we give all the bounds in the one-disk model. In the general D-disk model they should all be divided by D.

Initial theoretical work on I/O-complexity was done by Floyd [47] and by Hong and Kung [56] who studied matrix transposition and fast Fourier transformation in restricted I/O models. The general I/O model was introduced by Aggarwal and Vitter [6] and the notion of parallel disks was introduced by Vitter and Shriver [96]. The latter papers also deal with fundamental problems such as permutation, sorting and matrix transposition, and a number of authors have considered the difficult problem of sorting optimally on parallel disks [5, 21, 72, 70]. The problem of implementing various classes of permutations has been addressed in [38, 39, 40]. More recently researchers have moved on to more specialized problems in the computational geometry [12, 14, 19, 34, 52, 98], graph theoretical [13, 14, 34, 33, 69, 65] and string processing areas [15, 35, 45, 46].

As already mentioned the number of I/O operations needed to read the entire input is N/B and for convenience we call this quotient n. One normally uses the term *scanning* to describe the fundamental primitive of reading (or writing) all elements in a set stored contiguously in external memory by reading (or writing) the blocks of the set in a sequential manner in $O(n)$ I/Os. Furthermore, one says that an algorithm uses a linear number of I/O operations if it uses $O(n)$ I/Os. Similarly, we introduce $m = M/B$ which is the number of blocks that fit in internal memory. Aggarwal and Vitter [6] showed that the number of I/O operations needed to sort N elements is $\Omega(n \log_m n)$, which is then the external-memory equivalent of the well-known $\Omega(N \log_2 N)$ internal-memory bound.[1] Furthermore, they showed that the number of I/Os needed to rearrange N elements according to a given permutation is $\Omega(\min\{N, n \log_m n\})$.

Taking a closer look at the above bounds for typical values of B and M reveals that because of the large base of the logarithm, $\log_m n$ is less than 3 or 4 for all realistic values of N and m. Thus in practice the important term is

---

[1] We define $\log_m n = \max\{1, \log n / \log m\}$. For extremely small values of M and B the comparison model is assumed in the sorting lower bound—see also [16, 17]

the B-term in the denominator of the $O(n \log_m n) = O(\frac{N}{B} \log_m n)$ bound, and an improvement from an $\Omega(N)$ bound (which we have seen is the worst case I/O performance of many internal-memory algorithms) to the sorting bound $O(n \log_m n)$ can be extremely significant in practice. Also the small value of $\log_m n$ in practice means that in all realistic cases the sorting term in the permutation bound will be smaller than N. Thus $\min\{N, n \log_m n\} = n \log_m n$ and the problem of permuting is as hard as the more general problem of sorting. This fact is one of the important facts distinguishing the parallel disk model from the RAM-model, as any permutation can be performed in $O(N)$ time in the latter. Actually, it turns out that the permutation bound is a lower bound on the list ranking problem discussed above [33], and as an $O(n \log_m n)$ I/O algorithm is known for the problem [12, 34, 33] we have an asymptotically optimal algorithm for all realistic systems. Even though the algorithm is more complicated than the simple RAM algorithm, Vengroff [89] has performed simulations showing that on large problem instances it has a much better performance than the simple internal-memory algorithm. In the parallel algorithm (PRAM) world list ranking is a very fundamental graph problem which extracts the essence in many other problems, and it is used as an important subroutine in many parallel algorithms [9]. This turns out also to be the case in external memory [34, 33].

3.2 Summary

- RAM algorithms typically use $\Omega(N)$ I/Os when analyzed in parallel disk model.
- Typical bounds in one-disk model (divide by D in D-disk model):
  - Scanning bound: $\Theta(\frac{N}{B}) = \Theta(n)$.
  - Sorting bound: $\Theta(\frac{N}{B} \log_{M/B} \frac{N}{B}) = \Theta(n \log_m n)$.
  - Permutation bound: $\Theta(\min\{N, n \log_m n\})$.
- In practice:
  - $\log_m n < 4$ and B is the important term in $O(\frac{N}{B} \log_m n)$ bound. Going from $\Omega(N)$ to $O(n \log_m n)$ algorithm extremely important.
  - Permutation bound equal to sorting bound.

4 Paradigms for Designing I/O-Efficient Algorithms

Originally Aggarwal and Vitter [6] presented two basic paradigms for designing I/O-efficient algorithms; the *merging* and the *distribution* paradigms. In Section 4.1 and 4.2 we demonstrate the main ideas in these paradigms by showing how to use them to sort N elements in the optimal number of I/Os. Another important paradigm is to construct I/O-efficient versions of commonly used data structures. This enables the transformation of efficient internal-memory algorithms to efficient I/O-algorithms by exchanging the data structures used in the internal algorithms with the external data structures. This approach has the extra benefit of isolating the I/O-specific parts of an algorithm in the data structures. We call the paradigm the *data structuring paradigm*, and in Section 4.3 we illustrate it by way of the so called *buffer tree* designed in [12]. As we shall see later I/O-efficient data structures turn out to be a very powerful tool in the development of efficient I/O algorithms. For simplicity we only discuss the paradigms in the one disk ($D = 1$) model. In Section 4.4 we then briefly discuss how to make the sorting algorithms work in the general model.

4.1 Merge Sort

External merge sort is a generalization of internal merge sort. First, in the "run formation phase", N/M ($= n/m$) sorted "runs" are formed by repeatedly filling up the internal memory, sorting the elements, and writing them back to disk. The run formation phase requires $2n$ I/Os as we read and write each block ones. Next we continually merge m runs together to a longer sorted run, until we end up with one sorted run containing all the elements—refer to Figure 6.

The crucial property is that we can merge m runs together in a linear number of I/Os. To do so we simply load a block from each of the m runs and collect and output the B smallest elements. We continue this process until we have processed all elements in all runs, loading a new block from a run every time a block becomes empty. Since there are $\log_m n/m$ levels in the merge process, and since we only use $2n$ I/O operations on each level, we in total use $2n + 2n \cdot \log_m n/m = 2n + 2n(\log_m n - 1) = 2n\log_m n$ I/Os and have thus obtained an optimal $O(n\log_m n)$ algorithm.

4.2 Distribution Sort

External distribution sort is in a sense the reverse of merge sort and the external-memory equivalent of quick sort. Like in merge sort the distribution sort algorithm consists of a number of levels each using a linear number of I/Os. However,

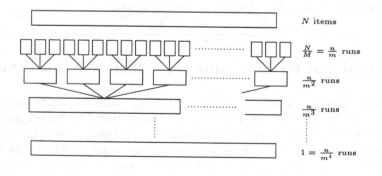

N items

$\frac{N}{M} = \frac{n}{m}$ runs

$\frac{n}{m^2}$ runs

$\frac{n}{m^3}$ runs

$1 = \frac{n}{m^i}$ runs

Fig. 6. Merge sort.

220

instead of repeatedly merging m run together, we repeatedly distribute the elements in a "bucket" into m smaller "buckets" of equal size. All elements in the first of these smaller buckets are smaller than all elements in the second bucket and so on. The process continues until the elements in a bucket fit in internal memory, in which case the bucket is sorted using an internal-memory sorting algorithm. The sorted sequence of elements is then obtained by appending the small sorted buckets—refer to Figure 7.

Like m-way merge, m-way distribution can also be performed in a linear number of I/Os, by just keeping a block in internal memory for each of the buckets we are distributing elements into—writing it to disk when it becomes full. However, in order to distribute the N elements in a bucket into m smaller buckets we need to find m "pivot" elements among the N elements, such that the buckets each defined by two such pivot elements are of equal size (corresponding to finding the median in the quicksort algorithm). In order to do so I/O-efficiently we need to decrease the distribution factor and distribute the elements in a bucket into $\sqrt{m}$ instead of m smaller buckets. If the elements are divided perfectly among the buckets this will only double the number of levels as $\log_{\sqrt{m}} n = \log_m n / \log_m \sqrt{m} = 2\log_m n$. Thus we will still obtain an $O(n\log_m n)$ algorithm if we can process each level in a linear number of I/Os.

The obvious way to find the $\sqrt{m}$ pivot elements would be to find every $N/\sqrt{m}$'th element using the k-selection algorithm $\sqrt{m}$ times. The k-selection algorithm [26] finds the k'th smallest element in $O(N)$ time in internal memory and it can easily be modified to work in $O(n)$ I/Os in external memory. The $\sqrt{m}$ elements found this way would give us buckets of exact equal size, but unfortunately we would use $O(n \cdot \sqrt{m})$ and not $O(n)$ I/Os to compute them. Therefore we first choose $4N/\sqrt{m}$ of the N elements by sorting the N/M memory loads individually and choosing every $\sqrt{m}/4$'the element from each of them. Then we can use k-selection $\sqrt{m}$ times on these elements to obtain the pivot elements using only $\sqrt{m} \cdot O(n/4\sqrt{m}) = O(n)$ I/Os. However, now we cannot be sure that the $\sqrt{m}$ buckets have equal size $N/\sqrt{m}$, but fortunately one can prove that they are of approximately equal size, namely that no bucket contains more

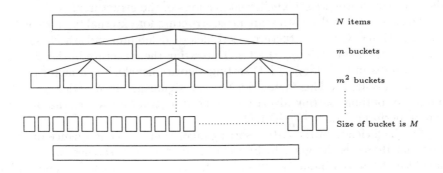

Fig. 7. Distribution sort.

than $5N/4\sqrt{m}$ elements [6, 62]. Thus the number of levels in the distribution is less than $\log_{4/5\sqrt{m}} n/m = O(\log_m n)$ and the overall complexity remains the optimal $O(n \log_m n)$ I/Os.

Even though merge sort is a lot simpler and efficient in practice than distribution sort [63], the distribution paradigm is the most frequently used of the two paradigms. Mainly two factors make distribution sort less efficient in practice than merge sort, namely the larger number of levels in the recursion and the computation of the pivot elements. Especially the last factor (the k-selection algorithm) can be very I/O expensive. However, in many applications it turns out that one can compute all the pivot elements used during the whole algorithm before the actual distribution starts, and thus *both* avoid the expensive k-selection algorithm and obtain an m distribution factor. We will see an example of this in Section 5.1.

Both the distribution sort and merge sort algorithm demonstrates two of the most fundamental and useful features of the I/O-model, which is used repeatedly when designing I/O algorithms. First the fact that we can do m-way merging or distribution in a linear number of I/O operations, and secondly that we can solve a complicated problem in a linear number of I/Os *if* it fits in internal memory. In the two algorithms the sorting of a memory load is an example of the last feature, which is also connected with what is normally referred to as "locality of reference"—one should try to work on data in chunks of the block (or internal memory) size, and do as much work as possible on data once it is loaded into internal memory.

4.3 Buffer Tree Sort

In internal memory we can sort N elements in $O(N \log_2 N)$ time using a balanced search tree; we simply insert all N elements in the tree one by one using $O(\log_2 N)$ time on each insertion, and then we can easily obtain the sorted set of element in $O(N)$ time. Similarly, we can use a priority queue to sort optimally; first we insert all N elements in the queue and then we perform N deletemin operations. Why not use the same algorithms in external memory, exchanging the data structures with I/O-efficient versions of the structures?

The standard well-known search tree structure for external memory is the B-tree [22, 41, 64]. On this structure insert, delete, deletemin and search operations can be performed in $O(\log_B n)$ I/Os. Thus using the structure in the algorithms above results in $O(N \log_B n)$ I/O sorting algorithms which is a factor of $B \frac{\log m}{\log B}$ away from optimal. This factor can be very significant in practice. In order to obtain an optimal sorting algorithm we need a structure where the operations can be performed in $O(\frac{\log_m n}{B})$ I/Os.

The inefficiency of the B-tree sorting algorithm is a consequence of the fact that the B-tree is designed to be used in an "on-line" setting, where queries should be answered immediately and within a good worst-case number of I/Os, and thus updates and queries are handled on an individual basis. This way one is not able to take full advantage of the large internal memory. Actually, using a decision tree like argument as in [60], one can show that the search bound is indeed optimal in such an "on-line" setting (assuming the comparison model). However, in an "off-line" environment where we are only interested in the overall I/O use of a series of operations on the involved data structure, and where we are willing to relax the demands on the search operations, we could hope to develop data structures on which a series of N operations could be performed in $O(n \log_m n)$ I/Os in total. Below we sketch such a basic tree structure developed using what is called the *buffer tree* technique [12]. The structure can be used in the normal tree sort algorithm. We also sketch how the structure can be used to develop an I/O-efficient external priority queue.

Basically the buffer tree is a fan-out $m/2$ tree structure built on top of n leaves, each containing B of the N elements stored in the structure. Thus the tree has height $O(\log_m n)$—refer to Figure 8. A *buffer* of size $m/2$ *blocks* is assigned to each internal node and operations on the structure are done in a "lazy" manner. In order to insert a new element in the structure we do not (like in a normal tree) search all the way down the tree to find the place among the leaves to insert the element. Instead, we wait until we have collected a block of insertions, and then we insert this block in the buffer of the root (which is stored on disk). When a buffer "runs full" its element are then "pushed" one level down to buffers on the next level. We call this a *buffer-emptying process*, and it is basically performed as a $m/2$-way distribution step; we load the $M/2$ elements from the buffer and the $m/2$ partition elements into internal memory, sort the elements from the buffer, and write them back to the appropriate buffers on the next level. If the buffer of any of the nodes on the next level becomes full by this process the buffer-emptying process is applied recursively.

The main point is now that we can perform the buffer-emptying process in $O(m)$ I/Os, basically because the elements in a buffer fit in memory and because the fan-out of the structure is $\Theta(m)$. We use $O(m)$ I/Os to read and write all the elements, plus at most one I/O for each of the $O(m)$ children to distribute elements in non-full blocks. Thus we push $m/2$ blocks one level down the tree using $O(m)$ I/Os, or put another way, we use a constant number of I/Os to push

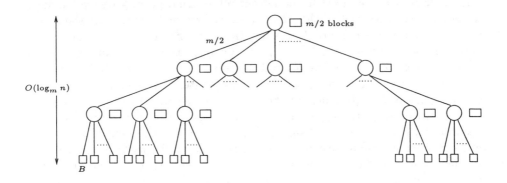

Fig. 8. Buffer tree.

223

one block one level down. In this way we can argue that every *block* is touched a constant number of times on each of the $O(\log_m n)$ levels of the tree, and thus inserting N elements (or n blocks) in the structure requires $O(n \log_m n)$ I/Os in total. Of course one also has to consider how to empty a buffer of a node on the last level in the tree, that is, how to insert elements among the leaves of the tree and perform rebalancing. In [12] it is shown that by using an (a, b)-tree [57] as the basic tree structure this can also be handled in the mentioned I/O bound. An (a, b)-tree is a generalization of the B-tree. Deletions (and queries) can be handled using similar ideas [12].

Note that while N insertions (and deletions) in total take $O(n \log_m n)$ I/Os, a single insertion (or deletion) can take a lot more than $O(\frac{\log_m n}{B})$ I/Os, as a single operation can result in a lot of buffer-emptying processes. Thus we do not as in the B-tree case have a *worst-case* I/O bound on performing an update. Instead we say that each operation can be performed in $O(\frac{\log_m n}{B})$ I/Os *amortized* [84].

In order to use the structure in a sorting algorithm we also need an operation that reports all the elements in the structure in sorted order. To do so we first empty all buffers in the structure using the buffer-emptying process from the root of the tree and down. As the number of internal nodes is $O(n/m)$ this can easily be done in $O(n)$ I/Os. After this all the elements are stored in the leaves and can be reported in sorted order using a simple scan. Using the buffer idea we have thus obtained a structure with the operations needed to sort N elements in $O(n \log_m n)$ I/O with precisely the same tree sort algorithm as can be used in internal memory. The algorithm has the extra benefit of isolating the I/O-specific parts in the data structure. Preliminary simulation results suggests that in practice the algorithm has a bit worse performance than merge sort, but that it performs much better than distribution sort.

External-memory priority queue. Normally, we can use a search tree structure to implement a priority queue because we know that the smallest element in a search tree is in the leftmost leaf. Thus when we want to perform a deletemin operation we simply delete and return the element in the leftmost leaf. The same general strategy can be used to implement an external priority queue based on the buffer tree. However, in the buffer tree we cannot be sure that the smallest element is in the leftmost leaf, since there can be smaller elements in the buffers of the nodes on the leftmost path. There is, however, a simple strategy for performing a deletemin operation in the desired *amortized* I/O bound.

When we want to perform a deletemin operation, we simply do a buffer-emptying process on all nodes on the path from the root to the leftmost leaf. This requires $O(m) \cdot O(\log_m n)$ I/Os. Now we can be sure not only that the leftmost leaf consists of the B smallest elements in the tree, but also that the $m/2 \cdot B = M/2$ smallest elements in the tree are in the $m/2$ leftmost leaves. As these elements fit in internal memory we can delete them all and hold them in internal memory in order to be able to answer future deletemin operations without having to do any I/Os at all. In this way we have used $O(m \log_m n)$ I/Os to answer $M/2$ deletemin operations which means that we amortized use

$O(\frac{\log_m n}{B})$ I/Os on one such operation. There is one complication, however, as insertions of small elements may be performed before we have performed $M/2$ deletemin operations. Therefore we also on each insertion check if the element to be inserted is smaller than the largest of the minimal elements we currently hold in internal memory. If this is the case we keep the new element in memory as one of the minimal elements and insert the largest of the smallest elements in the buffer tree instead. Note that this extra check do not require any extra I/Os.

To summarize we have sketched an external priority queue on which insertions and deletemin operations can be performed in $O(\frac{\log_m n}{B})$ I/Os amortized and thus we are able to sort optimally with yet another well-known algorithm. It should be noted that recently an alternative priority queue was developed in [65].

4.4 Sorting on Parallel Disks

In the previous sections we have discussed a number of sorting algorithms working in the one-disk model. As mentioned in the introduction, one approach to increase the throughput of I/O systems is to use a number of disks in parallel. In this section we briefly survey results on sorting optimally using D independent disks. We assume that by the start of the algorithm the N elements to be sorted are spread among the D disks with n/D blocks on each disk.

One very simple method of using D disks in parallel is *disk striping*, in which the heads of the disks are moving synchronously, so that in a single I/O operation each disk reads or writes a block in the same location as each of the others. In terms of performance, disk striping has the effect of using a single large disk with block size $B' = DB$. Even though disk striping does not in theory achieve asymptotic optimality when D is very large, it is often the method of choice in practice for using parallel disks [91]. The non-optimality of disk striping can be demonstrated via the sorting bound. While sorting N elements using disk striping and one of the previously described one-disk sorting algorithms requires $O(\frac{n}{D} \log_{m/D} n)$ I/Os, the optimal bound is $O(\frac{n}{D} \log_m n)$ I/Os [6]. Note that the optimal bound gives a linear speedup in the number of disk.

In order to use the D disks optimally in merge sort we should be able to merge $\Theta(m)$ sorted runs containing N elements in $O(n/D)$ I/Os, that is, every time we do an I/O we should load $\Theta(D)$ useful blocks from the disks. However, as we only have room in internal memory for a constant number of blocks from each input run, we cannot hold D blocks from each run and just load the next D blocks once the old ones expire. Instead, every time we want to read the next block of a run, we have to predict which $\Theta(D)$ block we will need to load next and "prefetch" them together with the desired block. The prediction can be done with a technique due to Knuth called *forecasting* [64]. However, in order to prefetch the blocks efficiently they must reside on different disks, and that is the main reason why merge sorting on parallel disk is difficult—during one merge pass one has to store the output blocks on disks in such a way that they can be efficiently prefetched in the next merge pass. But the way the blocks should be assigned to disks depends on the merge steps forming the other $m - 1$ runs which will participate in the next merging pass, and therefore it seems hard to figure out how to assign the blocks to disks. Nevertheless, Nodine and Vitter [72] managed to developed a (rather complicated) D disk sorting algorithm based on merge sort. Very recently Barve et al. [21] develop a very simple and practical randomize D-disk merge sort algorithm.

Intuitively, it seems easier to make distribution sort work optimally on parallel disks. During one distribution pass we should "just" make sure to distribute the blocks belonging to the same bucket evenly among the D disks, such that we can read them efficiently in the next pass. Vitter and Shriver [96] used randomization to ensure this and developed an algorithm which performs optimally with high probability. Later Nodine and Vitter [70] managed to develop a deterministic version of D disk distribution sort. An alternative distribution-like algorithm is develop by Aggarwal and Plaxton [5].

The buffer tree sorting algorithm can also be modified to work on D disks. Recall that the buffer-emptying process basically was performed like a distribution step, where a memory load of elements were distribute to $m/2$ buffers one level down the tree. Thus using the techniques developed in [70] the buffer-emptying algorithm can be modified to work on D disks, and we obtain an optimal D-disk sorting algorithm. As already mentioned, distribution sort is rather inefficient in practice, mainly because of the overhead used to compute the pivot elements. Also the deterministic D-disk merge sorting algorithms is rather complicated. As the computation of the pivot elements is avoided in the buffer tree, the D-disk buffer tree sorting algorithm could be very efficient in practice. In the future we hope to investigate this experimentally.

4.5 Summary

- Three main paradigms: Merging, distributing, and data structuring.
- Main features used:
  - m-way merging/distribution possible in linear number of I/Os.
  - Complicated *small* problems can be solved in linear number of I/Os.
  - Buffered data structures. Using B-trees typically yields algorithms a factor B away from optimal.
- In practice:
  - All three paradigms can be used to develop optimal sorting algorithms. One-disk merge sort fastest, followed by buffer and distribution sort.

---

5 External-Memory Computational Geometry Algorithms

Most GIS systems at some level store map data as a number of layers. Each layer is a thematic map, that is, it stores only one type of information. Examples are a layer storing all roads, a layer storing all cities, and so on. The theme of a layer can also be more abstract, as for example a layer of population density or land utilization (farmland, forest, residential). Even though the information stored in different layers can be very different, it is typically stored as geometric information like line segments or points. A layer for a road map typically stores the roads as line segments, a layer for cities typically contains points labeled with city names, and a layer for land utilization could store a subdivision of the map into regions labeled with the use of a particular region.

One of most fundamental operations in many GIS systems is map overlaying—the computation of new scenes or maps from a number of existing maps. Some existing software packages are completely based on this operation [10, 11, 74, 86]. Given two thematic maps the problem is to compute a new map in which the thematic attributes of each location is a function of the thematic attributes of the corresponding locations in the two input maps. For example, the input maps could be a map of land utilization and a map of pollution levels. The map overlay operation could then be used to produce a new map of agricultural land where the degree of pollution is above a certain level. One of the main problems in overlaying of maps stored as line segments is "line-breaking"—the problem of computing the intersections between the line segments making up the maps. This problem can be abstracted as the in computational geometry well-known problem of red/blue line segment intersection. In this problem one is given a set of non-intersecting red segments and a set of non-intersecting blue segments and should compute all intersection red-blue segment pairs.

In general many important problems from computational geometry are abstractions of important GIS operations [43, 86]. Examples are range searching which e.g. can be used in finding all objects inside a certain region, planar point location which e.g. can be used when locating the region a given city lies in, and region decomposition problems such as trapezoid decomposition, (Voronoi or Delaunay) triangulation, and convex hull computation. The latter problems are useful for rendering and modeling. Furthermore, as mentioned in the introduction, GIS systems frequently store and manipulate enormous amounts of data, and they are thus a rich source of problems that require good use of external-memory techniques. In this section we therefore consider external-memory algorithms for computational geometry problems. Like we in the previous section focused on the fundamental paradigms for designing efficient sorting algorithms, we will present the fundamental paradigms and techniques for designing computational geometry algorithms, and at the same time present some of the algorithms for problems with applications in GIS systems. In order to do so we define two additional parameters in our model:

K = number of queries in the problem instance;

T = number of elements in the problem solution.

In analogy with the definition of n and m we define $k - K/B$ and $t = T/B$ to be respectively the number of query blocks and number of solution element blocks.

In internal memory one can prove what might be called sorting lower bounds $O(N \log_2 N + T)$ on a large number of important computational geometry problems. The corresponding bound $O(n \log_m n + t)$ can be obtained for the external versions of the problems either by redoing standard proofs [17, 52], or by using a conversion result from [16].

Computational geometry problems in external memory were first considered by Goodrich et al. [52], who developed a number of techniques for designing I/O-efficient algorithms for such problems. They used their techniques to develop I/O algorithms for a large number of important problems. In internal memory the *plane-sweep* paradigm [76] is a very powerful technique for designing computational geometry algorithms, and in [52] an external-memory version of this technique called *distribution sweeping* is developed. As the name suggests the technique relies on the distribution paradigm. In [12] it is shown how the data structuring paradigm can also be use to solve computational geometry problems. It is shown how data structures based on the buffer tree can be used in the standard internal-memory plane-sweep algorithm for a number of problems. In [52] two techniques called *batched construction of persistent B-trees* and *batched filtering* are also discussed. In [18] some results from [52, 12] are extended and generalized, and some external-memory computational geometry results are also reported in [49, 98]. In [19] efficient I/O algorithms for a large number of problems involving line segments in the plane are designed by combining the ideas of distribution sweeping, batched filtering, buffer trees and a new technique, which can be regarded as an external-memory version of *fractional cascading* [31]. Most of these problems have important applications in GIS systems. In [32, 34, 18] some experimental results on the practical performance of external-memory algorithms for computational geometry problems are reported.

We divide our survey of external-memory computational geometry into four main parts. In the next section we illustrate the distribution sweeping and the data structure paradigm using the orthogonal line segment intersection problem. We also present some experimental results. In Section 5.2 we then use the batched range searching problem to introduce the external segment tree data structure. Section 5.3 is then devoted to a discussion of the red/blue line segment intersection problem. In that section we also discuss batched filtering and external fractional cascading. Finally, we in Section 5.4 survey some other important external-memory computational geometry results.

For simplicity we restrict the discussion to the one-disk model. Some of the algorithms can be modified to work optimally in the general model and we refer the interested reader to the research papers for a discussion of this. For completeness it should be mentioned that recently a number of researchers have considered the design of worst-case efficient external-memory "on-line" data structures, mainly for (special cases of) two and three dimensional range searching [20, 25, 58, 60, 77, 83, 92]. While B-trees [22, 41, 64] efficiently support range searching in one dimension they are inefficient in higher dimensions. Similarly the many sophisticated internal-memory data structures for range searching are not efficient when mapped to external memory. This has lead to the development of a large number of structures that do not have good theoretical worst-case update and query I/O bounds, but do have good average-case behavior for common problems—see Chapter 6. Range searching is also considered in [73, 81, 82] where the problem of maintaining range trees in external memory is considered. However, the model used in this work is different from the one considered here. In [27] an external on-line version of the topology tree is developed and this structure is used to obtain structures for a number of dynamic problems, including approximate nearest neighbor searching and closest pair maintenance. Very recently, an algorithm has been given [1] for preprocessing a TIN into an external data structure such that the contour lines of a query elevation can be computed I/O optimally.

5.1 The Orthogonal Line Segment Intersection Problem

The orthogonal line segment intersection problem is that of reporting all intersecting orthogonal pairs in a set of N line segment in the plane parallel to the axis. In internal memory a simple optimal solution to the problem based on the plane-sweep paradigm [76] works as follows (refer to Figure 9): We imagine that we sweep with a horizontal sweep line from the top to the bottom of the plane and every time we meet a horizontal segment we report all vertical segments intersecting the segment. To do so we maintain a balanced search tree containing the vertical segments currently crossing the sweep line, ordered according to x-coordinate. This way we can report the relevant segments by performing a range query on the search tree with the x-coordinates of the endpoints of the horizontal segment. To be more precise we start the algorithm by sorting all the segment endpoints by y-coordinate. We use the sorted sequence of points to perform the sweep, that is, we process the segments in endpoint y order. When the top endpoint of a *vertical* segment is reached the segment is inserted in the search tree. The segment is removed again when its bottom endpoint is reached. This way the tree at all times contains the segments intersection the sweep line. When a *horizontal* segment is reached a range query is made on the search tree. As inserts and deletes can be performed in $O(\log_2 N)$ time and range querying in $O(\log_2 N + T')$ time, where T' is the number of reported segments, we obtain the optimal $O(N \log_2 N + T)$ solution.

As discussed in Section 4.3 a simple natural external-memory modification of the plane-sweep algorithm would be to use a B-tree as the tree data structure, but this would lead to an $O(N \log_B n + t)$ I/O solution, while we are looking for an $O(n \log_m n + t)$ I/O solution. In the next two subsections we discuss I/O-optimal solutions to the problem using the distribution sweeping and buffer tree techniques.

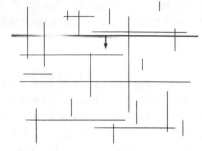 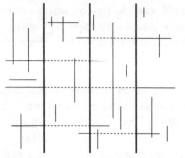

Fig. 9. Solution to the orthogonal line segment intersection problem using plane-sweep.

Fig. 10. Solution to the orthogonal line segment intersection problem using distribution sweeping.

Distribution sweeping. Distribution sweeping [52] is a powerful technique obtained by combining the distribution and the plane-sweep paradigms. Let us briefly sketch how it works in general. To solve a given problem we divide the plane into m vertical *slabs*, each of which contains $\Theta(n/m)$ input objects, for example points or line segment endpoints. We then perform a vertical top to bottom sweep over all the slabs in order to locate components of the solution that involve interaction between objects in different slabs or objects (such as line segments) that completely span one or more slabs. The choice of m slabs ensures that one block of data from each slab fits in main memory. To find components of the solution involving interaction between objects residing in the same slab, the problem is then solved recursively in each slab. The recursion stops after $O(\log_m n/m) = O(\log_m n)$ levels when the subproblems are small enough to fit in internal memory. In order to get an $O(n \log_m n)$ algorithm we thus need to be able to perform one sweep in $O(n)$ I/Os.

To use this general technique to solve the orthogonal line segment intersection problem we first sort the endpoints of all the segments twice and create two lists, one with the endpoints sorted according to x-coordinate and the other by y-coordinate. The list sorted by y-coordinate is used to perform sweeps from top to bottom as in the plane-sweep algorithm. The list sorted according to x-coordinate is used to locate the pivot elements used throughout the algorithm to distribute the input into m vertical slabs. In this way we avoid using the complicated k-selection algorithm as discussed Section 4.2.

The algorithm now proceeds as follows (refer to Figure 10): We divide the plane into m slabs and sweep from top to bottom. When a top endpoint of a vertical segment is reached, we insert the segment in an *active list* (a stack where we keep the last block in internal memory) associated with the slab containing the segment. When a horizontal segment is reached we scan through all the active lists associated with the slabs it completely spans. During this scan we know that every vertical segment in an active list is either intersected by the horizontal segment, or will not be intersected by any of the following hori-

230

zontal segments and can therefore be removed from the list. The process finds all intersections except those between vertical segments and horizontal segments (or portions of horizontal segments) that do not completely span vertical slabs (the solid parts of the horizontal segments in Figure 10). These are found after distributing the segments to the slabs, when the problem is solved recursively for each slab. A horizontal segment may be distributed to two slabs, namely the slabs containing its endpoints, but it will at most be represented twice on each level of the recursion. It is easy to realize that if T' is the number of intersections reported, one sweep can be performed in $O(n + t')$ I/Os—every vertical segment is only touched twice where an intersection is not discovered, namely when it is distributed to an active list and when it is removed again. Also blocks can be used efficiently because of the distribution factor of m. Thus by the general discussion of distribution sweeping above we report all intersections in the optimal $O(n \log_m n + t)$ I/O operations.

Using the buffer tree. As discussed previously, the idea in the data structuring paradigm is to develop efficient external data structures and use them in the standard internal-memory algorithms. In order to make the plane-sweep algorithm for the orthogonal line segment intersection problem work in external memory, we thus need to extend the basic buffer tree with a rangesearch operation.

Basically a rangesearch operation on the buffer tree is done in the same way as insertions and deletions. When we want to perform a rangesearch we create a special element which is pushed down the tree in a lazy way during buffer-emptying processes, just as all other elements. However, we now have to modify the buffer-emptying process. The basic idea in the modification is the following (see [12, 14] for details). When we meet a rangesearch element in a buffer-emptying process, instructing us to report elements in the tree between x_1 and x_2, we first determine whether x_1 and x_2 are contained in the same subtree among the subtrees rooted at the children of the node in question. If this is the case we just insert the rangesearch element in the corresponding buffer. Otherwise we "split" the element in two, one for x_1 and one for x_2, and report the elements in those subtrees where *all* elements are contained in the interval $[x_1, x_2]$—refer to Figure 11. The splitting only occurs once and after that the rangesearch element is treated like inserts and deletes in buffer-emptying processes, except that we report the elements in the sub-trees for which all elements are contained in the interval. In [12, 14] it is show how we can report all elements in a subtree (now containing other rangesearch elements) in a linear number of I/Os. Using the normal argument it then follows that a rangesearch operation requires $O(\frac{\log_m n}{B} + t')$ I/Os amortized.

Note that the above procedure means that the rangesearch operation gets batched in the sense that we do not obtain the result of a query immediately. Actually, parts of the result will be reported at different times as the query element is pushed down the tree. However, this suffices in the plane-sweep algorithm in question, since the updates performed on the data structure do not

231

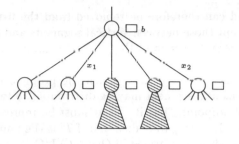

Fig. 11. Buffer-emptying process with rangesearch-elements. Elements in marked sub-trees are reported when buffer b is emptied

depend on the results of the queries. This is the crucial property that has to be fulfilled in order to used the buffer tree structure. Actually, in the plane-sweep algorithm the entire sequence of updates and queries on the data structure is known in advance, and the only requirement on the queries is that they must all eventually be answered. In general such problems are known as *batched dynamic problems* [44].

To summarize, the buffer tree, extended with a rangesearch operation, can be used in the normal internal-memory plane-sweep algorithm for the orthogonal segment intersection problem, and doing so we obtain an optimal $O(n \log_m n + t)$ I/O solution to the problem.

Experimental results. One main reason why we choose the orthogonal line segment intersection problem as our initial computational geometry problem is that Chiang [32, 34] has performed experiments on the practical performance of several of the described algorithms for the problem.

Chiang considered four algorithms, namely the distribution sweeping algorithm, denoted `Distribution`, and three variants of the plane-sweep algorithm, denoted `B-tree`, `234-Tree`, and `234-Tree-Core`. As discussed the theoretical I/O cost of the distribution sweeping algorithm is the optimal $O(n \log_m n + t)$. The plane-sweep algorithms differ by the sorting method used in the preprocessing step and in the dynamic data structure used in the sweep. The first variation, `B-tree`, uses external merge sort and a B-tree as search tree structure. As discussed previously this is the simple natural way to modify the plane-sweep algorithm to external memory. It uses $O(n \log_m n)$ I/Os in the preprocessing phase and $O(N \log_B n + t)$ I/Os to do the sweep. The second variation, `234-Tree`, also uses external merge sort but uses a 2-3-4 Tree [37] (a generic search tree structure equivalent to a red-black tree) as sweep structure, viewing the internal memory as having an infinite size and letting the virtual memory feature of the operating systems handle the page faults during the sweep. This way $O(N \log_2 N + t)$ I/Os is used to do the sweep. Finally, the third variation, `234-Tree-Core`, uses internal merge sort and a 2-3-4 tree, letting the operating system handle page faults at all times. The last variant is the most commonly used algorithm in practice, as viewing the internal memory as virtually having infinite size and letting the operating system handle the swapping is conceptually simplest. The I/O cost is $O(N \log_2 N)$ and $O(N \log_2 N + t)$ in the two phases, respectively.

In order to compare the I/O performance of the four algorithms, Chiang generated test data with particular interesting properties. One can prove that if we just randomly generate segments with lengths uniformly distributed over $[0,N]$, place them randomly in a square with side length N, and make horizontal and vertical segments equally likely to occur, then the expected number of intersections is $\Theta(N^2)$. In this case any algorithm must use $O(N^2/B)$ I/Os to report these intersections and thus the reporting cost will dominate in all four algorithms. Thus Chiang generated test data with only a linear number of intersections. Also, it is conceivable that the number of vertical overlaps among vertical segments at a given time decides the tree size at that moment of the plane-sweep and also the total size of the active lists at that time of the distribution sweep. Thus we would like the vertical overlap to be relatively large in order to study I/O issues. In the three data sets generated by Chiang the average number of vertical overlaps among vertical segments, that is, the average number of vertical segments intersected by the horizontal sweep line when it passes through an event, is $\frac{1}{4}\sqrt{N}$, $\frac{1}{8}N$ and $\frac{1}{4.8}N$, respectively. The average is taken over all sweeping events.

Chiang experimented on a Sun Sparc-10 workstation with a main memory size of 32Mb and with a page size of 4Kb. The performance measures used was total running time (wall not cpu), number of I/O operations performed (i.e. number of blocks read and written by the program), and the number of page faults occurred (I/Os controlled by the operating system)—see [32] for a precise description of the experimental setting. The first surprising result of the experiments was that the main memory available for use is typically much smaller than what would be expected. The algorithms were implemented such that the amount of main memory used could be parameterized, and Distribution was run on a fixed data set with various sizes of main memory. In theory one would expect that the performance would increase with main memory size up close to the actual 32Mb of main memory, but it turned out that 4Mb gave the best performance—refer to Figure 12 where the number of I/Os and page faults is plotted as a function of the memory used. Going from 1Mb to 4Mb the same number of page faults occur and the number of I/Os decrease slightly, so that the actual running time is decreasing slightly. Going from 4Mb to 20Mb the number of I/Os again decreases only slightly while the number of page faults increase significantly, thus resulting in a much worse overall performance. The reason is that a lot of daemon programs are also taking up memory in the machine. Chiang thereafter performed all experiments with the parameter of the main memory size set to 4Mb

Experiments were performed with the four algorithms on data sets of sizes ranging from 250 thousand segments to 2.5 million segments. The overall conclusion made by Chiang is that while the performance of the tree variations of plane-sweep algorithms depends heavily on the average number of vertical over-

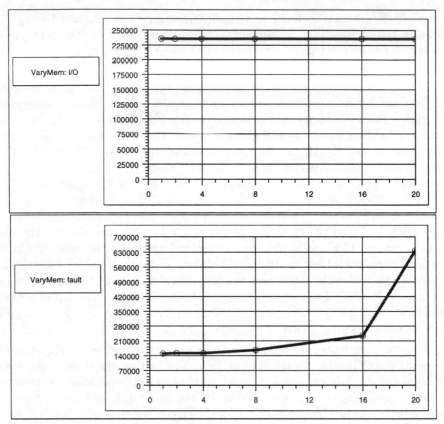

VaryMem: I/O

VaryMem: fault

* X-axis: size of the main memory used (Mb)

Fig. 12. Number of I/Os and page faults when running `Distribution` on a data set of $1.5 \cdot 10^6$ segments with various sizes of main memory.

laps, the performance of distribution sweeping is both steady and efficient. As could be expected `234-Tree-Core` performs the best for very small inputs in all experiments, but as input size grows the performance quickly becomes considerably worse than that of the other algorithms. Excluding `234-Tree-Core`, `234-Tree` always runs the fastest and `Distribution` always the slowest for the data set with a small average number of vertical overlaps ($\frac{1}{4}\sqrt{N}$)—because the search tree structure is small enough to fit into internal memory. Another reason is that `Distribution` sorts the data twice in the preprocessing step, while `234-Tree` only sorts ones. However, for the data set with a large average number of vertical overlaps ($\frac{1}{4.8}N$) `Distribution` runs much faster than the other algorithms for just moderately large data sets. For example, for $N = 1.37 \cdot 10^6$ `Distribution` runs for 45.29 minutes, `B-Tree` for 74.54 minutes, but `234-Tree` for more than 10.5 hours. Also, for $N = 2.5 \cdot 10^6$ `Distribution` runs for less than

234

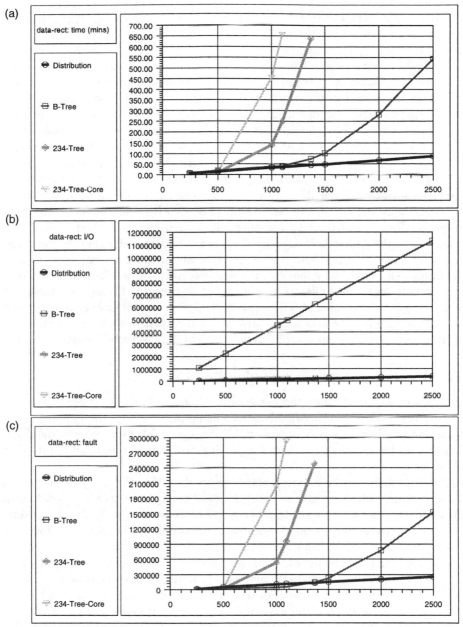

Fig. 13. Experimental results for the algorithms running on the data set with an average number of vertical overlaps of $\frac{1}{4.8}N$. (a) running time in minutes, (b) number of I/O operations, (c) number of page faults.

235

1.5 hours, while B-Tree runs for more than 8.5 hours. The full result of Chiang's experiments with this data set is shown in Figure 13. Note that Distribution always performs less I/Os than B-Tree. Recall that the I/O cost of the two algorithms is $O(n \log_m n + t)$ and $O(N \log_B n + t)$, respectively. With the parameter of the main memory size set to 4Mb the two logarithmic terms in these bounds are almost the same, so it is the $1/B$ term that makes the difference significant.

¿From the experiments performed by Chiang one may conclude that explicitly considering the I/O cost of solving a problem can be very important, and that algorithms developed for the theoretical parallel disk model can very well have a good practical performance and can lead to the solution of problems one would not be able to solve in practice with algorithms developed for main memory. Chiang did not perform experiments with the buffer tree solution described in the last subsection. Even though the constants in the I/O bounds of the buffer tree operations are small, the buffer emptying algorithm for buffers containing rangesearch elements is quite complicated, so a worse performance could be expected of the buffer tree algorithm compared to the distribution sweeping algorithm. On the other hand the buffer tree algorithm does not need to sort the input twice as the distribution sweeping algorithm does. We plan to perform experiments with the buffer tree solution in the future. Finally, it should be mentioned that Vengroff and Vitter [91] and Arge et al. [18] have also performed some experiments with I/O algorithms. We will return to these experiments in Section 6 when we discuss the TPIE environment developed by Vengroff [88, 93].

5.2 The Batched Range Searching Problem

In this section we consider another computational geometry problem with applications in GIS which is normally solved using plane-sweep; the batched range searching problem. Given N points and N (axis-parallel) rectangles in the plane (Figure 14) the problem consists of reporting for each rectangle all points that lie inside it.

The optimal internal-memory plane-sweep algorithm for the problem uses a data structure called a segment tree [23, 76]. The segment tree is a well-known dynamic data structure used to store a set of N segments in one dimension, such that given a query point all segments containing the point can be found in $O(\log_2 N + T)$ time. Such queries are normally called *stabbing queries*—refer to Figure 15. Using a segment tree the algorithm works as follows: A vertical sweep with a horizontal line is made. When the top horizontal segment of a rectangle is reached it is inserted in a segment tree. The segment is deleted again when the corresponding bottom horizontal segment is reached. When a point is reached in the sweep, a stabbing query is performed with it on the segment tree and in this way all rectangles containing the point are found. As insertions and deletions can be performed in $O(\log_2 N)$ time on a segment tree the algorithm runs in the optimal $O(N \log_2 N + T)$ time.

In external memory the batched range searching problem can be solved optimally using distribution sweeping [52] or an external buffered version of the segment tree [12]. As we will also use the external segment tree structure to solve

236

the red/blue line segment intersection problem in Section 5.3, we will sketch the structure below.

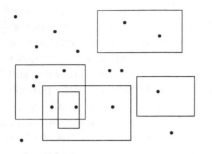

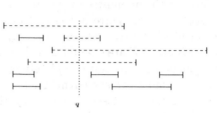

Fig. 14. The batched range searching problem.

Fig. 15. Stabbing query with q. Dotted segments are reported.

The external segment tree. In internal memory a static segment tree consists of a binary base tree storing the endpoints of the segments, and a given segment is stored in up to two nodes on each level of the tree. More precisely a segment is stored in all nodes v where it contains the interval consisting of all endpoints below v, but not the interval associated with $parent(v)$. The segments stored in a node are just stored in an unordered list. A stabbing query can be answered efficiently on such a structure, simply by searching down the base tree for the query value, reporting all segments stored in the nodes encountered.

When we want to "externalize" the segment tree and obtain a structure with height $O(\log_m n)$, we need to increase the fan-out of the nodes in the base tree. This creates a number of problems when we want to store segments space-efficiently in secondary structures such that queries can be answered efficiently. Therefore we make the nodes have fan-out $\Theta(\sqrt{m})$ instead of the normal $\Theta(m)$. As discussed in Section 4.2 this smaller branching factor at most doubles the height of the tree, but as we will see it allows us to efficiently store segments in a number of secondary structures of each node.

The external segment tree [12] is sketched in Figure 16. The base structure is a perfectly balanced tree with branching factor $\sqrt{m}$ over the endpoints. A buffer of size $m/2$ blocks and $m/2 - \sqrt{m}/2$ lists of segments are associated with each node in the tree. A list (block) of segments is also associated with each leaf. A set of segments is stored in this structure as follows: The first level of the tree partitions the data into $\sqrt{m}$ intervals σ_i—for illustrative reasons we call them *slabs*—separated by dotted lines in Figure 16. *Multislabs* are then defined as contiguous ranges of slabs, such as for example $[\sigma_1, \sigma_4]$. There are $m/2 - \sqrt{m}/2$ multislabs and the lists associated with a node are precisely a list for each multislab. The key point is that the number of multislabs is a quadratic function of the branching

237

factor. Thus by choosing the branching factor to be $\Theta(\sqrt{m})$ rather than $\Theta(m)$ we have room in internal memory for a constant number of blocks for each of the $\Theta(m)$ multislabs. Segments such as $\overline{CD}$ in Figure 16 that spans at least one slab completely are called *long segments*. A copy of each long segment is stored in the list of the largest multislab it spans. Thus, $\overline{CD}$ is stored in the list associated with the multislab $[\sigma_1, \sigma_3]$. All segments that are not long are called *short segments*. They are not stored in any multislab, but are passed down to lower levels of the tree where they may span recursively defined slabs and be stored. $\overline{AB}$ and $\overline{EF}$ are examples of short segments. Additionally, the portions of long segments that do not completely span slabs are treated as small segments. There are at most two such synthetically generated short segments for each long segment. Segments passed down to a leaf are just stored in one list. Note that we at most store one block of segments in each leaf. A segment is thus at most stored in two lists on each level of the tree and hence total space utilization is $O(n \log_m n)$ blocks.

Given an external segment tree *with empty buffers* a stabbing query can in analogy with the internal case be answered by searching down the tree for the query value, and at every node encountered report all the long segments associated with each of the multislabs that spans the query value. However, because of the size of the nodes and the auxiliary multislab data, the external segment tree is inefficient for answering single queries. But by using the general idea from Section 4.3 and 5.1 and make updates and queries buffered, we can perform the whole batch of operations needed to solve the batched range searching problem in the optimal $O(n \log_m n + t)$ I/Os. When we want to perform an update or a query we thus just keep an element for the operation in internal memory, and when we have collected a block of such operations we insert it in the buffer of the root. If this buffer now contains more than $m/2$ blocks we perform a buffer-emptying process on it as follows: First we load the elements in the buffer into

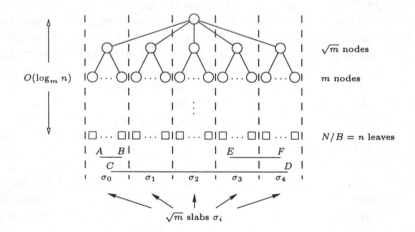

Fig. 16. An external-memory segment tree over a set of N segments, three of which, $\overline{AB}$, $\overline{CD}$, and $\overline{EF}$, are shown.

238

internal memory. Then we in internal memory collect all the segment that need to be stored in the node and distribute a copy of them to the relevant multislab lists. After that we report the relevant stabbings, by for every multislab list in turn decide if the segments in the list are stabbed by any of the query points from the buffer. Finally, we distribute the segments and queries to the buffers of the nodes on the next level of the tree. As previously we obtain the $O(\frac{\log_m n}{B})$ and $O(\frac{\log_m n}{B} + t)$ amortized I/O bounds for the update and query operation, respectively, if the buffer-emptying process can be performed in $O(m + t')$ I/Os. That this is indeed the case can easily be realized by observing that we use $O(m)$ I/Os to load the element in the buffer and to distribute them back to the buffers one level down, and that we only use $O(m)$ I/Os extra to manage the multislab lists basically because the number of such lists is $\Theta(m)$. Note that like the range searching operation on the buffer tree, the stabbing queries become batched. Thus, as discussed previously, the external segment tree can in general only be used to solve batched dynamic problems.

As mentioned in the beginning of this section distribution sweeping can also be used to solve the batched range searching problem. Also having obtained an algorithm for this problem, as well as for the orthogonal line segment intersection problem, one can also obtain an optimal external algorithm for the pairwise rectangle intersection problem—the problem of given N rectangles in the plane (with sides parallel to the axes) to report all intersecting pairs of rectangles [24]. Also the external segment tree approach for solving the batched range searching problem can be extended from 2 to d dimensions. A couple of new ideas is needed and the resulting algorithm uses $O(n \log^{d-1} n + t)$ I/Os [18]. The same technique works for a general class of problems called *colorable external-decomposable problems*.

5.3   The Red/Blue Line Segment Intersection Problem

After having presented the basic paradigms for designing I/O-efficient computational geometry algorithms through the development of optimal algorithms for the relatively simple problems of orthogonal line segment intersection and batched range searching, we now turn to the more complicated red/blue line segment intersection reporting problem: Given two internally non-intersecting sets of line segments, the problem is to report all intersections between segments in the two sets.

As previously discuss the red/blue line segment intersection problem is at the core of the important GIS problem of map overlaying. Unfortunately, it turns out that distribution sweeping and buffer trees are inadequate for solving the problem, as well as other problems involving line segments which are not axis-parallel. In the next subsection we try to illustrate this before we in the following two subsections sketch how to actually solve the problem in the optimal number of I/Os.

The endpoint dominance problem. Let us consider the endpoint dominance (EPD) problem defined as follows [19]: Given N non-intersecting line segments in the plane, find the segment directly above each endpoint of each segment—refer to Figure 17.

Even though EPD seems to be a rather simple problem, it is a powerful tool for solving other important problems. As an example EPD can be used to sort non-intersecting segments in the plane, an important subproblem in the algorithm for the red/blue line segment intersection problem. A segment $\overline{AB}$ in the plane is *above* another segment $\overline{CD}$ if we can intersect both $\overline{AB}$ and $\overline{CD}$ with the same vertical line l, such that the intersection between l and $\overline{AB}$ is above the intersection between l and $\overline{CD}$. Two segments are incomparable if they cannot be intersected with the same vertical line. The problem of sorting N non-intersecting segments is to extend the partial order defined in this way to a total order.

Figure 18 demonstrates that if two segments are comparable then it is sufficient to consider vertical lines through the four endpoints to obtain their relation. Thus one way to sort N segments [19] is to add two "extreme" segments as indicated in Figure 19, and use EPD twice to find for each endpoint the segments immediately above and below it. Using this information we create a (planar s, t-) graph where nodes correspond to segments and where the relations between the segments define the edges. Then the sorted order can be obtained by topologically sorting this graph in $O(n \log_m n)$ I/Os using an algorithm developed in [33]. This means that if EPD can be solved in $O(n \log_m n)$ I/Os then N segments can be sorted in the same number of I/Os.

In internal memory EPD can be solved optimally with a simple plane-sweep algorithm. We sweep the plane from left to right with a vertical line, inserting a segment in a search tree when its left endpoint is reached and removing it again when the right endpoint is reached. For every endpoint we encounter, we also perform a search in the tree to identify the segment immediately above the point (refer to Figure 17). One might think that it is equally easy to solve EPD in external memory, using distribution sweeping or buffer trees. Unfortunately, this is not the case.

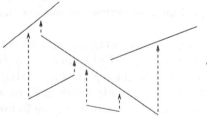
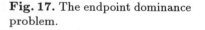
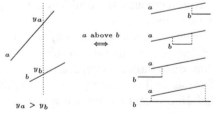

Fig. 17. The endpoint dominance problem.

Fig. 18. Comparing segments. Two segments can be related in four different ways.

240

One important property of the internal-memory plane-sweep algorithm for EPD is that only segments that actually cross the sweep-line are stored in the search tree at any given time during the sweep. This means that all segments in the tree are comparable and that we can easily compute their order. However, if we try to store the segments in a buffer tree during the sweep, the tree can (because of the "laziness" in the structure) also contain "old" segments which do not cross the sweep-line. This means that we can end up in a situation where we try to compare two incomparable segments. In general the buffer tree only works if we know a total order on the elements inserted in it or if we can compare all pair of elements. Thus we cannot directly use the buffer tree in the plane-sweep algorithm. We could try to compute a total order on the segments before solving EPD, but as discussed above the solution to EPD is one of the major steps towards finding such an order so this seems infeasible.

For similar reasons using distribution sweeping seems infeasible as well. Recall that in distribution sweeping we need to perform one sweep in a linear number of I/Os to obtain an efficient solution. Normally this is accomplished by sorting the objects by y-coordinate in a preprocessing phase. This e.g. allows one to sweep over the objects in y order without sorting on each level of recursion, because as the objects are distributed to recursive subproblems their y ordering is retained. In the orthogonal line segment intersection case we presorted the segments by endpoint in order to sweep across them in endpoint y order. In order to use distribution sweeping to solve EPD it seems that we need to presort the segments and not the endpoints.

External-memory fractional cascading. As attempts to solve EPD optimally using the buffer tree or distribution sweeping fail we are led to other approaches. It is possible to come close to solving EPD by first constructing an external-memory segment tree over the projections of the segments onto the x-axis and then performing stabbing queries at the x-coordinates of the endpoints of the segments. However, what we want is the single segment directly above each query point in the y dimension, as opposed to all segments it stabs. This segment could be found if we were able to compute the segment directly above a query point among the segments stored in a given node of the external segment tree. We call such a segment a *dominating segment*. Then we could examine each node on the path from the root to the leaf containing the query point, and

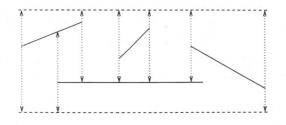

Fig. 19. Algorithm for the segment sorting problem.

241

in each such node find the dominating segment and compare it to the segment found to be closest to the query so far. When the leaf is reached we would then know the "global" dominating segment.

However, there are a number of problems that have to be dealt with in order to find the dominating segment of a query point among the segments stored in a node. The main problems are that the dominating segment could be stored in a number of multislab lists, namely in all lists containing segments that contain the query point, and that a lot of segments can be stored in a multislab list. Both of these facts seem to suggest that we need a lot of I/Os to find the dominating segment. However, as we are looking for an $O(n \log_m n)$ solution, and as the segment tree has $O(\log_m n)$ levels, we are only allowed to use a linear number of I/Os to find the positions of *all* the N query points among the segments stored in one level of the tree. This gives us less than one I/O per query point per node!

Fortunately, it is possible to modify the external segment tree and the query algorithm to overcome these difficulties [19]. To do so we first strengthen the definition of the external segment tree and require that the segments in the multislab lists are sorted. Note that all pairs of segments in the same multislab list can be compared just by comparing the order of their endpoints on one of the boundaries of the multislab, and that a multislab list thus can be sorted using a standard sorting algorithm. In [19] it is shown how to build an external segment tree with sorted multislab lists on N non-intersecting segments in $O(n \log_m n)$ I/Os. The construction is basically done using distribution sweeping.

The sorting of the multislab lists makes it easier to search for the dominating segment in a given multislab list but it may still require a lot of I/Os. We also needs to be able to look for the dominating segment in many of the multislabs lists. However, one can overcome these problems using *batched filtering* [52] and a technique similar to what in internal memory is called *fractional cascading* [30, 31, 85]. The idea in batched filtering is to process all the queries at the same time and level by level, such that the dominating segments in nodes on one level of the structure are found for all the queries, before continuing to consider nodes on the next level. In internal memory the idea in fractional cascading is that instead of e.g. searching for the same element individually in S sorted lists containing N elements each, each of the lists are in a preprocessing step augmented with sample elements from the other lists in a controlled way, and with "bridges" between different occurrences of the same element in different lists. These bridges obviate the need for full searches in each of the lists. To perform a search one only searches in one of the lists and uses the bridges to find the correct position in the other lists. This results in a $O(\log_2 N + S)$ time algorithm instead of an $O(S \log_2 N)$ time algorithm.

In the implementation of what could be called *external fractional cascading*, we do not explicitly build bridges but we still use the idea of augmenting some lists with elements from other lists. The construction is rather technical, but the general idea is the following (the interested reader is referred to [19] for details): First a preprocessing step is used (like in fractional cascading) to sample a set of segments from each slab in each node and the multislab lists of the corresponding

242

child are augmented with these segments. The sampling is done in $O(n \log_m n)$ I/Os using the distribution paradigm. Having preprocessed the structure the N queries are filtered through it. In order to do so in the optimal number of I/Os the filtering is done in a rather untraditional way—from the leaves towards the root. First the query points are sorted and distributed to the leaves to which they belong. Then for each leaf in turn the dominating segment among the segments stored in the leaf is found for all query points assigned to the leaf. This can be done efficiently using an internal-memory algorithm, because the segments stored in a leaf easily fit in internal memory. This is also the reason for the untraditional search direction—one cannot in the same way efficiently find the dominating segments among the segments stored in the root of the tree, because more than a memory load of segments can be stored there. Next the actual filtering up the $O(\log_m n)$ levels is performed, and on each level the dominating segment is found for all the query points. This is done I/O efficiently using the merging paradigm and the sampled segments from the preprocessing phase. In [19] it is shown that one filtering step can be performed in $O(n)$ I/Os, and thus EPD can be solved in $O(n \log_m n)$ I/O operations.

External red/blue line segment intersection algorithm. Using the solution to the EPD problem, or rather the ability to sort non-intersecting segments, we can now solve the red/blue line segment intersection problem with (a variant of) distribution sweeping. Recall that we in the solution to the orthogonal line segment intersection problem presorted the endpoints of the segments by y-coordinate, and used the sorted sequence throughout the algorithm to perform vertical sweeps. The key to solving the red/blue problem is to presort the red and the blue *segments* (not endpoints) individually, and perform the sweep in segment order rather than in y order of the endpoints. Thus given input sets S_r of non-intersecting red segments and S_b of non-intersecting blue segments, we construct two intermediate sets

$$T_r = S_r \cup \bigcup_{(p,q) \in S_b} \{(p,p),(q,q)\}$$

$$T_b = S_b \cup \bigcup_{(p,q) \in S_r} \{(p,p),(q,q)\}$$

Each new set is the union of the input segments of one color and the endpoints of the segments of the other color (or rather zero length segments located at the endpoints). Both T_r and T_b are of size $O(|S_r| + |S_b|) = O(N)$ and can thus be sorted in $O(n \log_m n)$ I/Os.

We now locate intersections with distribution sweeping with a branching factor of $\sqrt{m}$. Recall that the structure of distribution sweeping is that we divide the plane into $\sqrt{m}$ slabs, and that we then find intersections involving parts of segments that completely span one or more slabs, before we solve the problem recursively in each slab. The recursion continues through $O(\log_m n)$ levels until the subproblems are small enough to be solved in internal memory. If we can do a sweep in $O(n)$ I/Os plus the number of I/O's used to report intersections, we obtain the optimal $O(n \log_m n + t)$ I/O solution.

So let us consider the sweep algorithm. In one sweep we define *long segments* as those crossing one or more slabs and *short segments* as those completely contained in a slab. Furthermore, we shorten the long segments by "cutting" them at the right boundary of the slab that contains their left endpoint, and at the left boundary of the slab containing their right endpoint. Thus our task in a sweep is to report all intersections between long segments of one color and long and short segments of the other color—refer to Figure 20. To find the intersections between long and short segments we use the sweep algorithm used in Section 5.1 to solve the orthogonal line segment intersection problem—except that we sweep in the order of T_r and T_b. We use the algorithm twice, treating long segments of one color as horizontal segments and short segments of the other color as vertical segments. For long red and short blue segments we proceed as follows: We sweep from top to bottom by scanning through the sorted list T_r of red segments and blue endpoints. When a top endpoint of a small blue segment is encountered we insert the segment in the active list associated with the slab containing the segment. When a long red segment is encountered we then scan through all the active lists associated with the slabs it completely spans. During this scan we know that every small blue segment is either intersected by the red segment, or will not be intersected by any of the following red segments and can therefore be removed from the list. As previously we use $O(n + t')$ I/Os to do the sweep.

Note that whereas the important property that a small segment not intersected by a long segment is not intersected by any of the following long segments in the orthogonal case followed from the fact that we were working on the segments in y order, it now follows from the fact that we are sweeping the segments and endpoints in sorted order. As an illustration of this refer to Figure 21. In the sweep we will meet the segments in order a, b, c, d, e, whereas we would meet them in order a, b, d, c, e if we were sweeping in endpoint y order. In the latter case the important property would not hold as segment b actually intersect segment c, even though it does not intersect d which is encountered after b but before c.

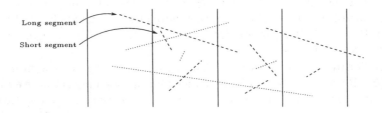

Fig. 20. Long and short segments (red segments dotted, blue segments dashed).

In order to report intersections between long segments of different colors the notion of multislab (as in Section 5.2) is used. First we scan through T_r and distribute the long red segments to the $O(m)$ multislabs. Next, we scan through the blue set T_b, and for each long blue segment we report intersections with the relevant long red segments. This is the same as reporting intersections with the appropriate red segments in each of the multislab lists. As each of the multislabs lists are sorted, and as we also process the blue segments in sorted order, it turns out that this can be done in a simple and efficient way using a merging idea, where it again is crucial that the number of multislab lists is $O(m)$ (that the distribution factor is $\sqrt{m}$). Details in the algorithm appear in [19], where it is also proved that the sweep can be performed in $O(n + t')$ I/Os as required.

To summarize, the red/blue line segment intersection problem can be solved in the optimal $O(n \log_m n + t)$ I/Os. However, as the segment sorting algorithm used in the solution is relatively complicated, its practical importance may be limited. It would be interesting to experimentally compare the algorithms performance to that of other (internal-memory) algorithms for the problem [28, 29, 30, 67, 74]. An experimental comparison of internal-memory algorithms for the problem is already reported in [10].

5.4 Other External-Memory Computational Geometry Algorithms

In the previous sections we have discussed the basic techniques for designing efficient external-memory computational geometry algorithms. We have illustrated the powerful distribution sweeping technique using the orthogonal line segment intersection problem. We have also already mentioned that the technique can be used to solve the batched range searching problem. In [52] it is discussed how it can be used to develop optimal algorithms for a number of other important problems, including for the problems of finding the pairwise intersection of N rectangles, finding all nearest neighbors for a set of N points in the plane, computing the measure (area) of the union of N rectangles in the plane, and for several geometric dominance problems. Several of these problems have applications in GIS systems. In [18] some of the algorithms are extended to work in d dimensions.

Goodrich et al. [52] also discussed external-memory algorithms for the convex hull problem, that is, the problem of computing the smallest convex poly-

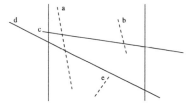

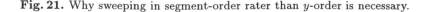

Fig. 21. Why sweeping in segment-order rater than y-order is necessary.

245

tope completely enclosing a set of N points in d-dimensional space. In the two-dimensional case the internal-memory algorithm known as Graham's scan [53, 76] can be modified in a simple way to obtain an $O(n \log_m n)$ I/O external algorithm. They also discussed how to obtain an output-sensitive algorithm based upon an external version of the marriage-before conquest technique [61]. The algorithm uses $O(n \log_m t)$ I/Os. Finally, they developed $O(n \log_m n)$ algorithms for the three-dimensional case which is particularly interesting because of the close relation to the two-dimensional versions of the problems of computing the Voronoi diagram and the Delaunay triangulation of a set of N points. Using the reduction described in [54] the 3-d convex hull algorithm immediately gives algorithms for the two latter problems with the same I/O performance.

The $O(n \log_m n)$ solution to the EPD problem discussed in the last section, which lead to the segment sorting and the red/blue line segment intersection algorithms, has several other almost immediate consequences [19]. If one takes a closer look at the algorithm for EPD one realizes that it works in general with K query points, which are not necessarily endpoints of the segments. Therefore the result leads to an $O((n + k) \log_m n)$ I/O solution to the batched planar point location problem, that is, the problem in which one are given a planar decomposition by N line segments and wants for each of K query points to locate the region in which it lies. Similarly the EPD algorithm leads to algorithms for a couple of region decomposition problems. First it leads to an algorithm for trapezoid decomposition of a set of N segments [68, 76], as the core of this problem precisely is to find for each segment endpoint the segment immediately above it. Using a slightly modified version of an internal-memory algorithm [48], the ability to compute a trapezoid decomposition of a simple polygon then leads to an $O(n \log_m n)$ polygon triangulation algorithm. Finally, using a complicated integration of all the ideas in the red/blue line segment intersection algorithm with the external priority queue discussed in Section 4.3 [12], one can obtain an $O((n + t) \log_m n)$ I/O algorithm for the general line segment intersection problem, where one is just given N segments in the plane and should report their pairwise intersections.

- Main paradigms for developing external computational geometry algorithms:
  - Distribution sweeping.
  - Batched dynamic data structures (external buffered one-dimensional range tree and segment tree).
  - Batched filtering.
  - External fractional cascading.
- Optimal algorithms developed for a large number of problems.
- In practice:
  - Experiments with orthogonal line segment intersection algorithms suggest that algorithms devcloped for the parallel disk model perform well in practice.
  - Much larger problems instances seem to be solvable with I/O algorithms than with main memory algorithms.

6 TPIE — A Transparent Parallel I/O Environment

In Section 5.1 we discussed the experiments with orthogonal line segment intersection algorithms carried out by Chiang [32, 34]. As discussed these experiments suggest that algorithms developed for the parallel disk model perform well in practice, and that they can very well lead to the solution of problem instances one would not be able to solve in practice with algorithms developed for main memory. Unfortunately, existing systems tend not to adequately support the functionality required in order to implement algorithms designed for the parallel disk model directly. Most operating systems basically lets the programmer program a virtual machine with (practically) unlimited main memory, and control I/O "behind the back" of the programmer. However, in order to implement algorithms developed for the parallel disk model one needs to be able to explicitly control I/O, and thus it seems that one has to try to bypass the operating system and write very low level code in order to implement the algorithms we have discussed. Doing so would be a very complicated task and would probably lead to very inflexible code, which would not be portable across different platforms.

On the other hand we have seen how a large number of problems can be solved using a relatively small number of paradigms, such as merging, distribution (and distribution sweeping), and buffered external data structures. The Transparent Parallel I/O Environment (TPIE) proposed by Vengroff [88, 90, 93] tries to take advantage of this. While Chiang [32, 34] performed experiments in order to compare the efficiency of algorithms designed for internal and external memory and to validate the I/O-model, TPIE is designed to assist programmers in the development of I/O-efficient (and easily portable) programs. TPIE implements a set of high-level paradigms (access methods) which lets the programmers specify the functional details of the computation they wish to perform within a given paradigm, without explicitly having to worry about doing I/O or managing internal memory. The paradigms supported by the current prototype of TPIE includes scanning, distribution, merging, sorting, permuting, and matrix arithmetic [90, 93].

In order to allow programmers to abstract away I/O, TPIE uses a stream approach. A computation is viewed as a continuous process in which a program is fed streams of data from an outside source and leave trails (in form of other streams of data) behind it. In this way programmers only need to specify the functional details in the computation they wish to perform within a given paradigm. TPIE then choreograph an appropriate sequence of I/Os in order to keep the computation fed. To realize that the stream approach is indeed natural, just consider a simple version of merge sort. Here a stream of data is first read and divided into a number of (sorted) main memory sized streams, which are then continually read m at a time and merged into a longer stream. Having implemented basic stream handling routines, the programmers only need to specify how to compare objects in order to sort a given set of objects using the external merge sort paradigm—without having to worry about I/O. Note that the programmers do not even need to worry if the streams are stored on a single disk or if a number of parallel disks are used.

TPIE is implemented in C++ as a set of template classes and functions and a run-time library. The current implementation supports access to data stored on one or more disks attached to a workstation. In the future, it is the plan that TPIE will support multiprocessors and/or collections of workstations. TPIE is a modular system with three components; a block transfer engine (BTE), a memory manager (MM) and an access method interface (AMI). The BTE is responsible for moving blocks of data to and from disk, that is, it is intended to bridge the gap between the I/O hardware and the rest of the system. If the system consists of several processors, every processor has its own BTE. The MM running on top of one or more BTEs is responsible for managing main memory resources. All memory allocated by application programs or other parts of TPIE is handled by the MM. Finally, the AMI provides the high-level interface to the programmer and is the only component with which most programmers will need to interact directly. As mentioned the access methods supported by the AMI currently include scanning, distribution, merging, sorting, permuting, and matrix arithmetic. The interested reader is refereed to [88, 90, 93] for details. In [90] implementations of algorithms such as convex hull and list ranking are also discussed. Finally, it is discussed how to obtain the prototype version of TPIE. Currently, TPIE does not support external buffered data structures but we hope in the future to include such structures in the environment. Similarly, it is the plan to extend TPIE with support for more application controlled I/O in order to allow implementation of the in GIS commonly used indexing structures.

In [91] Vengroff and Vitter discuss applications of TPIE to problems in scientific computing, and report some performance results of programs written to solve certain benchmark problems. The TPIE paradigms used in these experiments are scanning, sorting, and matrix arithmetic. The main conclusions made are that TPIE is indeed practical and efficient, and that algorithms for the theoretical parallel disk model perform well in practice. Actually, Vengroff and Vitter show that using TPIE results in a small CPU overhead compared to entirely main memory implementation, but allows much larger data sets to be used. Also, for the implemented benchmarks, the time spent on I/O range from being negligible to being of the same order of magnitude as internal computation time, showing that using TPIE a large degree of overlap between computation and I/O can be accomplished. Very recently, Arge et al. [18] reported similar encouraging experiences with a TPIE implementation of an algorithms for the pairwise rectangle intersection problem.

7 Conclusions

As GIS systems frequently handle huge amounts of data it is getting increasingly important to design algorithms with good I/O performance for problems arising in such systems. Many important computational geometry problems are abstractions of important GIS operations, and in recent years a number of basic techniques for designing I/O-efficient algorithms for such problems have been developed. In this chapter we have surveyed these techniques and the algorithms developed using them. However, the young field of I/O-efficient computation is to a large extend still wide open. Even though the experimental results reported so far are encouraging, a major future goal is to investigate the practical merits of the developed I/O algorithms.

Acknowledgments

I would like to thank Yi-Jen Chiang and Pavan Kumar Desikan for reading earlier drafts of this chapter and Yi-Jen Chiang for providing Figure 12 and 13. The work in this chapter was supported in part by the U.S. Army Research Office MURI grant DAAH04–96–1–0013. Part of the work was done while I was with BRICS, Department of Computer Science, University of Aarhus, Denmark, supported in part by the ESPRIT Long Term Research Programme of the EU under project number 20244 (ALCOM–IT).

References

1. P. K. Agarwal, L. Arge, T. M. Murali, K. Varadarajan, and J. S. Vitter. I/O-efficient algorithms for contour line extraction and planar graph blocking. In *Proc. ACM-SIAM Symp. on Discrete Algorithms*, 1998.
2. A. Aggarwal, B. Alpern, A. K. Chandra, and M. Snir. A model for hierarchical memory. In *Proc. ACM Symp. on Theory of Computation*, pages 305–314, 1987.
3. A. Aggarwal and A. K. Chandra. Virtual memory algorithms. In *Proc. ACM Symp. on Theory of Computation*, pages 173–185, 1988.
4. A. Aggarwal, A. K. Chandra, and M. Snir. Hierarchical memory with block transfer. In *Proc. IEEE Symp. on Foundations of Comp. Sci.*, pages 204–216, 1987.

5. A. Aggarwal and G. Plaxton. Optimal parallel sorting in multi-level storage. *Proc. ACM-SIAM Symp. on Discrete Algorithms*, pages 659–668, 1994.

6. A. Aggarwal and J. S. Vitter. The Input/Output complexity of sorting and related problems. *Communications of the ACM*, 31(9):1116–1127, 1988.

7. A. V. Aho, J. E. Hopcroft, and J. D. Ullman. *The Design and Analysis of Computer Algorithms*. Addison-Wesley, Reading, MA, 1974.

8. B. Alpern, L. Carter, and E. Feig. Uniform memory hierarchies. In *Proc. IEEE Symp. on Foundations of Comp. Sci.*, pages 600–608, 1990.

9. R. J. Anderson and G. L. Miller. A simple randomized parallel algorithm for list-ranking. *Information Processing Letters*, 33:269–273, 1990.

10. D. S. Andrews, J. Snoeyink, J. Boritz, T. Chan, G. Denham, J. Harrison, and C. Zhu. Further comparisons of algorithms for geometric intersection problems. In *Proc. 6th Int'l. Symp. on Spatial Data Handling*, 1994.

11. ARC/INFO. *Understanding GIS—the ARC/INFO method*. ARC/INFO, 1993. Rev. 6 for workstations.

12. L. Arge. The buffer tree: A new technique for optimal I/O-algorithms. In *Proc. Workshop on Algorithms and Data Structures, LNCS 955*, pages 334–345, 1995. A complete version appears as BRICS technical report RS-96-28, University of Aarhus.

13. L. Arge. The I/O-complexity of ordered binary-decision diagram manipulation. In *Proc. Int. Symp. on Algorithms and Computation, LNCS 1004*, pages 82–91, 1995. A complete version appears as BRICS technical report RS-96-29, University of Aarhus.

14. L. Arge. *Efficient External-Memory Data Structures and Applications*. PhD thesis, University of Aarhus, February/August 1996.

15. L. Arge, P. Ferragina, R. Grossi, and J. Vitter. On sorting strings in external memory. In *Proc. ACM Symp. on Theory of Computation*, pages 540–548, 1997.

16. L. Arge, M. Knudsen, and K. Larsen. A general lower bound on the I/O-complexity of comparison-based algorithms. In *Proc. Workshop on Algorithms and Data Structures, LNCS 709*, pages 83–94, 1993.

17. L. Arge and P. B. Miltersen. On showing lower bounds for external-memory computational geometry. In preparation.

18. L. Arge, O. Procopiuc, S. Ramaswamy, T. Suel, and J. S. Vitter. Theory and practice of I/O-efficient algorithms for multidimensional batched searching problems. In *Proc. ACM-SIAM Symp. on Discrete Algorithms*, 1998.

19. L. Arge, D. E. Vengroff, and J. S. Vitter. External-memory algorithms for processing line segments in geographic information systems. In *Proc. Annual European Symposium on Algorithms, LNCS 979*, pages 295–310, 1995. A complete version (to appear in special issue of Algorithmica) appears as BRICS technical report RS-96-12, University of Aarhus.

20. L. Arge and J. S. Vitter. Optimal dynamic interval management in external memory. In *Proc. IEEE Symp. on Foundations of Comp. Sci.*, pages 560–569, 1996.

21. R. D. Barve, E. F. Grove, and J. S. Vitter. Simple randomized mergesort on parallel disks. In *Proc. ACM Symp. on Parallel Algorithms and Architectures*, 1996.

22. R. Bayer and E. McCreight. Organization and maintenance of large ordered indizes. *Acta Informatica*, 1:173–189, 1972.

23. J. L. Bentley. Algorithms for klee's rectangle problems. Dept. of Computer Science, Carnegie Mellon Univ., unpublished notes, 1977.

24. J. L. Bentley and D. Wood. An optimal worst case algorithm for reporting intersections of rectangles. *IEEE Transactions on Computers*, 29:571–577, 1980.

25. G. Blankenagel and R. H. Güting. XP-trees—External priority search trees. Technical report, FernUniversität Hagen, Informatik-Bericht Nr. 92, 1990.
26. M. Blum, R. W. Floyd, V. Pratt, R. L. Rievest, and R. E. Tarjan. Time bounds for selection. *Journal of Computer and System Sciences*, 7:448–461, 1973.
27. P. Callahan, M. T. Goodrich, and K. Ramaiyer. Topology B-trees and their applications. In *Proc. Workshop on Algorithms and Data Structures, LNCS 955*, pages 381–392, 1995.
28. T. M. Chan. A simple trapezoid sweep algorithm for reporting red/blue segment intersections. In *Proc. of 6th Canadian Conference on Computational Geometry*, 1994.
29. B. Chazelle and H. Edelsbrunner. An optimal algorithm for intersecting line segments in the plane. *Journal of the ACM*, 39:1–54, 1992.
30. B. Chazelle, H. Edelsbrunner, L. J. Guibas, and M. Sharir. Algorithms for bichromatic line-segment problems and polyhedral terrains. *Algorithmica*, 11:116–132, 1994.
31. B. Chazelle and L. J. Guibas. Fractional cascading: I. A data structuring technique. *Algorithmica*, 1:133–162, 1986.
32. Y.-J. Chiang. Experiments on the practical I/O efficiency of geometric algorithms: Distribution sweep vs. plane sweep. In *Proc. Workshop on Algorithms and Data Structures, LNCS 955*, pages 346–357, 1995.
33. Y.-J. Chiang, M. T. Goodrich, E. F. Grove, R. Tamassia, D. E. Vengroff, and J. S. Vitter. External-memory graph algorithms. In *Proc. ACM-SIAM Symp. on Discrete Algorithms*, pages 139–149, 1995.
34. Yi-Jen Chiang. *Dynamic and I/O-Efficient Algorithms for Computational Geometry and Graph Problems: Theoretical and Experimental Results*. PhD thesis, Brown University, August 1995.
35. D. R. Clark and J. I. Munro. Efficient suffix trees on secondary storage. In *Proc. ACM-SIAM Symp. on Discrete Algorithms*, pages 383–391, 1996.
36. A. Cockcroft. *Sun Performance and Tuning. SPARC & Solaris*. Sun Microsystems Inc., 1995.
37. T. H. Cormen, C. E. Leiserson, and R. L. Rivest. *Introduction to Algorithms*. The MIT Press, Cambridge, Mass., 1990.
38. Thomas H. Cormen. *Virtual Memory for Data Parallel Computing*. PhD thesis, Department of Electrical Engineering and Computer Science, Massachusetts Institute of Technology, 1992.
39. Thomas H. Cormen. Fast permuting in disk arrays. *Journal of Parallel and Distributed Computing*, 17(1-2):41–57, 1993.
40. Thomas H. Cormen and Leonard F. Wisniewski. Asymptotically tight bounds for performing BMMC permutations on parallel disk systems. In *Proc. ACM Symp. on Parallel Algorithms and Architectures*, pages 130–139, 1993.
41. D. Cormer. The ubiquitous B-tree. *ACM Computing Surveys*, 11(2):121–137, 1979.
42. R. F. Cromp. An intellegent information fusion system for handling the archiving and querying of terabyte-sized spatial databases. In *S. R. Tate ed., Report on the Workshop on Data and Image Compression Needs and Uses in the Scientific Community, CESDIS Technical Report Series, TR-93-99*, pages 75–84, 1993.
43. M. de Berg, M. van Kreveld, M. Overmars, and O. Schwarzkopf. *Computational Geometry – Algorithms and Applications*. Springer Verlag, Berlin, 1997.
44. H. Edelsbrunner and M. Overmars. Batched dynamic solutions to decomposable searching problems. *Journal of Algorithms*, 6:515–542, 1985.
45. P. Ferragina and R. Grossi. A fully-dynamic data structure for external substring search. In *Proc. ACM Symp. on Theory of Computation*, pages 693–702, 1995.

251

46. P. Ferragina and R. Grossi. Fast string searching in secondary storage: Theoretical developments and experimental results. In *Proc. ACM-SIAM Symp. on Discrete Algorithms*, pages 373–382, 1996.

47. R. W. Floyd. Permuting information in idealized two-level storage. In *Complexity of Computer Calculations*, pages 105–109, 1972. R. Miller and J. Thatcher, Eds. Plenum, New York.

48. A. Fournier and D. Y. Montuno. Triangulating simple polygons and equivalent problems. *ACM Trans. on Graphics*, 3(2):153–174, 1984.

49. P. G. Franciosa and M. Talamo. Orders, implicit k-sets representation and fast halfplane searching. In *Proc. Workshop on Orders, Algorithms and Applications (ORDAL'94)*, pages 117–127, 1994.

50. G. R. Ganger, B. L. Worthington, R. Y. Hou, and Y. N. Patt. Disk arrays. High-performance, high-reliability storage subsystems. *IEEE Computer*, 27(3):30–46, 1994.

51. D. Gifford and A. Spector. The TWA reservation system. *Communications of the ACM*, 27:650–665, 1984.

52. M. T. Goodrich, J.-J. Tsay, D. E. Vengroff, and J. S. Vitter. External-memory computational geometry. In *Proc. IEEE Symp. on Foundations of Comp. Sci.*, pages 714–723, 1993.

53. R. L. Graham. An efficient algorithm for determining the convex hull of a finite planar set. *Information Processing Letters*, 1:132–133, 1972.

54. L. J. Guibas and J. Stolfi. Primitives for the manipulation of general subdivisions and the computation of voronoi diagrams. *ACM Trans. on Graphics*, 4:74–123, 1985.

55. Laura M. Haas and William F. Cody. Exploiting extensible dbms in integrated geographic information systems. In *Proc. of Advances in Spatial Databases, LNCS 525*, 1991.

56. J. W. Hong and H. T. Kung. I/O complexity: The red-blue pebble game. In *Proc. ACM Symp. on Theory of Computation*, pages 326–333, 1981.

57. S. Huddleston and K. Mehlhorn. A new data structure for representing sorted lists. *Acta Informatica*, 17:157–184, 1982.

58. Ch. Icking, R. Klein, and Th. Ottmann. Priority search trees in secondary memory. In *Proc. Graph-Theoretic Concepts in Computer Science, LNCS 314*, pages 84–93, 1987.

59. B. H. H. Juurlink and H. A. G. Wijshoff. The parallel hierarchical memory model. In *Proc. Scandinavian Workshop on Algorithms Theory, LNCS 824*, pages 240–251, 1993.

60. P. C. Kanellakis, S. Ramaswamy, D. E. Vengroff, and J. S. Vitter. Indexing for data models with constraints and classes. In *Proc. ACM Symp. Principles of Database Systems*, 1993. A complete version (to appear in special issue of JCSS on principles of database systems) appears as technical report 90-31, Brown University.

61. D. G. Kirkpatrick and R. Seidel. The ultimate planar convex hull algorithm? *SIAM Journal of Computing*, 15:287–299, 1986.

62. M. Knudsen and K. Larsen. I/O-complexity of comparison and permutation problems. Master's thesis, University of Aarhus, November 1992.

63. M. Knudsen and K. Larsen. Simulating I/O-algorithms. Master student project, University of Aarhus, August 1993.

64. D. Knuth. *The Art of Computer Programming, Vol. 3 Sorting and Searching.* Addison-Wesley, 1973.

65. V. Kumar and E. Schwabe. Improved algorithms and data structures for solving graph problems in external memory. In *Proc. IEEE Symp. on Parallel and Distributed Processing*, 1996.

66. R. Laurini and A. D. Thompson. *Fundamentals of Spatial Information Systems*. A.P.I.C. Series, Academic Press, New York, NY, 1992.

67. H. G. Mairson and J. Stolfi. Reporting and counting intersections between two sets of line segments. In *R. Earnshaw (ed.), Theoretical Foundation of Computer Graphics and CAD, NATO ASI Series, Vol. F40*, pages 307–326, 1988.

68. K. Mulmuley. *Computational Geometry. An introduction through randomized algorithms*. Prentice-Hall, 1994.

69. M. H. Nodine, M. T. Goodrich, and J. S. Vitter. Blocking for external graph searching. *Algorithmica*, 16(2):181–214, 1996.

70. M. H. Nodine and J. S. Vitter. Deterministic distribution sort in shared and distributed memory multiprocessors. In *Proc. ACM Symp. on Parallel Algorithms and Architectures*, pages 120–129, 1993.

71. M. H. Nodine and J. S. Vitter. Paradigms for optimal sorting with multiple disks. In *Proc. of the 26th Hawaii Int. Conf. on Systems Sciences*, 1993.

72. M. H. Nodine and J. S. Vitter. Greed sort: An optimal sorting algorithm for multiple disks. *Journal of the ACM*, pages 919–933, 1995.

73. M. Overmars, M. Smid, M. de Berg, and M. van Kreveld. Maintaining range trees in secondary memory. Part I: Partitions. *Acta Informatica*, 27:423–452, 1990.

74. L. Palazzi and J. Snoeyink. Counting and reporting red/blue segment intersections. In *Proc. Workshop on Algorithms and Data Structures, LNCS 709*, pages 530–540, 1993.

75. Yale N. Patt. The I/O subsystem—a candidate for improvement. *Guest Editor's Introduction in IEEE Computer*, 27(3):15–16, 1994.

76. F. P. Preparata and M. I. Shamos. *Computational Geometry. An Introduction*. Springer-Verlag, 1985.

77. S. Ramaswamy and S. Subramanian. Path caching: A technique for optimal external searching. In *Proc. ACM Symp. Principles of Database Systems*, 1994.

78. Chris Ruemmler and John Wilkes. An introduction to disk drive modeling. *IEEE Computer*, 27(3):17–28, 1994.

79. H. Samet. *Applications of Spatial Data Structures: Computer Graphics, Image Processing, and GIS*. Addison Wesley, MA, 1989.

80. J. E. Savage. Space-time tradeoffs in memory hierarchies. Technical Report CS-93-08, Brown University, 1993.

81. M. Smid. *Dynamic Data Structures on Multiple Storage Media*. PhD thesis, University of Amsterdam, 1989.

82. M. Smid and M. Overmars. Maintaining range trees in secundary memory. Part II: Lower bounds. *Acta Informatica*, 27:453–480, 1990.

83. S. Subramanian and S. Ramaswamy. The p-range tree: A new data structure for range searching in secondary memory. In *Proc. ACM-SIAM Symp. on Discrete Algorithms*, pages 378–387, 1995.

84. R. E. Tarjan. Amortized computational complexity. *SIAM J. Alg. Disc. Meth.*, 6(2):306–318, 1985.

85. V. K. Vaishnavi and D. Wood. Rectilinear line segment intersection, layered segment trees, and dynamization. *Journal of Algorithms*, 3:160–176, 1982.

86. M. van Kreveld. Geographic information systems. Utrecht University, INF/DOC–95–01, 1995.

87. J. van Leeuwen. *Handbook of Theoretical Computer Science, Volume A: Algorithms and Complexity*. Elsevier, 1990.

88. D. E. Vengroff. A transparent parallel I/O environment. In *Proc. 1994 DAGS Symposium on Parallel Computation*, 1994.

89. D. E. Vengroff. Private communication, 1995.

90. D. E. Vengroff. *TPIE User Manual and Reference*. Duke University, 1995. Available via WWW at http://www.cs.duke.edu/~dev.

91. D. E. Vengroff and J. S. Vitter. Supporting I/O-efficient scientific computation in TPIE. In *Proc. IEEE Symp. on Parallel and Distributed Computing*, 1995. Appears also as Duke University Dept. of Computer Science technical report CS-1995-18.

92. D. E. Vengroff and J. S. Vitter. Efficient 3-d range searching in external memory. In *Proc. ACM Symp. on Theory of Computation*, pages 192–201, 1996.

93. D. E. Vengroff and J. S. Vitter. I/O-efficient computation: The TPIE approach. In *Proceedings of the Goddard Conference on Mass Storage Systems and Technologies*, NASA Conference Publication 3340, Volume II, pages 553–570, College Park, MD, September 1996.

94. J. S. Vitter. Efficient memory access in large-scale computation (invited paper). In *Symposium on Theoretical Aspects of Computer Science, LNCS 480*, pages 26–41, 1991.

95. J. S. Vitter and M. H. Nodine. Large-scale sorting in uniform memory hierarchies. *Journal of Parallel and Distributed Computing*, 17:107–114, 1993.

96. J. S. Vitter and E. A. M. Shriver. Algorithms for parallel memory, I: Two-level memories. *Algorithmica*, 12(2–3):110–147, 1994.

97. J. S. Vitter and E. A. M. Shriver. Algorithms for parallel memory, II: Hierarchical multilevel memories. *Algorithmica*, 12(2–3):148–169, 1994.

98. B. Zhu. Further computational geometry in secondary memory. In *Proc. Int. Symp. on Algorithms and Computation*, pages 514–522, 1994.

Chapter 9. Precision and Robustness in Geometric Computations

Stefan Schirra

Max-Planck-Institut für Informatik
Saarbrücken
Germany

stschirr@mpi-sb.mpg.de

1 Introduction

This part gives a concise overview of techniques that have been proposed and successfully used to attack precision problems in the implementation of geometric algorithms.

In reference to issues of quality of spatial data in GIS as well as in reference to implementation issues of geometric data structures and algorithms, the terms precision and accuracy are often used interchangeably. We adopt the terminology used in [47]. *Accuracy* refers to the relationship between reality and the measured data modelling it. *Precision* refers to the level of detail with which (numerical) data are represented in a model or in (arithmetic) calculations with the model.

Here, our attention is directed to precision, more precisely, to how to deal with the notorious problems that imprecise geometric calculations can cause. Inaccuracy in GIS data is not our main objective. Basically, we assume that the geometric data to be processed are accurate. Precision problems can make implementing geometric algorithms very unpleasant [27, 72] even under the assumption of perfectly accurate data, if no appropriate techniques are used to deal with imprecision. A quite sketchy discussion of dealing with inaccurate data is given in Section 5.2.

1.1 Precision and Correctness

Geometric algorithms are usually designed and proven to be correct in a computational model that assumes exact computation over the real numbers. In implementations of geometric algorithms, exact real arithmetic is mostly replaced by fast finite precision floating-point arithmetic provided by the hardware of a computer system. For some problems and restricted sets of input data, this approach works well, but in many implementations the effects of squeezing the infinite set of real numbers into the finite set of floating-point numbers can cause catastrophic errors in practice. Due to (accumulated) rounding errors many implementations of geometric algorithms crash, loop forever, or in the best case, simply compute wrong results for some of the inputs for which they are supposed to work. Figure 1 gives an example.

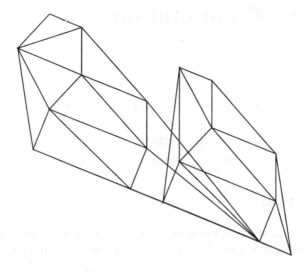

Fig. 1. Incorrect Delaunay triangulation. The error was caused by precision problems. The correct Delaunay triangulation is given in Figure 2. *Courtesy of J. R. Shewchuk* [102].

Conditional tests are critical parts of an implementation, because they determine the control flow. If in every test the same decision is made as if all computations would have been done over the reals, the algorithm is always in a state equivalent to that of its theoretical counterpart. In this case, the combinatorial part of the geometric output of the algorithm will be correct. Numerical data, however, computed by the algorithm might nevertheless be imprecise.

Rounding and cancellation errors may cause wrong decisions and hence lead to errors in the combinatorial part of the geometric output as well. Thereby imprecise calculations can destroy the correctness of the implementation of an otherwise correct algorithm.

1.2 Robustness and Stability

Along with the substitution of real arithmetic by floating-point arithmetic, correctness is often replaced by robustness. Robustness is a measure of the ability to recover from error conditions, e.g., tolerance of failures of internal components or errors in input data.

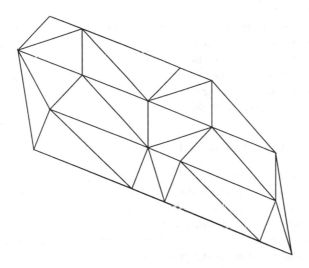

Fig. 2. Correct Delaunay triangulation. *Courtesy of J. R. Shewchuk*[102].

Often an implementation of an algorithm is considered to be *robust* if it produces the correct result for some perturbation of the input. It is called *stable* if the perturbation is small. This terminology has been adopted from numerical analysis where backward error analysis is used to get bounds on the sizes of the perturbations. Geometric computation, however, goes beyond numerical computation. Since geometric problems involve not only numerical but also combinatorial data it is not always clear what perturbation of the input, especially of the combinatorial part, means. Perturbation of the input is justified by the fact that in many geometric problems the numerical data are real world data obtained by measuring and hence known to be inaccurate. This is certainly true for most of the geometric problems in GIS.

1.3 Degeneracy

A related problem in the implementation of geometric algorithms is degeneracies. Theoretical papers on computational geometry often assume the input in general position and leave the "straightforward" handling of special cases to the reader. This might make the presentation of an algorithm more readable, but it can put a huge burden on the implementor, because the handling of degeneracies is often less straightforward than claimed. Since precision problems are caused by degenerate and nearly degenerate configurations in the input, degeneracy is closely related to precision and robustness. Symbolic perturbation schemes [31, 32, 33, 112, 113] have been proposed to abolish the handling of degeneracies. Exact computation is a prerequisite for applying these techniques [111]. The handling of degeneracies and the use of symbolic perturbation schemes are a point of controversy in the computational geometry literature [15, 99, 100]. For a discussion of degeneracy we refer the reader to [15] and [99].

Sometimes, the term robustness is also used with respect to degeneracies. Dey et al. [26] define robustness as the ability of a geometric algorithm to deal with degeneracies and "inaccuracies" during various numerical computations. The definition of robustness in [97] is similar.

1.4 Attacks on the Precision Problem

There are two obvious approaches for solving the precision problem. The first is to change the model of computation: design algorithms that can deal with imprecise computation. For a small number of basic problems this approach has been applied successfully but a general theory of how to design algorithms with imprecise primitives or how to adopt algorithms designed for exact computation with real numbers is still a distant goal. The second approach is exact computation: compute with a precision that is sufficient to keep the theoretical correctness of an algorithm designed for real arithmetic alive. This is basically possible, at least theoretically, in almost all cases arising in practical geometric computing. The second approach is very promising, because it allows exact implementations of numerous geometric algorithms developed for real arithmetic without modifications of these algorithms.

1.5 Floating-Point Arithmetic

Floating-point numbers are the standard substitution for real numbers in scientific computation. In some programming languages the floating-point number type is even called `real` [59]. Since most geometric computations are executed with floating-point arithmetic, it is worth taking a closer look at floating-point computation. Goldberg [46] gives an excellent overview.

A finite-precision floating-point system has a base B, a fixed mantissa length l and and an exponent range $[e_{min}..e_{max}]$.

$$\pm d_0.d_1 d_2 \cdots d_{p-1} * B^e$$

$0 \leq d_i < B$, represents the number

$$\pm(d_0 + d_1 \cdot B^{-1} + d_2 \cdot B^{-2} + \cdots + d_{p-1}B^{-p+1}) \cdot B^e.$$

A representation of a floating point number is called normalized iff $d_0 \neq 0$. For example, the rational number $1/2$ has representations $0.500 * 10^0$ or $5.000 * 10^{-1}$ in a floating-point system with base 10 and mantissa length 4 and normalized representation $1.00 * 2^{-1}$ in a floating-point system with base 2 and mantissa length 3.

Since an infinite set of numbers is represented by finitely many floating-point numbers, rounding errors occur. A real number is called representable if it is zero or its absolute value is in the interval $[B^{e_{min}}, B^{e_{max}+1}]$. Let r be some real number and f_r be a floating-point representation for r. Then $|r - f_r|$ is called *absolute error* and $|r - f_r|/|r|$ is called *relative error*. The relative error of rounding a representable real toward the nearest floating-point number in a floating-point system with base B and mantissa length l is bounded by $1/2 \cdot \text{B}^{-l}$, which is called *machine epsilon*. Calculations can underflow or overflow, i.e., leave the range of representable numbers.

Fortunately, the times where the results of floating-point computations could drastically differ from one machine to another, depending on the accuracy of the floating-point machinery, seem be coming to an end. The IEEE standard 754 for binary floating-point computation [104] is becoming widely accepted by hardware-manufacturers. The IEEE standard 754 requires that the results of $+, -, \cdot, /$ and $\sqrt{}$ are exactly rounded, i.e., the result is the exact result rounded according to the chosen rounding mode. The default rounding mode is round to nearest. Ties in round to nearest are broken such that the least significant bit becomes 0. Besides rounding toward nearest, rounding toward zero, rounding toward ∞, and rounding toward $-\infty$ are rounding modes that have to be supported according to IEEE standard 754.

The standard makes reasoning about correctness of a floating-point computation machine-independent. The result of the basic operations will be the same on different machines if both support IEEE standard and the same precision is used. Thereby code becomes portable.

The IEEE standard 754 specifies floating-point computation in single, single extended, double, and double extended precision. Single precision is specified for a 32 bit word, double precision for two consecutive 32 bit words. In single precision the mantissa length is $l = 24$ and the exponent range is $[-126..127]$. Double precision has mantissa length $l = 53$ and exponent range $[-1022..1023]$. Hence the relative errors are bounded by 2^{-23} and 2^{-52}. The single and double precision formats usually correspond to the number types **float** and **double** in C++.

Floating-point numbers are represented in normalized representation. Since the zeroth bit is always 1 in normalized representation with base 2, it is not stored. There are exceptions to this rule. *Denormalized* numbers are added to let the floating-point numbers underflow nicely and preserve the property $x - y = 0$ iff $x = y$. Zero and the denormalized numbers are represented with exponent $e_{\min}$. Besides these floating point numbers there are special quantities $+\infty$, $-\infty$ and NaN (Not a Number) to handle exceptional situations. For example $-1.0/0.0 = -\infty$, NaN is the result of $\sqrt{-1}$ and ∞ is the result of overflow in positive range.

Due to the unavoidable rounding errors, floating-point arithmetic is inherently imprecise. Basic laws of arithmetic like associativity and distributivity are not satisfied by floating-point arithmetic. Section 13.2 in [83] gives some examples. Since the standard fixes the layout of bits for mantissa and exponent in the representation of floating-point numbers, bit-operations can be used to extract information.

2 Geometric Computation

Geometric computing is a combination of numerical and combinatorial computation.

2.1 Geometric Problems

A geometric problem can be seen as a mapping from a set of permitted input data, consisting of a combinatorial and a numerical part, to a set of valid output data, again consisting of a combinatorial and a numerical part. A geometric algorithm solves a problem if it computes the output specified by the problem mapping for a given input. For some geometric problems the numerical data of the output are a subset of the data of the input. Those geometric problems are called *selective*. In other geometric problems new geometric objects are created which involve new numerical data that have to be computed from the input data. Such problems are called *constructive*. Geometric problems might have various facets, even basic geometric problems appear in different variants.

We use two classical geometric problems for illustration, convex hull and intersection of line segments in two dimensions. In the two-dimensional *convex hull problem* the input is a set of points. The numerical part might consist of the coordinates of the input points; the combinatorial part is simply the assignment of the coordinate values to the points in the plane. The output might be the convex hull of the set of points, i.e., the smallest convex polygon containing all the input points. The combinatorial part of the output might be the sorted cyclic sequence of the points on the convex hull, given in counterclockwise order. The point coordinates form the numerical part of the output. In a variant of the problem only the extreme points among the input points have to be computed, where a point is called is extreme if its deletion from the input set would change the convex hull. Note that the problem is selective according to our definition even if a convex polygon and hence a new geometric object is constructed.

In the *line segment intersection problem* the intersections among a set of line segments are computed. The numerical input data are the coordinates of the segment endpoints, the combinatorial part of the input just pairs them together. The combinatorial part of the output might be a combinatorial embedding of a graph whose vertices are the endpoints of the segments and the points of intersection between the segments. Edges connect two vertices if they belong to the same line segment l and no other vertex lies between them on l. Combinatorial embedding means that the set of edges incident to a vertex are given in cyclic order. The numerical part is formed by the coordinates of the points assigned to the vertices in the graph. Since the intersection points are in general not part of the input, the problem is constructive. A variant might ask only for all pairs of segments that have a point in common. This version is selective.

Line simplification problems in cartography can be selective or constructive as well, depending on whether only input points are allowed as vertices of the simplified polyline or not.

2.2 Geometric Predicates

Geometric primitives are the basic operations in geometric algorithms. There is a fairly small set of such basic operations that cover most of the computations in a geometric algorithm. Geometric primitives subsume constructions of basic geometric objects, like line segments or circles, and predicates. Geometric predicates test properties of basic geometric objects. They are used in conditional tests that direct the control flow in geometric algorithms. Well-known examples are: testing whether two line segments intersect, testing whether a sequence of points defines a right turn, or testing whether a point is inside or on the circle defined by three other points.

Geometric predicates involve the comparison of numbers which are given by arithmetic expressions. The operands of the expressions are constants, in practical problems mainly integers, and numerical data of the geometric objects that are tested. Expressions differ by the operations used, but many geometric predicates involve arithmetic expressions over $+, -, *$ only, or can at least be reformulated in such a way.

2.3 Arithmetic Expressions in Geometric Predicates

One can think of an arithmetic expression as a labeled binary tree. Each inner node is labeled with a binary or unary operation. It has pointers to trees defining its operands. The pointers are ordered corresponding to the order of the operands. The leaves are labeled with constants or variables which are place-holders for numerical input values. Such a representation is called an *expression tree*.

The numerical data that form the operands in an expression evaluated in a geometric predicate in the execution of a a geometric algorithm might be again defined by previously evaluated expressions. Tracing these expressions backwards we finally get expressions on numerical input data whose values for concrete problem instances have to be compared in the predicates. Since intermediate results are used in several places in an expression we get a directed acyclic graph (dag) rather than a tree.

Without loss of generality we may assume that the comparison of numerical values in predicates is a comparison of the value of some arithmetic expression with zero. The *depth of an expression tree* is the length of the longest root-to-leaf path in the tree. For many geometric problems the depth of the expressions appearing in the predicates is bounded by some constant [111]. Expressions over input variables involving operations $+, -, *$ only are called *polynomial*, because they define multivariate polynomials in the variables. If all constants in the expression are integral, a polynomial expression is called *integral*. The *degree* of a polynomial expression is the total degree of the resulting multivariate polynomial. In [11, 69] the notion of the degree of an expression is extended to expressions involving square roots. An expression involving operations $+, -, *, /$ only is called *rational*.

2.4 Geometric Computation with Floating-Point Numbers

In a branching step of a geometric algorithm, numerical values of some expression given by an expression dag are compared. In the theoretical model of computation a real-valued expression is evaluated correctly for all real input data, but in practice only an approximation is computed. The accumulated error in the numerical calculation might be so large that the truth value of the predicate with the expressions evaluated with inherently imprecise floating-point computation is different from the truth value of the predicate with an exact evaluation of the predicate.

Naively applied floating-point arithmetic can set axioms of geometry out of order. A classical example is Ramshaw's braided lines (see Figure 3 and [83, 84]).

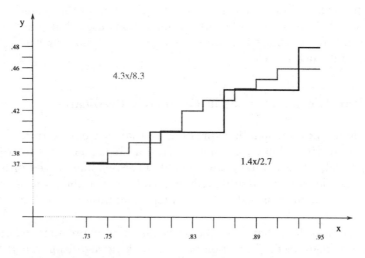

Fig. 3. Evaluation of the line equations $y = 4.3 \cdot x/8.3$ and $y = 1.4 \cdot x/2.7$ in a floating-point system with base 10 and mantissa length 2 and rounding to nearest suggests that the lines have several intersection points besides the true intersection point at the origin.

Rewriting an expression to an expression dag that leads to a numerically more stable evaluation order can help a lot. Goldberg [46] gives the following example due to Kahan. Consider a triangle with sides of length $a \geq b \geq c$ respectively. The area of a such a triangle is

$$\sqrt{s(s-a)(s-b)(s-c)}$$

where $s = (a + b + c)/2$. For $a = 9.0$, $b = c = 4.53$ the correct value of s in a floating-point system with base 10, mantissa length 3 and exact rounding is 9.03 while the computed value $\tilde{s}$ is 9.05. The area is 2.34, the computed area, however, is 3.04, an error of nearly 30%. Using the expression

$$\sqrt{(a + (b + c)) \cdot (c - (a - b)) \cdot (c + (a - b)) \cdot (a + (b - c))}/4$$

one gets 2.35, an error of less than 1%. For a less needle-like triangle with $a = 6.9$, $b = 3.68$, and $c = 3.48$ the improvement is not so drastic. Using the first expression, the result computed by a floating-point system with base 10, mantissa length 3 and exact rounding is 3.36. The second expression gives 3.3. The exact area is approximately 3.11. One can show that the relative error of the second expression is at most 11 times machine precision [46].

As the example above shows, the way a numerical value is computed can highly influence its precision. Summation of floating-point numbers is another classical example. Rearranging the summands helps to reduce imprecision due to extinction.

2.5 Heuristic Epsilons

A widely used method to deal with numerical inaccuracies is based on the rule of thumb

If something is close to zero it is zero.

Some trigger-value $\varepsilon_{\text{magic}}$ is added to a conditional test where a numerical value is compared to zero. If the computed approximation is smaller than $\varepsilon_{\text{magic}}$ it is treated as zero. Adding such epsilons is popular folklore. What should the $\varepsilon_{\text{magic}}$ be? In practice, $\varepsilon_{\text{magic}}$ is usually chosen as some fixed tiny constant and hence not sensitive to the actual sizes of the operands in a concrete expression. Furthermore, the same epsilon is often taken for all comparisons, no matter which expression or which predicate is being evaluated. Normally, no proof is given that the chosen $\varepsilon_{\text{magic}}$ makes sense. $\varepsilon_{\text{magic}}$ is guessed and adjusted by trial and error until the current value works for the considered inputs, i.e., until no catastrophic errors occur anymore. Yap [114] suggests calling this procedure *epsilon-tweaking*.

Adding epsilon is justified by the following reasoning: If something is so close to zero, then a small modification of the input, i.e., a perturbation of the numerical data by a small amount, would lead to value zero in the evaluated expression. There are, however, severe problems with that reasoning. The size of the perturbation causes a problem. The justification for adding epsilons assumes that the perturbation of the (numerical) input is small. Even if such a small perturbation exists for each predicate, the existence of a global small perturbation of the input data is not guaranteed. Figure 4 shows a polyline, where every three consecutive vertices are collinear under the "close to zero is zero" rule. In each

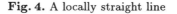

Fig. 4. A locally straight line

case, a fairly small perturbation of the points exists that makes them collinear.

There is, however, no small perturbation that makes the whole polyline straight. The example indicates that collinearity is not transitive. Generally, equality is not transitive under epsilon-tweaking. This might be the most serious problem with this approach. Another problem is that different tests might require different perturbations, e.g., predicate Γ_1 might require a larger value for input variable x_{56} while test P_2 requires a smaller value, such that both expressions evaluate to zero. There might be no perturbation of the input data that leads to the decisions made by the "close to zero is zero" rule. Finally, a result computed with "close to zero is zero" is not the exact result for the input data but only for a perturbation of it. For some geometric problems that might cause trouble, since the computed output and the exact output can be combinatorially very different [15].

3 Exact Geometric Computation

An obvious approach to the precision problem is to compute "exactly". In this approach the computation model over the reals is mimiced in order to preserve the theoretical correctness proof. Exact computation means to ensure that all decisions made by the algorithm are correct decisions for the actual input, not only for some perturbation of it. As we shall see, it does not mean that in all calculations exact representations for all numerical values have to be computed. Approximations that are sufficiently close to the exact value can often be used to guarantee the correctness of a decision. Empirically it turns out to be true for most of the decisions made by a geometric algorithm that approximations are sufficient. Only degenerate and nearly degenerate situations cause problems. That is why most implementations based on floating-point numbers work very well for the majority of the considered problem instances and fail only occasionally.

If an implementation of an algorithm does all branchings the same way as its theoretical counterpart, the control flow in the implementation corresponds to the control flow of the algorithm proved to be correct under the assumption of exact computation over the reals, and hence the validity of the combinatorial part of the computed output follows. Thus, for selective geometric problems, it is sufficient to guarantee correct decisions, since all numerical data are already part of the input.

For constructive geometric problems, new numerical data have to be computed "exactly". A representation of a real number r should be called exact only if it allows one to compute an approximation of r to whatever precision, i.e. no information has been lost. According to Yap [114] a representation of a subset of the reals is exact if it allows the exact comparison of any two real numbers in that representation. This reflects the necessity for correct comparisons in branchings steps in the exact geometric computation approach. Examples of exact representations are the representation of rationals by numerator and denominator, where both are arbitrary precision integers, and the representation of algebraic numbers by an integral polynomial P having root α and an interval

264

that isolates α from the other roots of P. Further examples are symbolic and implicit representations. For example, rather than compute the coordinates of an intersection point of line segments explicitly, one can represent them implicitly by maintaining the intersecting segments. Another similar example is the representation of a number by an expression dag, which reflects the computation history. Allowing symbolic or implicit representation can be seen as turning a constructive geometric problem into a selective one.

As suggested in the discussion above, there are different flavours of exact geometric computation. Franklin's survey [44] already discusses the basics of many approaches to exact computation. Since the publication of his paper much progress has been made in improving the efficiency of exact computation (see [111] for an overview). Thus some of his conclusions have to be revisited.

3.1 Exact Integer and Rational Arithmetic

A number of geometric predicates in basic geometric problems include only integral expressions in their tests. Thus, if all numerical input data are integers, the evaluation of these predicates involves integers only. With the integer arithmetic provided by the hardware only overflow may occur, but no rounding errors. The problem with overflow in integral computation is abolished if arbitrary precision integer arithmetic is used. There are several software packages for arbitrary or multiple precision integers, e.g., BigNum [101], GNU MP [49], LiDIA [68], or the number type `integer` in LEDA [74]. Fortune and Van Wyk [41, 43] report on experiments with such packages.

Since the integral input data are usually bounded in size, e.g., by the maximal representable `int`, there is not really a need for *arbitrary precision* integers. Integer arithmetic with a fixed precision adjusted to the maximum possible integer size in the input and to the degree of the integral polynomial expression arising in the computation is adequate. If the input integers have binary representation with at most b-bits and if d is the maximum degree and m the maximum number of monomials of the integral polynomial expressions, then an integer arithmetic for integers with $db + \log m + O(1)$ bits suffices. Usually, m is in $O(1)$. The degree of polynomial expressions in geometric predicates has recently gained more attention in the design of geometric algorithms. Liotta et al. [69] investigate the degree involved in some proximity problems in 2- and 3-dimensional space.

Many predicates include only expressions involving operations $+, -, *, /$. All the predicates arising in problems like map overlay in cartography and in most of the problems discussed in textbooks on computational geometry [70, 91, 30, 82, 86, 66, 62, 23, 8] are of this type. Such problems are called *rational* [111].

A rational number can be exactly stored as a pair of arbitrary precision integers representing numerator and denominator respectively. Let us call this *exact rational arithmetic*. The intermediate values computed in rational problems are often solutions to systems of linear equations like the coordinates of the intersection point of two straight lines.

Division can be avoided in rational predicates, e.g., exact rational arithmetic postpones division. With exact rational arithmetic, numerator and denominator of the result of the evaluation of a rational expression are integral polynomial expressions in the numerators and denominators of the rational operands. A sign test for a rational expression can be done by two sign tests for integral polynomial expressions. Hence rational expressions in conditional tests in geometric predicates can be replaced by tests involving integral polynomial expressions.

Homogeneous coordinates known from projective geometry and computer graphics can be used to avoid division, too. In homogeneous representation, a point in d-dimensional affine space with Cartesian coordinates $(x_0, x_1, \ldots, x_{d-1})$ is represented by a vector $(hx_0, hx_1, \ldots, hx_{d-1}, hx_d)$ such that $x_i = hx_i/hx_d$ for all $0 \leq i \leq d - 1$. Note that the homogeneous representation of a point is not unique; multiplication of the homogeneous representation vector with any $\lambda \neq 0$ gives a representation of the same point. The homogenizing coordinate hx_d is the common denominator of the coordinates. Homogeneous representation allows division-free representation of the intersection point of two straight lines given by $a \cdot X + b \cdot Y - c = 0$ and $d \cdot X + e \cdot Y + f = 0$. The intersection point can be represented by homogeneous coordinates $(b \cdot f - c \cdot e, a \cdot f - c \cdot d, a \cdot e - b \cdot d)$.

A test including rational expressions in Cartesian coordinates transforms into a test including only polynomial expressions in homogeneous coordinates after multiplication with an appropriate product of homogenizing coordinates. Since all monomials appearing in the resulting expressions have the same degree in the homogeneous coordinates, the resulting polynomial is a homogeneous polynomial. For example, the test $a \cdot x_0 + b \cdot x_1 + c = 0$?, which tests whether point (x_0, x_1) is on the line given by the equation $a \cdot X + b \cdot Y + c = 0$, transforms into $a \cdot hx_0 + b \cdot hx_1 + c \cdot hx_2 = 0$?.

Many geometric predicates that do not obviously involve only integral polynomial expressions can be rewritten so that they do. Above, we have illustrated this for rational problems. In principal, even sign tests for expressions involving square roots can be turned into a sequence of sign tests of polynomial expressions by repeated squaring [14, 69]. Therefore, arbitrary or multiple precision integer arithmetic is a powerful tool for exact geometric computation, but arbitrary precision integer arithmetic has to be supplied by software and is therefore much slower than the hardware-supported fixed precision integer arithmetic. The actual cost of an operation on arbitrary precision integers depends on the size of the operands, more precisely on the length of their binary representation. If expressions of large depth are involved in the geometric calculations the size of the operands can increase drastically. In the literature huge slow down factors are reported if floating-point arithmetic is simply replaced by exact rational arithmetic. Karasick, Lieber, and Nackman [61] report slow-down factors of about 10 000.

While in most rational problems the depth of the involved rational expressions is a small constant, there are problems where the size of the numbers has a linear dependence on the problem size. An example is computing minimum link paths inside simple polygons [60]. Numerator and denominator of the knickpoints on a minimum link path can have superquadratic bitlength with respect to the number of polygon vertices [60]. This is by the way a good example of how strange the assumption of constant time arithmetic operations in theory may be in practice.

Fortune and Van Wyk [41, 43] noticed that in geometric computations the sizes of the integers are small to medium compared to those arising in computer algebra and number theory. Multiple precision integer packages are mainly used in these areas and hence tuned for good performance with larger integers. Consequently Fortune and Van Wyk developed LN [42], a system that generates efficient code for integer arithmetic with fairly "little" numbers. LN takes an expression and a bound on the size of the integral operands as input. The generated code is very efficient if all operands are of the same order of magnitude as the bound. For much smaller operands the generated code is clearly not optimal. LN can be used to trim integer arithmetic in an implementation of a geometric algorithm for special applications. On the other hand, LN is not useful for generating general code.

For integral polynomial expressions, modular arithmetic [1, 64] is an alternative to arbitrary precision integer arithmetic. Let $p_0, p_1, \ldots, p_{k-1}$ be a set of integers that are pairwise relatively prime and let p be the product of the p_i. By the Chinese remainder theorem there is a one-to-one correspondence between the integers r with $\lfloor \frac{p}{2} \rfloor \leq r < \lceil \frac{p}{2} \rceil$ and the k-tupels $(r_0, r_1, \ldots, r_{k-1})$ with $-\lfloor \frac{p_i}{2} \rfloor \leq r_i < \lceil \frac{p_i}{2} \rceil$. By the integer analog of the Lagrangian interpolation formula for polynomials [1], we have

$$r = \sum_{i=0}^{k-1} r_i s_i q_i \mod p$$

where $r_i = r \mod p_i$, $q_i = p/p_i$, and $s_i = q_i^{-1} \mod p_i$. Note that s_i exists because of the relative primality and can be computed with an extended Euclidean gcd algorithm [64]. To evaluate an expression, a set of relatively prime integers is chosen such that the product of the primes is at least twice the absolute value of the integral value of the expression. Then the expression is evaluated modulo each p_i. Finally Chinese remaindering is used to reconstruct the value of the expression.

Modular arithmetic is frequently used in number theory, but not much is known about its application to exact geometric computation. Fortune and Van Wyk [41, 43] compared modular arithmetic with multiple precision integers provided by software packages for a few basic geometric problems without observing much of a difference in the performance. Recently, Brönnimann et al. reported on promising results concerning the use of modular arithmetic in combination with single precision floating-point arithmetic for sign evaluation of determinants [9].

Modular arithmetic is particularly useful if intermediate results can be very large, but the final result is known to be relatively small. The drawback is that a good bound on the size of the final result must be known in order to choose sufficiently many relatively prime integers, but not too many.

3.2 Lazy Evaluation

The LEA system [7] favors the rule

Why compute something that is never used,

so why compute numbers to high precision, before you know that this precision is actually needed. Since it is hard to know in advance which precision will be needed in later decisions, numbers have to be presented in a way that allows for recomputation with higher precision if the currently available precision is not sufficient. In the LEA system, numbers are represented by intervals and expression dags that reflects their creation history. Initially only a low precision representation is calculated, representations with repeatedly increased precision are computed only if decisions can't be made with the current precision.

In LEA, interval arithmetic [7] with floating-point numbers is used to compute rough representations of a number. The interval is then repeatedly refined by redoing the computation along the expression dag with refined intervals for the operands. If the interval representation can't be refined anymore with floating-point evaluation, exact rational arithmetic is used to solve the decision problem.

Another approach based on expression trees is described by Yap and Dubé [29, 111, 114]. In this approach the precision used to evaluate the operands is not systematically increased, but the increase is demanded by the intended increase in the precision of the result. The data type `real` in LEDA [16] also stores the creation history in expression dags and uses floating-point approximations and errors bounds as first approximations. The strategy of repeatedly increasing the precision is similar to [29, 111, 114]. In both approaches software-based multiple precision floating-point arithmetic with a mantissa length that can be arbitrarily chosen and an unbounded exponent is used to compute representations with higher precision. Furthermore, both approaches include square root operations besides $+, -, *, /$.

The C++-programming language is well suited for using number types that provide exact computation in a packed form like lazy numbers. Since arithmetic operators can be overloaded, software-based number types can be used exactly like `int` and `double`. Thereby lazy numbers can be used by a programmer exactly like the built-in number types. The user does not notice that his numbers are lazy-evaluated.

Lazy evaluation has to detect whether the precision of a computation is sufficient or not. How this can be done is described in the following subsections.

3.3 Floating-Point Filter

Replacing exact arithmetic, on which the correctness of a geometric algorithm was based, by imprecise finite-precision arithmetic works in practice for most of the given input data and fails only occasionally. Thus always computing exact values would put a burden on the algorithm that is rarely really needed. The idea of floating-point filters is to filter out those branching steps where a floating-point computation gives the correct result. Only if it is not certified that the floating-point evaluation leads to a correct decision is the branching step reevaluated at a higher cost by calculating the exact value or a better approximation.

Filter techniques allow the use of high speed floating-point arithmetic. A filter simply computes a bound on the error of the floating-point computation and compares the absolute value of the computed result to the computed error bound. If the error bound is smaller, the computed approximation and the exact value have the same sign. Error bounds can be computed a priori if specific information on the input data is available, e.g., if all input data are integers from a bounded range, e.g., the range of integers representable in a computer word. Such so-called static filters require only little additional effort at run time, just one additional test per branching, plus the refined reevaluation in the worst case. Dynamic filters compute an error bound on the fly parallel to the evaluation in floating point arithmetic. Since they take the actual values of the operands into account and not only bounds derived from the bounds on the input data, the estimates for the error involved in the floating-point computation can be much tighter than in a static filter. Thus dynamic filters can let more floating-point calculations pass the filter but at the cost of the online error computation. In the error computation one can put emphasis on speed or on precision. The former makes arithmetic operations more efficient while the latter lets more floating-point computations pass a test.

Note the difference between static filters and heuristic epsilons. If the computed approximate value is larger than the error bound or $\varepsilon_{\text{magic}}$ respectively, the behavior is identical. The program continues based on the assumption that the computed floating-point value has the correct sign. If, however, the computed approximate value is too small, the behavior is completely different. Epsilon-tweaking assumes that the actual value is zero, which might be wrong, while a floating-point filter invokes a more expensive computation, which finally leads to a correct decision.

Mehlhorn and Näher use the following easily computable error bounds for integral expressions evaluated in floating-point arithmetic in their implementation of the Bentley-Ottmann plane sweep algorithm for computing the intersections among a set of line segments in the plane [71]. It assumes that neither overflow nor underflow occurs. Let E be an integral expression. E is also used to denote the value of E while $\tilde{E}$ is used to denote the value of the expression when evaluated with floating point arithmetic, i.e., all operations are replaced by their floating point counterparts.

Mehlhorn and Näher [71] define the *measure mes(E)* and the *index ind(E)* of a polynomial expression E such that

$$|\tilde{E} - E| \leq ind(E) \cdot \varepsilon_{\text{prec}} \cdot mes(E).$$

where ε_{prec} is the machine precision of the floating-point system used. Both the index and the measure are easily computable by the following rules.

	$mes(E)$	$ind(E)$
float $f \neq 0$	$2^{\lceil \log |f| \rceil}$	0
float 0	0	0
$E_1 \pm E_2$	$2 \cdot \max(mes(E_1), mes(E_2))$	$(1 + ind(E_1) + ind(E_2))/2$
$E_1 \cdot E_2$	$mes(E_1) \cdot mes(E_2)$	$1/2 + ind(E_1) + ind(E_2)$

If a filter fails, a refined filter can be used. A refined filter might compute a tighter error bound or use floating-point arithmetic with higher precision and thereby get better approximations and smaller error bounds. This step can be iterated. Composition of more and more refined filters leads to a lazy evaluation strategy. Finally, if necessary, exact arithmetic can be used. Such lazy evaluation strategies are called *adaptive*, because they do not compute more precisely than needed.

For orientation predicates and incircle tests in two- and three-dimensional space Shewchuk [102, 103] presents such a lazy evaluation strategy. It uses an (exact, if neither underflow nor overflow occurs) representation of sums and products of floating-point numbers as a symbolic sum of double precision floating-point numbers. Computation with numbers in this representation, called expanded doubles in [102], is based on the interesting results of Priest [92, 93] and Dekker [25] on extending the precision of floating-point computation. An adapted combination of these techniques allows one to reuse values computed in previous filtering steps in later filtering steps.

For integral expressions scalar products delivering exactly rounded results can be used in filters to get best possible floating-point approximations, as suggested by Ottmann et al. [87].

3.4 Interval Arithmetic

Approximation and error bound define an interval that contains the exact value. Interval arithmetic [2, 79, 80] is another method to get an interval with this property. In interval arithmetic real numbers are represented by intervals, whose endpoints are floating-point numbers. The interval representing the result of an operation is computed by floating-point operations on the endpoints of the intervals representing the operands. For example, the lower endpoint of the interval representing the result of an addition is the sum of the lower endpoints of the intervals of the summands. Since this floating-point addition might be inexact, either the rounding mode is changed to rounding toward $-\infty$ before addition or a correction term is subtracted. For interval arithmetic, rounding modes toward ∞ and toward $-\infty$ are very useful. See, for example, [81, 105] for applications of interval methods to geometric computing. The combination of exact rational arithmetic with interval arithmetic based on fast floating-point computation has been pioneered by Karasick, Lieber and Nackman [61] to geometric computing.

A refinement of standard interval arithmetic is so-called affine arithmetic proposed by Comba and Stolfi [22]. While standard interval arithmetic assumes that the unknown values of operands and subexpressions can vary independently, affine arithmetic keeps track of first-order dependencies and takes these into account. Thereby error explosion can often be avoided and tighter bounds on the computed quantities can be achieved. An extreme example is computing $x - x$ where for x some interval $[x.lo, x.hi]$ is given. Standard interval arithmetic would compute the interval $[x.lo - x.hi, x.hi - x.lo]$, while affine arithmetic gives the true range $[0, 0]$.

3.5 Exact Sign of Determinant

Many geometric primitives can be formulated as sign computations of determinants. The classical example of such a primitive is the orientation test, which in two-dimensional space determines whether a given sequence of three points is a clockwise or a counterclockwise turn or whether they are collinear. Another example is the incircle test used in the construction of Voronoi diagrams of points.

Recently some effort has been focused on exact sign determination. Clarkson [21] gives an algorithm to evaluate the sign of a determinant of a $d \times d$ matrix with integer entries using floating-point arithmetic. His algorithm is a variant of the modified Graham-Schmidt orthogonalization. In his variant, scaling is used to improve the conditioning of the matrix. Since only positive scaling factors are used, the sign of the determinant does not change. Clarkson shows that only $b + O(d)$ bits are required, if all entries are b-bit integers. Hence, for small dimensional matrices his algorithm can be used to evaluate the sign of its determinant with fast hardware floating-point arithmetic.

Avnaim et al. [4] consider determinants of small matrices with integer entries, too. They present algorithms to compute the sign of 2×2 and 3×3 matrices with b-bit integer entries using precision b and $b + 1$ only, respectively. Brönnimann and Yvinec [10] extend the method of [4] to $d \times d$ matrices and compare it with a variant of Clarkson's method.

3.6 Certified Epsilons

While the order of two different numbers can be found by computing sufficiently close approximations, it is not so straightforward to determine whether two numbers are equal or, equivalently, whether the value of an expression is zero. ¿From a theoretical point of view arithmetic expressions arising in geometric predicates are expressions over the reals. Hence the value of an expression can in general get arbitrarily close to zero if the variable operands are replaced by arbitrary real numbers. In practice the numerical input data originate from a finite, discrete subset of the reals, namely a finite subset of the integers or a finite set of floating-point numbers, i.e., a finite subset of the rational numbers. The finiteness of such input excludes arbitrarily small absolute non-zero values for expressions of bounded depth. There is a gap between zero and other values that a parameterized expression can take on. A separation bound for an arithmetic expression E is a lower bound on the size of this gap. Besides the finiteness of the number of possible numerical inputs, the coarseness of the input data can generate a gap between zero and other values taken on. A straightforward example is integral expressions. If all operands are integers the number 1 is clearly a separation bound.

Once a separation bound is available it is clear how to decide whether the value of an expression is zero or not. Representations with repeatedly increased precision are computed until either the error bound on the current approximation is less than the absolute value of the approximation or their sum is less than the separation bound. In the phrasing of interval arithmetic, it means to refine the interval until either 0 or the separation bound are not contained in the interval.

How can we get separation bounds without computing the exact value or an approximation and an error bound? Most geometric computations are on linear objects and involve only basic arithmetic operations over the rational numbers. In distance computations and operations on nonlinear objects like circles and parabolas, square root operations are used as well. For the rational numerical input data arising in practice, expressions over the operations $+, -, *, /, \sqrt{}$ take on only algebraic values.

Let E be an expression involving square roots. Furthermore we assume that all operands are integers. We use $\alpha(E)$ to denote the algebraic value of expression E. Computer algebra provides bounds for the size of the roots of polynomials with integral coefficients. These bounds involve quantities used to describe the complexity of an integral polynomial, e.g., degree, maximum coefficient size, or less well-known quantities like height or measure of a polynomial. Once an integral polynomial with root $\alpha(E)$ is known the root bounds from computer algebra give us separation bounds. In general, however, we don't have a polynomial having root $\alpha(E)$ at hand. Fortunately, all we need to apply the root bounds are bounds on the quantities involved in the root bounds. Upper bounds on these quantities for some polynomial having root $\alpha(E)$ can be derived automatically from an expression E. Mignotte discusses identification of algebraic numbers given by expressions involving square roots in [75].

The measure of a polynomial [76] can be used for automatic derivation of a root bound. Table 1 gives the rules for (over)estimating measure and degree of an integral polynomial having root $\alpha(E)$. We have $\alpha(E) = 0$ or $|\alpha(E)| \geq M(E)^{-1}$. This bound is easily computable but very weak [14].

Other recursive formulas for an expression involving square root operations leading to separation bounds are given in [111]. Here, a bound on the maximum absolute value of the coefficients of an integral polynomial is used. The rules are given in Table 2. By a result of Cauchy, $(h(E)+1)^{-1}$ is a separation bound, i.e., $\alpha(E) = 0$ or $\alpha(E) \geq (h(E)+1)^{-1}$.

In [17] Canny considers isolated solutions of systems of polynomial equations in several variables with integral coefficients. He gives bounds on the absolute values of the non-zero components of an isolated solution vector. The bound depends on the number of variables, the maximum total degree d of the multivariate integral polynomials in the system and their maximum coefficient size c. Although Canny solves a much more general problem, his bounds can be used to

	$M(E)$	$deg(E)$
integer n	$\lvert n\rvert$	1
$E_1 + E_2$	$2^{deg(E_1)deg(E_2)}M(E_1)^{deg(E_2)}M(E_2)^{deg(E_1)}$	$deg(E_1)\cdot deg(E_2)$
$E_1 - E_2$	$2^{deg(E_1)deg(E_2)}M(E_1)^{deg(E_2)}M(E_2)^{deg(E_1)}$	$deg(E_1)\cdot deg(E_2)$
$E_1 \cdot E_2$	$M(E_1)^{deg(E_2)}M(E_2)^{deg(E_1)}$	$deg(E_1)\cdot deg(E_2)$
E_1/E_2	$M(E_1)^{deg(E_2)}M(E_2)^{deg(E_1)}$	$deg(E_1)\cdot deg(E_2)$
$\sqrt{E_1}$	$M(E_1)$	$2\cdot deg(E_1)$

Table 1. Automatic derivation of separation bounds for expressions involving square roots based on the measure of a polynomial

	$h(E)$	$d(E)$
integer n	$\lvert n\rvert$	1
$E_1 + E_2$	$(h(E_1)2^{1+d(E_1)})^{d(E_2)}(h(E_2)\sqrt{1+d(E_2)})^{d(E_1)}$	$d(E_1)\cdot d(E_2)$
$E_1 - E_2$	$(h(E_1)2^{1+d(E_1)})^{d(E_2)}(h(E_2)\sqrt{1+d(E_2)})^{d(E_1)}$	$d(E_1)\cdot d(E_2)$
$E_1 \cdot E_2$	$(h(E_1)\sqrt{1+d(E_1)})^{d(E_2)}(h(E_2)\sqrt{1+d(E_2)})^{d(E_1)}$	$d(E_1)\cdot d(E_2)$
E_1/E_2	$(h(E_1)\sqrt{1+d(E_1)})^{d(E_2)}(h(E_2)\sqrt{1+d(E_2)})^{d(E_1)}$	$d(E_1)\cdot d(E_2)$
$\sqrt{E_1}$	$h(E_1)$	$2\cdot d(E_1)$

Table 2. Recursive formulas for quantities $h(E)$ and $d(E)$ of an arithmetic expression involving square roots.

get fairly good separation bounds for expressions involving square roots. Canny shows that the absolute value of a component of an isolated solution of a system of n integral polynomial equations in n variables is either zero or at least $(3dc)^{-nd^n}$ [17, 18].

Based on the structure of an expression E given by an expression tree, a system of polynomial equations can be built which has an isolated solution vector with $\alpha(E)$ as a component. The system of polynomial equations consists of a system $\mathcal{P}(E)$ in n_E variables $X_1,\ldots,X_{n_E}$ and a distinct equation of the form $X_E = P_E(X_1,\ldots,X_{n_E})$. The variables correspond to subexpressions of E, the variable X_E represents the value of E.

At the basis of recursion we have the distinct polynomial only. If $E = E_1 \pm E_2$ then $\mathcal{P}(E)$ is the union of the systems $\mathcal{P}(E_1)$ and $\mathcal{P}(E_2)$ and the distinct equation becomes $X_E = P_{E_1}(\ldots) \pm P_{E_2}(\ldots)$. Variables are renamed appropriately. The recursion step is completely analogous if $E = E_1 \cdot E_2$.

If $E = E_1/E_2$ the system $\mathcal{P}(E)$ contains the union of the systems $\mathcal{P}(E_1)$ and $\mathcal{P}(E_2)$. Furthermore the equation

$$X_{\text{new}} \cdot P_{E_2}(\ldots) = P_{E_1}(\ldots)$$

is added. It uses a new variable X_{new} and is based on the distinct equations for the subexpressions. The new distinct equation becomes $X_E = X_{\text{new}}$. If $E = \sqrt{E}$

the procedure is similar. The new equation is

$$X_{new}^2 = P_E(\dots).$$

The distinct equation is $X_E = X_{new}$ again.

If the system resulting from an expression tree has maximum degree d_E, maximum coefficient size c_E, and n_E equations, $(3d_E c_E)^{-n_E} d_E^{n_E}$ is a separation bound for E. Note that $n_E - 1$ is the number of square root and division operations involved in E. There are alternative ways to derive a system of polynomial equations for an expression E. One could also introduce a new variable and a new equation for each operation. That would guarantee degree at most 2 but result in a system with more equations and variables.

Recently Burnikel et al.[12] have shown that

$$\alpha(E) \geq \left(u(E)^{2^{2k(E)-1}} l(E) \right)^{-1}$$

where $k(E)$ is the number of (distinct) square root operations in E and the quantities $u(E)$ and $l(E)$ are defined as given in Table 3. Note that $u(E)$ and $l(E)$ are simply the numerator and denominator of an expression obtained by replacing in E all $+$ by $-$ and all integers by their absolute value. If E is division-free and $\alpha(E)$ is non-zero, then $\alpha(E) \geq u(E)^{1-2^{k(E)-1}}$.

	$u(E)$	$l(E)$
integer n	$\|n\|$	1
$E_1 \pm E_2$	$u(E_1) \cdot l(E_2) + l(E_1) \cdot u(E_2)$	$l(E_1) \cdot l(E_2)$
$E_1 \cdot E_2$	$u(E_1) \cdot u(E_2)$	$l(E_1) \cdot l(E_2)$
E_1 / E_2	$u(E_1) \cdot l(E_2)$	$l(E_1) \cdot u(E_2)$
$\sqrt{E_1}$	$\sqrt{u(E_1)}$	$\sqrt{l(E_1)}$

Table 3. Recursive formulas for quantities $u(E)$ and $l(E)$ of an arithmetic expression involving square roots.

This bound as well as the bound given in [111] involve square root operations. Hence they are not easily computable. In practice one computes ceilings of the results to get integers [111] or maintains integer bounds logarithmically [12, 16]. The `Real/Expr`-package [28, 88] and the number type `real` [16] in LEDA provide exact computation (in C++) for expressions with operations $+, -, \cdot, /$ and $\sqrt{\ }$ and initially integral operands, using techniques described above. In particular, the recent version of the `reals` in LEDA [74] uses the bounds given in [12].

Note the difference between separation bounds and ε_{magic}s in epsilon tweaking. In epsilon-tweaking a test for zero is replaced by the test $|\tilde{E}| < \varepsilon_{magic}$? . With separation bounds it becomes $|\tilde{E}| < sep(E) - E_{error}$? where $sep(E)$ is a separation bound and E_{error} is a bound on the error accumulated in the evaluation of E. The difference is that the latter term is self-adjusting, it is based on an error bound, and justified; it is guaranteed that the result is zero, if the condition is satisfied. While $\varepsilon_{\text{magic}}$ is always positive, it might happen that the accumulated error is so large that $sep(E) - E_{\text{error}}$ is negative. Last but not least, the conclusion is different if the test is not satisfied. Epsilon-tweaking concludes that the number is non-zero if it is larger than $\varepsilon_{\text{magic}}$ while the use of separation bounds allows this conclusion only if $|\tilde{E}| \geq sep(E) + E_{\text{error}}$.

4 Geometric Computation with Imprecision

In this section we briefly discuss the basic aspects of the design and implementation of geometric algorithms for calculations with imprecision.

4.1 Implementation with Imprecise Predicates

Imprecise arithmetic cannot guarantee correct evaluation of a geometric predicate. It can lead to wrong decisions and wrong results. But even if the result is not the exact result for the considered problem instance, it can be meaningful. An algorithm that computes the exact result for a very similar problem instance can be sufficient for an application, since the input data are known not to be exact either. This observation motivates the definition of robustness and stability given in Section 1.2. In addition to the existence of a perturbation of the input data, for which the computed result is correct, Fortune's definition of robustness and stability [37] requires in addition that the implementation of an algorithm would compute the exact result, if all computations were precise.

The output of an algorithm might be useful although it is not a correct output for any perturbation of the input. In some situations it might be feasible to allow perturbation of the output as well. For example, for some applications it might be sufficient that the output of a two-dimensional convex hull algorithm is a nearly convex polygon while other applications require convexity. Sometimes the requirements on the output are relaxed to allow "more general" perturbations of the input data. Robustness and stability are then defined with respect to the weaker problem formulation. For example, Fortune's and Milenkovic's line arrangement algorithm [40] computes a combinatorial arrangement that is realizable by pseudolines but not necessarily by straight lines. Shewchuk [102] suggests calling an algorithm *quasi-robust* if it computes useful information but not a correct output for any perturbation of the input.

For many implementations of geometric primitives it is easy to show that the computed result is correct for some perturbation of the input. The major problem in the implementation with imprecise predicates is their combination. The basic predicates evaluated in an execution of an algorithm operate on the same set of data and and hence they might be dependent. The results of dependent geometric predicates might be mutually exclusive, i.e., there might be no small perturbation leading to correctness for all predicates. Hence an algorithm might get into an inconsistent state, a state that could not be reached from any input with correct evaluation. A relaxation of the problem sometimes helps. An illegal state can be a legal state for a similar problem with weaker restrictions, e.g., a state illegal for an algorithm computing an arrangement of straight lines could be legal for arrangements of pseudolines. Although an inconsistent state cannot be reached from any legal input it can still contain useful information.

Avoiding inconsistencies among the decisions is a primary goal in achieving robustness in implementations with imprecise predicates. Consistency is a non-issue if an algorithm never evaluates a basic predicate whose outcome is implied by the results of previous evaluations of basic predicates. Such an algorithm is called *parsimonious* [37, 65].

It can be hard to achieve consistency with previous evaluations. For example, checking whether the outcome of an orientation test is implied by previous tests on the given set of points is as hard as the existential theory of the reals [37].

For the incremental construction of Voronoi diagrams of points Sugihara et al. show how consistency with previous decisions can be forced [107, 108]. Their algorithm is extremely (quasi-)robust. Some "meaningful" output is computed even if the results of all numerical comparisons are chosen at random. Meaningful means that the computed result is guaranteed to have some topological properties of a Voronoi diagram.

For some basic geometric problems there are stable, robust, or quasirobust implementations of geometric algorithms. Li and Milenkovic [67], Guibas et al. [53, 52], and Kawaguchi et al. [19] consider the convex hull problem in two dimensions, Barber [6] considers convex hulls and related problems, Hopcraft, Hoffmann, and Karasick [57] and Hopcroft and Kahn [58] consider intersection of polygons and convex polyhedra respectively. Fortune and Milenkovic [40] and Milenkovic [77] consider line arrangements. Fortune [39] considers the Delauney triangulation of point sets in two-dimensional space and Dey et al. [26] in three-dimensional space. For modelling polygonal regions in the plane Milenkovic [77] uses a technique called data normalization to modify the input such that it can be processed with imprecise arithmetic. Pullar [94] describes possible applications of these techniques to GIS. Sugihara and Iri present a solid modelling system free from topological errors [109].

The techniques used in these algorithms are fairly special and it seems unlikely that they can be easily transferred to other geometric problems. A general theory showing how to implement geometric algorithms with imprecise predicates is still a distant goal.

4.2 Epsilon Geometry

An interesting theoretical framework for the investigation of imprecision in geometric computation is epsilon geometry introduced by Guibas, Salesin, and Stolfi [52]. Instead of a boolean value, an epsilon predicates returns a real number that gives some information "how much" the input satisfies the predicate. In epsilon geometry the size of a perturbation is measured by a non-negative real number. Only the identity has size zero. If an input does not satisfy a predicate, the "truth value" of an epsilon predicate is the size of the smallest perturbation producing a perturbed input that satisfies the predicate. If the input satisfies a predicate, the "truth value" is the non-positive number ϱ if the predicate is still satisfied after perturbing with any perturbations of size at most $-\varrho$. In [52] epsilon predicates are combined with interval arithmetic. Imprecise evaluations of epsilon predicates compute a lower and an upper bound on the "truth value" of an epsilon predicate. Guibas, Salesin, and Stolfi compose basic epsilon predicates to less simple predicates. Unfortunately epsilon geometry has been applied successfully only to a few basic geometric primitives [52, 53]. Reasoning in the epsilon geometry framework seems to be difficult.

4.3 Axiomatic Approach

In [97, 98] Schorn proposes what he calls the *axiomatic approach*. The idea is to investigate which properties of primitive operations are essential for a correctness proof of an algorithm and to find algorithm invariants that are based on these properties only.

One of the algorithms considered in [97] is computing a closest pair of a set of points S by plane sweep [54]. Instead of a closest pair, the distance δ_S of a closest pair is computed. In his implementation Schorn uses distance functions $d(p,q)$, $d_x(p,q)$, $d_y(p,q)$, and $d'_y(p,q)$ on points $p = (p_x, p_y)$ and $q = (q_x, q_y)$ in the plane. In an exact implementation these functions would compute $\sqrt{(p_x - q_x)^2 + (p_y - q_y)^2}$, $p_x - q_x$, $p_y - q_y$, and $q_y - p_y$, respectively. Schorn lists properties for these functions that are essential for a correctness proof: First, they must have some monotonicity properties. d_x must be monotone with respect to the x-coordinate of its first argument, i.e., $[p_x \geq p'_x \Rightarrow d_x(p,q) \geq d_x(p',q)]$ holds, and inverse monotone in the x coordinate of its second argument, i.e. $[q_x \leq q'_x \Rightarrow d_x(p,q) \geq d_x(p,q')]$ holds. Similarly, $[q_y \leq q'_y \Rightarrow d_y(p,q) \geq d_y(p,q')]$ and $[q_y \geq q'_y \Rightarrow d'_y(p,q) \geq d'_y(p,q')]$ must hold for d_y and d'_y, respectively. Second, d_x, d_y, and d'_y must be "bounded by d", more precisely, $[p_x \geq q_x \Rightarrow d(p,q) \geq d_x(p,q)]$, $[p_y \geq q_y \Rightarrow d(p,q) \geq d_y(p,q)]$, and $[p_y \leq q_y \Rightarrow d(p,q) \geq d'_y(p,q)]$ must hold. Finally, d must be symmetric, i.e., $d(p,q) = d(q,p)$. These properties, called axioms in [97] are sufficient to prove that for the δ computed by Schorn's plane sweep implementation

$$\delta = \min_{s,t \in S} d(s,t)$$

holds. No matter what d, d_x, d_y, and d'_y are, as long as they satisfy all axioms, $\min_{s,t \in S} d(s,t)$ is computed by the sweep. In particular, if exact distance functions could be used, the correct distance of a closest pair would be computed. Schorn uses floating-point implementations of the distance functions d, d_x, d_y, and d'_y. He shows that they have the desired properties and that they guarantee a relative error of at most $8\varepsilon_{\text{prec}}$ in the computed approximation for δ_S, where $\varepsilon_{\text{prec}}$ is machine precision.

Further geometric problems to which the axiomatic approach is applied in [97] to achieve robustness are: finding pairs of intersecting line segments and computing the winding number of a point with respect to a not necessarily simple polygon. The latter involves point in polygon testing as a special case, which is also discussed in [36].

5 Related Issues

5.1 Rounding

The complexity, e.g., the bit-length of integers, of numerical data in the output of algorithms for constructive geometric problems is usually higher than that of the input data. Thus piping geometric computations can result in expensive arithmetic operations. If the cost caused by increased precision resulting from cascaded computation is not tolerable, precision must be decreased by rounding the geometric output data. The goal in rounding is not to deviate too much from the original data both with respect to geometry and topology while reducing the precision. Rounding geometric objects is related to simultaneous approximation of reals by rationals [106]. However, rounding geometric data is more complicated than rounding numbers and can be very difficult [78], because combinatorial and numerical data have to be kept consistent.

An intensively studied example is rounding an arrangement of line segments, the underlying geometric structure of cartographic maps. Greene and Yao [50] were the first to investigate rounding line segments consistently to a regular grid. Note that simply rounding each segment endpoint to its nearest grid point can introduce new intersections and hence significantly violate the original topology. Greene and Yao break line segments into polylines such that all endpoints lie on the grid and the topology is largely preserved. Largely means, incidences not present in the original arrangement might arise, but it can be shown that no additional crossings are generated. Currently the most promising structure is *"snap-rounding"*, also called *"hot-pixel"* rounding, introduced by Greene and Hobby. A pixel in the regular grid is called hot if it contains an endpoint of an original line segment or an intersection point of the original segments. In the rounding process all line segments are snapped to the pixel center, cf. Fig. 5. Snap-rounding is used in [55, 51, 48]. Rounding can be done as a postprocessing step after exact computation, but it can also be seen as part of the problem and be incorporated into the algorithmic solution, as e.g. in [51] and [48].

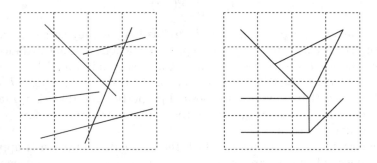

Fig. 5. Snap-rounding line segments

5.2 Inaccurate Data

Cartographic data are inherently inaccurate. Sometimes, they can nevertheless be treated as exact. In preprocessing and postprocessing steps, input and output data respectively might have to be "cleaned up". For example, in map overlay, spurious or sliver polygons [20] have to be removed that result from the overlay of objects which are identical in the real world but not in the overlaid maps. Treating inaccurate data as exact works (with exact geometric computation) as long as the input data are consistent. If not, we are in a situation similar to computation with imprecision. An algorithm might get into states it was not supposed to get in and which it therefore cannot handle. This similarity has led researchers to advocate imprecise computation and to attack both inconsistencies arising from imprecise computation and inconsistencies due to inaccurate data uniformly. In this approach, however, it is not clear whether errors in the output are caused by precision problems during computation or inaccuracies in the data. Source errors and processing errors become indistinguishable. Exact computation, on the other hand, assures that inconsistencies are due to faulty data. But knowing that an error was caused by a source error does not at all tell you how to proceed.

The alternative to treating possibly inaccurate data as exact is to incorporate uncertainty into the problem statement and to develop and use algorithms solving the resulting problems (exactly). Goodchild [47] gives an overview on approaches to incorporate inaccuracy and uncertainty in cartographic data in GIS. For example, tolerance regions can be added to geometric objects to model inaccuracy and uncertainty in the data, see e.g. [35]. Inaccuracies in the position of points can be modelled by epsilon circles, inaccuracies in lines by a Perkal epsilon band [90].

Pullar discusses consequences of using tolerance circles to point coincidence and point clustering problems [95]. Similar to point coincidence under the "close to zero is zero" rule, transitivity is a problem, cf. Fig. 6, if points are considered as coincident if their tolerance regions overlap. In [95] clustering of points is considered to solve the coincidence problem.

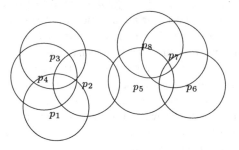

Fig. 6. Points with circular tolerance regions. An obvious clustering would be $\{\{p_1, \ldots, p_4\}, \{p_5, \ldots, p_8\}\}$.

Enhancing tolerance regions with a probability distribution leads to a better model of uncertainty. An example of this approach is modelling coordinates $x_1, \ldots, x_d$ of a point position which is known to be possibly inaccurate by probability distributions $X_1, \ldots, X_d$ such that the mean of X_i is at x_i. As with computation with imprecision, a lot of research on modelling and handling uncertainty in geometric data is still needed.

5.3 Geometric Algorithms in a Library

The purpose of a library is to provide reusable software components. Reusability requires generality. The components must be usable in or adaptable to various applications. Generality as such is not sufficient. The components must not only be adaptable, they must also lead to efficient solutions.

Library components should come with a precise description what they compute and for which inputs they are guaranteed to work. Correctness means that a component behaves according to such a specification. Clearly, correctness in the sense of reliability should be beyond question for geometric algorithms and primitives in a library. However, by far not all implementations of geometric algorithms are correct. Many implementations of geometric algorithms pretend to solve a geometric problem, but for a not-clearly-specified set of problems instances they don't. Due to missing or improper handling of special cases or just incorrect coding of complicated parts and especially to precision problems, many implementations of geometric algorithms disappoint the user occasionally by unexpected failures, break downs, or computing garbage.

Exactness should not be confused with correctness in the sense of reliability. There is nothing wrong with approximation algorithms or approximate solutions as long as they do what they profess to do. Correctness can have unlike appearances: An algorithm handling only non-degenerate cases can be correct in the above sense. Also, an algorithm that guarantees to compute the exact result only if the numerical input data are integral and smaller than some given bound can be correct as well as an algorithm that computes an approximation to the exact result with a guaranteed error bound. Correctness in the sense of reliability is a must for (re-)usability and hence for a geometric algorithms library.

A good library is more than just a collection of reusable software. It provides reliable, reusable components that can be combined in a fairly seamless way. Due to the composition problem with imprecise predicates described in Section 4.1, even stable, imprecise predicates are not very useful as library components. Building a library upon exact geometric predicates is much easier. With exact predicates, algorithms developed under the real computation model can be implemented in a straightforward way. A redesign that deals with imprecision in the predicates is not necessary. Exact basic predicates can simplify the task of implementing approximation algorithms as well. For input data that are known to be inaccurate, exactness is not so important. Correctness in the sense of reliability is then the primary goal, not exactness, but currently exact geometric computation seems to be the safest way to reach it.

Among the library and workbench efforts in computational geometry [3, 45, 63, 24, 74, 85] the XYZ-Geobench and LEDA deserve special attention concerning precision and robustness. In XYZ-Geobench [85, 96] the axiomatic approach to robustness, described in section 4.3, is used. In LEDA [73, 74] arbitrary precision integer arithmetic is combined with the floating-point filter technique to yield efficient exact components for rational problems. Recently, in Europe and the US, new projects called CGAL (Computational Geometry Algorithms Library) [34, 89] and GeomLib [5] have been started. The goal of both projects is to enhance the technology transfer from theory to practice in geometric computing by providing reliable, reusable implementations of geometric algorithms.

6 Conclusion

In his book on randomization and geometry [82] Mulmuley writes

Dealing with the finite nature of actual computers is an art that requires infinite patience.

Nevertheless, the precision problem is almost ignored and left to the implementor in the textbooks on computational geometry, for sake of simplicity and readability of presentation. The emphasis is on understanding an algorithm and their correctness over the reals rather than on implementation issues of these algorithms. More than a half page description of the precision problems is hardly given.

Despite a lot of research having been done on the precision and robust problem, no satisfactory general-purpose solution has been found. There is no consensus in the geometry literature on how to deal with precision problems. Some researchers want to use fast floating-point arithmetic exclusively and hence investigate design and implementation of algorithms with imprecise predicates. Others prefer exact geometric computation, because it allows fairly straightforward implementation of geometric algorithms designed for the real RAM model [91] and sometimes because they want to use perturbation schemes. Exact geometric computation seems to be the more practical approach to reach reliability, especially if number packages supporting exact geometric computation [29, 13] are available. However, there need not be a consensus. Both approaches have their merits.

Practitioners often ask for the impossible. Algorithms computing exact or at least highly accurate results are requested to be competitive in performance to algorithms that sometimes crash or exhibit otherwise unexpected behavior. Efficiency is compared for inputs that all of them handle. That is somewhat unfair. It should be clear that one has to pay for the detection of degenerate and nearly degenerate situations, but it should also be clear, that one gets much more.

Surely, this survey is incomplete and biased. Most of the presentation is devoted to exact geometric computation. Implementation with imprecise primitives has gained less attention here, because it lacks generality and its application is much less straightforward. Related surveys on the problem of precision and robustness in geometric computation are given by Fortune [38], Hoffmann [56], and Yap [110]. Franklin [44] especially discusses cartographic errors caused by precision problems.

Acknowledgements

Work on these notes was partially supported by the ESPRIT IV LTR Project No. 21957 (CGAL).

References

1. A.V. Aho, J.E. Hopcroft, and J.D. Ullman. *The Design and Analysis of Computer Algorithms.* Addison-Wesley, 1974.
2. G. Alefeld and J. Herzberger. *Introduction to Interval Computation.* Academic Press, New York, 1983.
3. F. Avnaim. *C++GAL: A C++ Library for Geometric Algorithms,* 1994.
4. F. Avnaim, J.D. Boissonnat, O. Devillers, F.P. Preparata, and M. Yvinec. Evaluating signs of determinants using single precision arithmetic. Technical Report 2306, INRIA Sophia-Antipolis, 1994.
5. J.E. Baker, R. Tamassia, and L. Vismara. GeomLib: Algorithm engineering for a geometric computing library, 1997. (Preliminary report).
6. J.L. Barber. Computational geometry with imprecise data and arithmetic : Phd Thesis. Technical Report CS-TR-377-92, Princeton University, 1992.
7. M.O. Benouamer, P. Jaillon, D. Michelucci, and J-M. Moreau. A "lazy" solution to imprecision in computational geometry. In *Proc. of the 5th Canad. Conf. on Comp. Geom.,* pages 73–78, 1993.
8. J.D. Boissonnat and M. Yvinec. *Algorithmic Geometry.* Cambridge University Press, Cambridge, UK, 1997.
9. H. Brönnimann, I.Z. Emiris, V.Y. Pan, and S. Pion. Computing exact geometric predicates using modular arithmetic with single precision. In *Proc. 13th Annu. ACM Sympos. Comput. Geom.,* pages 174–182, 1997.
10. H. Brönnimann and M. Yvinec. Efficient exact evaluation of signs of determinants. In *Proc. 13th Annu. ACM Sympos. Comput. Geom.,* pages 166–173, 1997.
11. C. Burnikel. *Exact Computation of Voronoi Diagrams and Line Segment Intersections.* PhD Thesis, Universität des Saarlandes, Saarbrücken, Germany, 1996.
12. C. Burnikel, R. Fleischer, K. Mehlhorn, and S. Schirra. A strong and easily computable separation bound for arithmetic expressions involving square roots. In *Proc. of the 8th ACM-SIAM Symp. on Discrete Algorithms,* pages 702–709, 1997.
13. C. Burnikel, J. Könemann, K. Mehlhorn, S. Näher, S. Schirra, and C. Uhrig. Exact geometric computation in LEDA. In *Proceedings of the 11th ACM Symposium on Computational Geometry,* pages C18–C19, 1995.
14. C. Burnikel, K. Mehlhorn, and S. Schirra. How to compute the Voronoi diagram of line segments: Theoretical and experimental results. In *ESA94,* pages 227–239, 1994.
15. C. Burnikel, K. Mehlhorn, and S. Schirra. On degeneracy in geometric computations. In *Proc. of the 5th ACM-SIAM Symp. on Discrete Algorithms,* pages 16–23, 1994.

16. C. Burnikel, K. Mehlhorn, and S. Schirra. The LEDA class **real** number. Technical Report MPI-I-96-1-001, Max-Planck-Institut für Informatik, 1996.

17. J.F. Canny. *The Complexity of Robot Motion Planning*. PhD Thesis, 1987.

18. J.F. Canny. Generalised characteristic polynomials. *J. Symbolic Computation*, 9:241–250, 1990.

19. Wei Chen, Koichi Wada, and Kimio Kawaguchi. Parallel robust algorithms for constructing strongly convex hulls. In *Proc. 12th Annu. ACM Sympos. Comput. Geom.*, pages 133–140, 1996.

20. N.R. Chrisman. The accuracy of map overlays: a reassessment. In D.J. Peuquet and D.F. Marble, editors, *Introductory Readings in Geographic Information Systems*, pages 308–320. Taylor & Francis, London, 1990.

21. K. L. Clarkson. Safe and effective determinant evaluation. In *Proc. 33rd Annu. IEEE Sympos. Found. Comput. Sci.*, pages 387–395, 1992.

22. J.L.D. Comba and J. Stolfi. Affine arithmetic and its applications to computer graphics, 1993. Presented at SIBGRAPI'93, Recife (Brazil), October 20-22.

23. M. de Berg, M. van Kreveld, M. Overmars, and O. Schwarzkopf. *Computational Geometry*. Springer Verlag, 1997.

24. P. de Rezende and W. Jacometti. Geolab: An environment for development of algorithms in computational geometry. In *Proc. 5th Canad. Conf. Comput. Geom.*, pages 175–180, Waterloo, Canada, 1993.

25. T.J. Dekker. A floating-point technique for extending the available precision. *Numerische Mathematik*, 18:224 – 242, 1971.

26. T.K. Dey, K. Sugihara, and C.L. Bajaj. Delaunay triangulations in three dimensions with finite precision arithmetic. *Computer Aided Geometric Design*, 9:457–470, 1992.

27. D. Douglas. It makes me so CROSS. In D.J. Peuquet and D.F. Marble, editors, *Introductory Readings in Geographic Information Systems*, pages 303–307. Taylor & Francis, London, 1990.

28. T. Dubé, K. Ouchi, and C.K. Yap. Tutorial for **Real/Expr** package. 1996.

29. T. Dubé and C.K. Yap. A basis for implementing exact computational geometry. extended abstract, 1993.

30. H. Edelsbrunner. *Algorithms in Combinatorial Geometry*. Springer Verlag, 1986.

31. H. Edelsbrunner and E. Mücke. Simulation of simplicity: A technique to cope with degenerate cases in geometric algorithms. *ACM Trans. on Graphics*, 9:66–104, 1990.

32. I. Emiris and J. Canny. A general approach to removing degeneracies. In *Proceedings of the 32nd IEEE Symposium on Foundations of Computer Sience*, pages 405–413, 1991.

33. I. Emiris and J. Canny. An efficient approach to removing geometric degeneracies. In *Proc. of the 8th ACM Symp. on Computational Geometry*, pages 74–82, 1992.

34. A. Fabri, G.-J. Giezeman, L. Kettner, S. Schirra, and S. Schönherr. The CGAL kernel : a basis for geometric computation. In Ming C. Lin and Dinesh Manocha, editors, *Applied Computational Geometry : Towards Geometric Engineering (WACG96)*, pages 191–202. Springer LNCS 1148, 1996.

35. S. Fang and B. Brüderlin. Robustness in geometric modeling - tolerance based methods. In *Proc. Workshop on Computational Geometry CG'91*, pages 85–102. Springer Verlag LNCS 553, 1991.

36. A. R. Forrest. Computational geometry in practice. In R. A. Earnshaw, editor, *Fundamental Algorithms for Computer Graphics*, volume F17 of *NATO ASI*, pages 707–724. Springer-Verlag, 1985.

37. S. Fortune. Stable maintenance of point-set triangulations in two dimensions. In *Proceedings of the 30th IEEE Symposium on Foundations of Computer Sience*, pages 494–499, 1989.

38. S. Fortune. Progress in computational geometry. In R. Martin, editor, *Directions in Geometric Computing*, pages 81 – 128. Information Geometers Ltd., 1993.

39. S. Fortune. Numerical stability of algorithms for 2D Delaunay triangulations and Voronoi diagrams. *Int. J. Computational Geometry and Applications*, 5:193–213, 1995.

40. S. Fortune and V. Milenkovic. Numerical stability of algorithms for line arrangements. In *Proc. of the 7th ACM Symp. on Computational Geometry*, pages 334–341, 1991.

41. S. Fortune and C. van Wyk. Efficient exact arithmetic for computational geometry. In *Proc. of the 9th ACM Symp. on Computational Geometry*, pages 163–172, 1993.

42. S. Fortune and C. van Wyk. *LN user manual*, 1993.

43. S. Fortune and C. Van Wyk. Static analysis yields efficient exact integer arithmetic for computational geometry. *ACM Transactions on Graphics*, 15(3):223–248, 1996.

44. W.R. Franklin. Cartographic errors symptomatic of underlying algebra problems. In *Proc. International Symposium on Spatial Data Handling*, volume 1, pages 190–208, Zürich, 20–24 August 1984.

45. G.-J. Giezeman. *PlaGeo, a library for planar geometry, and SpaGeo, a library for spatial geometry*, 1994.

46. D. Goldberg. What every computer scientist should know about floating-point arithmetic. *ACM Computing Surveys*, pages 5–48, 1991.

47. M.F. Goodchild. Issues of quality and uncertainty. In J.C. Muller, editor, *Advances in Cartography*, pages 113–139. Elsevier Applied Science, London, 1991.

48. M. Goodrich, L. Guibas, J. Hershberger, and P. Tanenbaum. Snap rounding line segments efficiently in two and three dimensions. In *Proc. 13th Annu. ACM Sympos. Comput. Geom.*, pages 284–293, 1997.

49. T. Granlund. *GNU MP, The GNU Multiple Precision Arithmetic Library*, 2.0.2 edition, June 1996.

50. D. Greene and F. Yao. Finite resolution computational geometry. In *Proc. of the 27th IEEE Symposium on Foundations of Computer Science*, pages 143–152, 1986.

51. L. Guibas and D. Marimont. Rounding arrangements dynamically. In *Proc. 11th Annu. ACM Sympos. Comput. Geom.*, pages 190–199, 1995.

52. L. Guibas, D. Salesin, and J. Stolfi. Epsilon geometry: Building robust algorithms from imprecise computations. In *Proc. of the 5th ACM Symp. on Computational Geometry*, pages 208–217, 1989.

53. L. Guibas, D. Salesin, and J. Stolfi. Constructing strongly convex approximate hulls with inaccurate primitives. In *Proc. SIGAL Symp. on Algorithms*, pages 261–270, Tokyo, 1990.

54. K. Hinrichs, J. Nievergelt, and P. Schorn. An all-round sweep algorithm for 2-dimensional nearest-neighbor problems. *Acta Informatica*, 29:383–394, 1992.

55. J.D. Hobby. Practical line segment interscetion with finite precision output. Technical Report 93/2-27, Bell Laboratories (Lucent Technologies), 1993.

56. C.M. Hoffmann. The problem of accuracy and robustness in geometric computation. *IEEE Computer*, pages 31–41, March 1989.

57. C.M. Hoffmann, J.E. Hopcroft, and M.S. Karasick. Towards implementing robust geometric computations. In *Proc. of the 4th ACM Symp. on Computational Geometry*, pages 106–117, 1988.

58. J.E. Hopcroft and P.J. Kahn. A paradigm for robust geometric algorithms. *Algorithmica*, 7:339–380, 1992.

59. K. Jensen and N. Wirth. *PASCAL- User Manual and Report. Revised for the ISO Pascal Standard*. Springer Verlag, 3rd edition, 1985.

60. S. Kahan and J. Snoeyink. On the bit complexity of minimum link paths: Superquadratic algorithms for problems solvable in linear time. In *Proc. 12th Annu. ACM Sympos. Comput. Geom.*, pages 151–158, 1996.

61. M. Karasick, D. Lieber, and L.R. Nackman. Efficient Delaunay triangulation using rational arithmetic. *ACM Transactions on Graphics*, 10(1):71–91, 1991.

62. R. Klein. *Algorithmische Geometrie*. Addison-Wesley, 1997. (in German).

63. A. Knight, J. May, M. McAffer, T. Nguyen, and J.-R. Sack. A computational geometry workbench. In *Proc. 6th Annu. ACM Sympos. Comput. Geom.*, page 370, 1990.

64. D.E. Knuth. *The Art of Computer Programming Vol. 2: Seminumerical Algorithms*. Addison-Wesley, 2nd edition, 1981.

65. Donald E. Knuth. *Axioms and Hulls*, volume 606 of *Lecture Notes in Computer Science*. Springer-Verlag, Heidelberg, Germany, 1992.

66. M.J. Laszlo. *Computational geometry and computer graphics in C++*. Prentice Hall, Upper Saddle River, NJ, 1996.

67. Z. Li and V. Milenkovic. Constructing strongly convex hulls using exact or rounded arithmetic. *Algorithmica*, 8:345–364, 1992.

68. LiDIA -Group, Fachbereich Informatik Institut für Theoretische Informatik TH Darmstadt. *LiDIA Manual A library for computational number theory*, 1.3 edition, April 1997.

69. G. Liotta, F. Preparata, and R. Tamassia. Robust proximity queries: An illustration of degree-driven algorithm design. In *Proc. 13th Annu. ACM Sympos. Comput. Geom.*, pages 156–165, 1997.

70. K. Mehlhorn. *Data Structures and Algorithms 3: Multi-dimensional Searching and Computational Geometry*. Springer Verlag, 1984.

71. K. Mehlhorn and S. Näher. Implementation of a sweep line algorithm for the straight line segment intersection problem. Technical Report MPI-I-94-160, Max-Planck-Institut für Informatik, 1994.

72. K. Mehlhorn and S. Näher. The implementation of geometric algorithms. In *13th World Computer Congress IFIP94*, volume 1, pages 223–231. Elsevier Science B.V. North-Holland, Amsterdam, 1994.

73. K. Mehlhorn and S. Näher. LEDA, a platform for combinatorial and geometric computing. *Communications of the ACM*, 38:96–102, 1995.

74. K. Mehlhorn, S. Näher, and C. Uhrig. *The LEDA User manual*, 3.5 edition, 1997. cf. http://www.mpi-sb.mpg.de/LEDA/leda.html.

75. M. Mignotte. Identification of algebraic numbers. *Journal of Algorithms*, 3:197–204, 1982.

76. M. Mignotte. *Mathematics for Computer Algebra*. Springer Verlag, 1992.

77. V. Milenkovic. Verifiable implementations of geometric algorithms using finite precision arithmetic. *Artificial Intelligence*, 37:377–401, 1988.

78. V. Milenkovic and L. R. Nackman. Finding compact coordinate representations for polygons and polyhedra. In *Proc. 6th Annu. ACM Sympos. Comput. Geom.*, pages 244–252, 1990.

79. R.E. Moore. *Interval Analysis.* Prentice-Hall, Englewood Cliffs, NJ, 1966.

80. R.E. Moore. *Methods and Applications of Interval Analysis.* SIAM, Philadelphia, 1979.

81. S.P. Mudur and P.A. Koparkar. Interval methods for processing geometric objects. *IEEE Computer Graphics and Applications*, 4(2):7–17, 1984.

82. K. Mulmuley. *Computational Geometry : An Introduction through Randomized Algorithms.* Prentice Hall, Englewood Cliffs, NJ, 1994.

83. J. Nievergelt and K. H. Hinrichs. *Algorithms and Data Structures : with Applications to Graphics and Geometry.* Prentice Hall, Englewood Cliffs, NJ, 1993.

84. J. Nievergelt and P. Schorn. Das Rätsel der verzopften Geraden. *Informatik Spektrum*, (11):163–165, 1988. (in German).

85. J. Nievergelt, P. Schorn, M. de Lorenzi, C. Ammann, and A. Brüngger. XYZ: Software for geometric computation. Technical Report 163, Institut für Theoretische Informatik, ETH, Zürich, Switzerland, 1991.

86. J. O'Rourke. *Computational geometry in C.* Cambridge University Press, Cambridge, 1994.

87. T. Ottmann, G. Thiemt, and C. Ullrich. Numerical stability of geometric algorithms. In *Proc. of the 3rd ACM Symp. on Computational Geometry*, pages 119–125, 1987.

88. K. Ouchi. Real/Expr: Implementation of exact computation, 1997.

89. M. Overmars. Designing the computational geometry algorithms library CGAL. In Ming C. Lin and Dinesh Manocha, editors, *Applied Computational Geometry : Towards Geometric Engineering (WACG96)*, pages 53–58. Springer LNCS 1148, 1996.

90. J. Perkal. On epsilon length. *Bulletin de l'Académie Polonaise des Sciences*, 4:399–403, 1956.

91. F. Preparata and M.I. Shamos. *Computational Geometry.* Springer Verlag, 1985.

92. D.M. Priest. Algorithms for arbitrary precision floating point arithmetic. In *10th Symposium on Computer Arithmetic*, pages 132 – 143. IEEE Computer Society Press, 1991.

93. D.M. Priest. *On Properties of Floating-Point Arithmetic: Numerical Stability and the Cost of Accurate Computations.* PhD Thesis, Department of Mathematics, University of California at Berkeley, 1992.

94. D. Pullar. Spatial overlay with inexact numerical data. In *Proc. of Auto-Carto 10*, pages 313–329, 1991.

95. D. Pullar. Consequences of using a tolerance paradigm in spatial overlay. In *Proc. of Auto-Carto 11*, pages 288–296, 1993.

96. P. Schorn. An object-oriented workbench for experimental geometric computation. In *Proc. 2nd Canad. Conf. Comput. Geom.*, pages 172–175, 1990.

97. P. Schorn. *Robust Algorithms in a Program Library for Geometric Algorithms.* PhD Thesis, Informatik-Dissertationen ETH Zürich, 1991.

98. P. Schorn. An axiomatic approach to robust geometric programs. *J. Symbolic Computation*, 16:155–165, 1993.

99. P. Schorn. Degeneracy in geometric computation and the perturbation approach. *The Computer Journal*, 37(1):35–42, 1994.

100. R. Seidel. The nature and meaning of perturbations in geometric computations. In *STACS94*, 1994.

101. B. Serpette, J. Vuillemin, and J.C. Hervé. BigNum, a portable and efficient package for arbitrary-precision arithmetic. Technical Report 2, Digital Paris Research Laboratory, 1989.

102. J. R. Shewchuk. Adaptive precision floating-point arithmetic and fast robust geometric predicates. Technical Report CMU-CS-96-140, School of Computer Science, Carnegie Mellon University, 1996.

103. J. R. Shewchuk. Triangle: Engineering a 2D quality mesh generator and delaunay triangulator. In Ming C. Lin and Dinesh Manocha, editors, *Applied Computational Geometry : Towards Geometric Engineering (WACG96)*, pages 203–222, 1996.

104. IEEE Standard. 754-1985 for binary floating-point arithmetic. *SIGPLAN*, 22:9–25, 1987.

105. K.G. Suffern and E.D. Fackerell. Interval methods in computer graphics. *Computers & Graphics*, 15(3):331–340, 1991.

106. K. Sugihara. On finite-precision representations of geometric objects. *J. Comput. Syst. Sci.*, 39:236–247, 1989.

107. K. Sugihara. A simple method for avoiding numerical errors and degeneracies in Voronoi diagram construction. *IEICE Trans. Fundamentals*, E75-A(4):468 477, 1992.

108. K. Sugihara and M. Iri. Construction of the Voronoi diagram for over 10^5 generators in single-precision arithmetic. In *Abstracts 1st Canad. Conf. Comput. Geom.*, page 42, 1989.

109. K. Sugihara and M. Iri. A solid modelling system free from topological inconsistency. *Journal of Information Processing*, 12(4):380–393, 1989.

110. C. K. Yap. Robust geometric computation. In J. E. Goodman and J. O'Rourke, editors, *CRC Handbook in Computational Geometry*. CRC Press. (to appear).

111. C. K. Yap and T. Dubé. The exact computation paradigm. In D.Z. Du and F. Hwang, editors, *Computing in Euclidean Geometry*, pages 452–492. World Scientific Press, 1995. 2nd edition.

112. C.K. Yap. A geometric consistency theorem for a symbolic perturbation scheme. In *Proc. of the 4th ACM Symp. on Computational Geometry*, pages 134–141, 1988.

113. C.K. Yap. Symbolic treatment of geometric degeneracies. *J. Symbolic Comput.*, 10:349–370, 1990.

114. C.K. Yap. Towards exact geometric computation. *Computational Geometry: Theory and Applications*, 7(1-2):3–23, 1997. Preliminary version appeared in Proc. of the 5th Canad. Conf. on Comp. Geom., pages 405-419, (1993).

Lecture Notes in Computer Science

For information about Vols. 1–1265

please contact your bookseller or Springer-Verlag

Vol. 1302: P. Van Hentenryck (Ed.), Static Analysis. Proceedings, 1997. X, 413 pages. 1997.

Vol. 1303: G. Brewka, C. Habel, B. Nebel (Eds.), KI-97: Advances in Artificial Intelligence. Proceedings, 1997. XI, 413 pages. 1997. (Subseries LNAI).

Vol. 1304: W. Luk, P.Y.K. Cheung, M. Glesner (Eds.), Field-Programmable Logic and Applications. Proceedings, 1997. XI, 503 pages. 1997.

Vol. 1305: D. Corne, J.L. Shapiro (Eds.), Evolutionary Computing. Proceedings, 1997. X, 307 pages. 1997.

Vol. 1306: C. Leung (Ed.), Visual Information Systems. X, 274 pages. 1997.

Vol. 1307: R. Kompe, Prosody in Speech Understanding Systems. XIX, 357 pages. 1997. (Subseries LNAI).

Vol. 1308: A. Hameurlain, A M. Tjoa (Eds.), Database and Expert Systems Applications. Proceedings, 1997. XVII, 688 pages. 1997.

Vol. 1309: R. Steinmetz, L.C. Wolf (Eds.), Interactive Distributed Multimedia Systems and Telecommunication Services. Proceedings, 1997. XIII, 466 pages. 1997.

Vol. 1310: A. Del Bimbo (Ed.), Image Analysis and Processing. Proceedings, 1997. Volume I. XXII, 722 pages. 1997.

Vol. 1311: A. Del Bimbo (Ed.), Image Analysis and Processing. Proceedings, 1997. Volume II. XXII, 794 pages. 1997.

Vol. 1312: A. Geppert, M. Berndtsson (Eds.), Rules in Database Systems. Proceedings, 1997. VII, 214 pages. 1997.

Vol. 1313: J. Fitzgerald, C.B. Jones, P. Lucas (Eds.), FME '97: Industrial Applications and Strengthened Foundations of Formal Methods. Proceedings, 1997. XIII, 685 pages. 1997.

Vol. 1314: S. Muggleton (Ed.), Inductive Logic Programming. Proceedings, 1996. VIII, 397 pages. 1997. (Subseries LNAI).

Vol. 1315: G. Sommer, J.J. Koenderink (Eds.), Algebraic Frames for the Perception-Action Cycle. Proceedings, 1997. VIII, 395 pages. 1997.

Vol. 1316: M. Li, A. Maruoka (Eds.), Algorithmic Learning Theory. Proceedings, 1997. XI, 461 pages. 1997. (Subseries LNAI).

Vol. 1317: M. Leman (Ed.), Music, Gestalt, and Computing. IX, 524 pages. 1997. (Subseries LNAI).

Vol. 1318: R. Hirschfeld (Ed.), Financial Cryptography. Proceedings, 1997. XI, 409 pages. 1997.

Vol. 1319: E. Plaza, R. Benjamins (Eds.), Knowledge Acquisition, Modeling and Management. Proceedings, 1997. XI, 389 pages. 1997. (Subseries LNAI).

Vol. 1320: M. Mavronicolas, P. Tsigas (Eds.), Distributed Algorithms. Proceedings, 1997. X, 333 pages. 1997.

Vol. 1321: M. Lenzerini (Ed.), AI*IA 97: Advances in Artificial Intelligence. Proceedings, 1997. XII, 459 pages. 1997. (Subseries LNAI).

Vol. 1322: H. Hußmann, Formal Foundations for Software Engineering Methods. X, 286 pages. 1997.

Vol. 1323: E. Costa, A. Cardoso (Eds.), Progress in Artificial Intelligence. Proceedings, 1997. XIV, 393 pages. 1997. (Subseries LNAI).

Vol. 1324: C. Peters, C. Thanos (Eds.), Research and Advanced Technology for Digital Libraries. Proceedings, 1997. X, 423 pages. 1997.

Vol. 1325: Z.W. Raś, A. Skowron (Eds.), Foundations of Intelligent Systems. Proceedings, 1997. XI, 630 pages. 1997. (Subseries LNAI).

Vol. 1326: C. Nicholas, J. Mayfield (Eds.), Intelligent Hypertext. XIV, 182 pages. 1997.

Vol. 1327: W. Gerstner, A. Germond, M. Hasler, J.-D. Nicoud (Eds.), Artificial Neural Networks – ICANN '97. Proceedings, 1997. XIX, 1274 pages. 1997.

Vol. 1328: C. Retoré (Ed.), Logical Aspects of Computational Linguistics. Proceedings, 1996. VIII, 435 pages. 1997. (Subseries LNAI).

Vol. 1329: S.C. Hirtle, A.U. Frank (Eds.), Spatial Information Theory. Proceedings, 1997. XIV, 511 pages. 1997.

Vol. 1330: G. Smolka (Ed.), Principles and Practice of Constraint Programming – CP 97. Proceedings, 1997. XII, 563 pages. 1997.

Vol. 1331: D. W. Embley, R. C. Goldstein (Eds.), Conceptual Modeling – ER '97. Proceedings, 1997. XV, 479 pages. 1997.

Vol. 1332: M. Bubak, J. Dongarra, J. Waśniewski (Eds.), Recent Advances in Parallel Virtual Machine and Message Passing Interface. Proceedings, 1997. XV, 518 pages. 1997.

Vol. 1333: F. Pichler. R.Moreno Díaz (Eds.), Computer Aided Systems Theory – EUROCAST'97. Proceedings, 1997. XI, 626 pages. 1997.

Vol. 1334: Y. Han, T. Okamoto, S. Qing (Eds.), Information and Communications Security. Proceedings, 1997. X, 484 pages. 1997.

Vol. 1335: R.H. Möhring (Ed.), Graph-Theoretic Concepts in Computer Science. Proceedings, 1997. X, 376 pages. 1997.

Vol. 1336: C. Polychronopoulos, K. Joe, K. Araki, M. Amamiya (Eds.), High Performance Computing. Proceedings, 1997. XII, 416 pages. 1997.

Vol. 1337: C. Freksa, M. Jantzen, R. Valk (Eds.), Foundations of Computer Science. XII, 515 pages. 1997.

Vol. 1338: F. Plášil, K.G. Jeffery (Eds.), SOFSEM'97: Theory and Practice of Informatics. Proceedings, 1997. XIV, 571 pages. 1997.

Vol. 1339: N.A. Murshed, F. Bortolozzi (Eds.), Advances in Document Image Analysis. Proceedings, 1997. IX, 345 pages. 1997.

Vol. 1340: M. van Kreveld, J. Nievergelt, T. Roos, P. Widmayer (Eds.), Algorithmic Foundations of Geographic Information Systems. XIV, 287 pages. 1997.

Vol. 1341: F. Bry, R. Ramakrishnan, K. Ramamohanarao (Eds.), Deductive and Object-Oriented Databases. Proceedings, 1997. XIV, 430 pages. 1997.

Vol. 1342: A. Sattar (Ed.), Advanced Topics in Artificial Intelligence. Proceedings, 1997. XVIII, 516 pages. 1997. (Subseries LNAI).

Vol. 1344: C. Ausnit-Hood, K.A. Johnson, R.G. Pettit, IV, S.B. Opdahl (Eds.), Ada 95 – Quality and Style. XV, 292 pages. 1997.